Bioeconomies

Vincenzo Pavone · Joanna Goven
Editors

Bioeconomies

Life, Technology, and Capital in the 21st Century

palgrave
macmillan

Editors
Vincenzo Pavone
Institute of Public Goods and Policies (IPP)
Consejo Superior de Investigaciones Científicas (CSIC)
Madrid, Spain

Joanna Goven
Kukupa Research Ltd.
Pigeon Bay, New Zealand

ISBN 978-3-319-55650-5 ISBN 978-3-319-55651-2 (eBook)
DOI 10.1007/978-3-319-55651-2

Library of Congress Control Number: 2017937125

Printed on acid-free paper

This Palgrave Macmillan imprint is published by Springer Nature
The registered company is Springer International Publishing AG
The registered company address is: Gewerbestrasse 11, 6330 Cham, Switzerland

To Livia, my daughter, and all those women who will show us the way out of the 21st century
V.P.

To all the beings, human and otherwise, who resist despoliation
J.G.

Preface

On a bright and warm day of winter in Buenos Aires, in 2014, some 15 panellists met under the rubric of *(Re)thinking the bioeconomy* to discuss through a reflexive and empirical lens, the meaning and significance of that increasingly ubiquitous concept. The conversations generated by that panel had a galvanizing effect on this book project; but they also mirrored and extended conversations between Vincenzo and Joanna that had been occurring since 2008. Those earlier conversations triggered our long-lasting interest in the bioeconomy. In November 2008, researchers from the New Zealand project "Constructive Conversations" and invited researchers from Europe and North America, met in sunny Nelson to learn, discuss, reflect and write together, drawing inspiration from the results of a project that had grappled with different ways of studying what would now be called the bioeconomy. Some of the contours of the bioeconomy as a political project began to emerge from those discussions of empirical studies of relationships among research institutions, government, industry, indigenous communities and biotechnologies in New Zealand. Although it was hardly known at the time, and few scholars had addressed it by then, we soon realized that the bioeconomy concept was here to stay. Since 2008, references to the bioeconomy in the academic literature have skyrocketed, and many countries around the world, as well as the EU and the OECD, now have a bioeconomy strategy. We have been concerned ever since with the politics of the bioeconomy concept: how it encodes power, how it enables and constrains political choices, how it redraws boundaries between the political and the technical, how it extends the principles of economic rationality and economic competitiveness to previously protected domains, and how it makes and unmakes communities and collectives.

We would like to thank the organizers of the Buenos Aires panel, Vincenzo, Eric Deibel, Sara Lafuente, and Luisa Reis-Castro, and the colleagues from those discussions in Buenos Aires and Nelson, with whom we began to develop the thinking that led to this book: Dragana Lassiter, Chih-hsing Ho, YouZhao Gou, Maya Fisher, Dani Filc, Lieta Vivaldi, Cathy Herbrand, Hugo Ferpozzi, Teun Zuiderent-Jerak, Ida-Elisabeth Andersen, Mike Burgess, Hugh Campbell, Fiona Cram, Richard LeHeron, Nick Lewis, Angela Mackenzie, Carolyn Morris, Hazel Philips, Two Crow, and Brian Wynne.

Madrid, Spain — Vincenzo Pavone
Pigeon Bay, New Zealand — Joanna Goven

Acknowledgements

We wish to acknowledge the funding to the project BIOARREME awarded by the National R&D Plan 2011 of the Spanish Ministry of Science of Innovation for the Grant CSO2011-26019, the New Zealand Foundation for Research, Science and Technology (Grant UOCX0221), and the Royal Society of New Zealand (COST Travel Grant). We also wish to acknowledge the comments and ideas shared with us by the panellists of the Panel *(Re)thinking the bioeconomy*, organized at the 4S-ESOCITE Conference, Buenos Aires, 2014.

Contents

Editors and Contributors

About the Editors

Vincenzo Pavone is tenured research fellow the Institute of Public Goods and Policies (IPP) of the Spanish National Research Council (CSIC). His research activities address the relationship between neoliberalism and emerging bioeconomies, with a special focus on reproductive bioeconomies.

Joanna Goven is currently the Director of Kukupa Research, Deputy Director of the Centre for Integrated Research in Biosafety, Evaluation Director for the Engagement Programme of the Deep South National Science Challenge and a researcher with the Centre for Integrated Biowaste Research (New Zealand). She has published on gender politics, antifeminism, transdisciplinarity and the politics and political economy of science and technology.

Contributors

Erik Aarden Department of Science and Technology Studies, University of Vienna, Vienna, Austria

J. Benjamin Hurlbut School for the Future of Innovation in Society & School of Life Sciences, Arizona State University, Tempe, USA

Pierre Delvenne SPIRAL Research Centre, University of Liège, Liege, Belgium

Tess Doezema School for the Future of Innovation in Society & School of Life Sciences, Arizona State University, Tempe, USA

John Gardner School of Social Sciences, Monash University, Melbourne, Australia

Joanna Goven Kukupa Research, Pigeon Bay, New Zealand

Emma King NMAHP-RU, University of Stirling, Stirling, UK

Sara Lafuente Funes Institute of Public Goods and Policies (IPP), Consejo Superior Investigaciones Científicas (CSIC), Madrid, Spain

Luca Marelli Department of Experimental Oncology, European Institute of Oncology (IEO), Milan, Italy

James Mittra Science, Technology and Innovation Studies, University of Edinburgh, Edinburgh, UK

Vincenzo Pavone Institute of Public Goods and Policies (IPP), Consejo Superior Investigaciones Científicas (CSIC), Madrid, Spain

Luísa Reis-Castro HASTS, MIT, Cambridge, MA, USA

Carmen Romero-Bachiller Complutense University of Madrid, Madrid, Spain

Pablo Santoro Complutense University of Madrid, Madrid, Spain

Giuseppe Testa Department of Experimental Oncology, European Institute of Oncology (IEO), Milan, Italy; Department of Oncology and Hemato-Oncology, University of Milan, Milan, Italy

Aaro Tupasela Department of Public Health, Centre for Medical Science and Technology Studies, University of Copenhagen, Copenhagen, Denmark

Andrew Webster Department of Sociology, University of York, York, UK

Sonja van Wichelen Department of Sociology and Social Policy, The University of Sydney, Sydney, Australia

Abbreviations

A*STAR	Agency for Science, Technology and Research (Singapore)
AA	AquaAdvantage (salmon)
AACREA	Asociación Argentina de Consorcios Regionales de Experimentación Agrícola
AAPRESID	Asociación Argentina de Productores en Siembra Directa (no-till farming association)
AI	Artificial insemination
ART	Assisted Reproductive Technology
ATMP	Advanced Therapies Medicinal Products
BBMRI	Biobanking and BioMolecular resources Research Infrastructure
BBMRI-ERIC	Biobanking and BioMolecular resources Research Infrastructure—European Reasearch Infrastructure Consortium
BE2030	*Bioeconomy to 2030*
BEB	*Bioeconomy Blueprint*
BRC	Biological Resource Center/Biomedical Research Centres
BU	Boston University
CCRM	Centre for Commercialization of Regenerative Medicine (Canada)
CGHR	Centre for Global Health Research (Canada)
CGTC	Cell and Gene Therapy Catapult
CIRM	California Institute for Regenerative Medicine
CNRE	Kuwait-MIT Center for Natural Resources and Environment
CNRHA	National Assisted Reproduction Committee (Comisión Nacional de Reproducción Humana Asistida) (Spain)
cRBC	Cultured red blood cell

EC	European Commission
ECHR	European Court of Human Rights
EFPIA	European Federation of Pharmaceutical Industries and Associations
EMA	European Medicines Agency
ERA	European Research Area
ETP	European Technology Platform
EUA	European University Association
EU	European Union
FDA	Food and Drug Administration (United States)
FHS	Framingham Heart Study
GE	Genetically engineered
GM	Genetically modified
GMO	Genetically modified organism
IMI	Innovative Medicines Initiative
iPSC	Induced pluripotent stem cell
iPS	Induced pluripotent stem (cell)
IVF	In vitro fertilization
JTI	Joint Technology Initiative
KBBE	Knowledge-based bio-economy
LERU	League of European Research Universities
LPC	Large prospective cohort
LSB	Life Sciences and Biotechnology (strategy)
MDS	Million Death Study
MINCYT	Ministry for Science, Technology and Productive Innovation (Argentina)
NHSBT	National Health Service Blood and Transplant
NHS	National Health Service (United Kingdom)
NIH	National Institutes of Health (United States)
NIHR	National Institute for Health Research (United States)
OECD	Organization for Economic Cooperation and Development
PGD	Pre-implantation genetic diagnosis
PGS	Pre-implantation genetic screening
PPP	Public-private partnership
RBC	Red blood cell
RGI	Office of the Registrar General of India
RM	Regenerative medicine
SEMM	School of Molecular Medicine
SFC	Scottish Funding Council
SMEs	Small–Medium Enterprises
SNBTS	Scottish National Blood Transfusion Service
SRS	Sample Registration System

SRT	Selective reproductive technology
STN	Singapore Tissue Network
STS	Science and Technology Studies
TM	Translational medicine
TRTs	Technologies for reproductive transference
TTI	Transfusion Transmitted Infection
UCB	Umbilical cord blood
UKRMP	UK Regenerative Medicine Platform

List of Figures

List of Tables

CHAPTER 1

Introduction

Vincenzo Pavone and Joanna Goven

A young woman in Barcelona, unemployed for the past two years, waits at a private fertility clinic to be screened as an egg donor; she calculates how far she can make the thousand-Euro payment go. A 10-year-old boy in Kakuma refugee camp reads a letter from his mother, who once thought him dead; now resettled in Australia, she writes that the Australian government requires that they prove their relationship through expensive DNA tests, which might take her years to save for. A scientist in Finland resists pressure to permit an unknown but well-funded researcher access to the collection of painstakingly obtained and documented tissue samples that form the basis of her research. An agricultural contractor in Argentina monitors the international market price of GM soy and considers whether investing in a more advanced harvester would improve his competitiveness.

These brief snapshots reveal moments in the expanding web of relations increasingly being conceptualized as a bioeconomy, or as multiple bioeconomies. While the subject of this book is "bioeconomies", we use

V. Pavone (✉)
Institute of Public Goods and Policies (IPP), Consejo Superior Investigaciones Científicas (CSIC), Madrid, Spain

J. Goven
Kukupa Research, Pigeon Bay, New Zealand

V. Pavone and J. Goven (eds.), *Bioeconomies*,
DOI 10.1007/978-3-319-55651-2_1

the word with considerable discomfort. Bioeconomy (singular or plural) is both polysemic and political. It is a neologism that has been coined more than once; at the same time, as a combination of two already familiar terms—bios or bio-, meaning "life", and economy—its meaning may seem self-evident. These attributes lead to confusion and facilitate rhetorical sleight of hand.

What might bioeconomy "self-evidently" mean? It could mean an economy that functions like a biological organism. It could mean an economy that respects biology or the bios, that is, an economy that is integrated with, and nurtures, webs of life. It could highlight the fact that the management or husbandry of that which is needed to meet material needs (that is, "economy") is intertwined with other forms of life: plants grown or gathered, animals hunted or herded, milked or shorn.

From this it becomes obvious that many previous societies have encompassed bioeconomies. They relied on other forms of life for their own survival. In many indigenous societies, we see an explicit understanding and acknowledgement of this that underpins practices that respect other forms of life. Agriculture is the harnessing of other life forms to human benefit, and some agricultural societies built empires based on agricultural surpluses. But humanity's long reliance on other forms of life for its own survival is not the reason for the sudden popularity of the term "bioeconomy".

Below we discuss the recent rise of the "notoriously elusive" (Styhre and Sundgren 2011) term bioeconomy from the perspective of three conceptually distinguishable coinages. The first two coinages we discuss—the bioeconomy as biotechnological-innovation economy and the bioeconomy as biomass economy—are associated strongly with the policy domain, which includes policy documents as well as policy-focused scholarly work. Among the most important of these documents for shaping discourse around the bioeconomy are bioeconomy *strategies*, which have been emanating from increasing numbers of national governments and supranational entities (e.g., OECD, EU) over approximately the past decade (German Bioeconomy Council 2015a, b). Bioeconomy strategies often combine elements of both coinages, but, in addition, they share a number of other, highly consequential attributes, which we discuss separately. We then move on to the third coinage, which is located in academic debates around theorizing the bioeconomy as a novel form of capitalism.[1]

The Bioeconomy as a Biotechnological-Innovation Economy

In this coinage, the essence of the bioeconomy is biotechnology. The OECD's Bioeconomy to 2030 project has played a major role in developing and disseminating this vision. It has defined the bioeconomy as "a world where biotechnology contributes to a significant share of economic output" (OECD 2009, 22); its emphasis is on the use of present and, especially, future biotechnologies to generate economic growth while also providing solutions to "global challenges" (OECD 2006, 2009). Similarly, according to the 2012 U.S. National Bioeconomy Blueprint, "[a] bioeconomy is one based on the use of research and innovation in the biological sciences to create economic activity and public benefit", building on three key "scientific and technological developments that transformed the practice and potential of biological research: genetic engineering, DNA sequencing, and robotic technologies that perform high throughput molecular operations rapidly and accurately" (The White House 2012, 7). According to the South African Bio-Economy Strategy, the bioeconomy "encompasses biotechnological activities and processes that translate into economic outputs, particularly those with industrial application" (The Republic of South Africa 2013).

The "bio" in this version is drawn from biotechnology, with "biotechnological knowledge" defined as "an understanding of DNA, RNA, proteins and enzymes at the molecular level; of ways to manipulate cells, tissues, organs or whole organisms; and of bioinformatics for analysis of genomes and proteins" (OECD 2009, 23). While government promotion of the economic potential of various biotechnologies predates these bioeconomy strategies, in the latter, biotechnology is given a more comprehensive, programmatic role. The strategies proclaim the enormous promise of biotechnologies across a wide range of sectors, and particularly their alleged ability to generate impressive economic growth while simultaneously addressing a variety of global and/or national "challenges". In this vision, climate change, environmental pollution, biodiversity loss, disease threats, hunger and malnutrition, and resource scarcity can all be addressed, through biotechnology, in a way that simultaneously also boosts national competitiveness and rates of return to capital.

This version of the bioeconomy has been criticized for promoting "an overtly neoliberal model of economic development" (Parry 2007), for "treat[ing] the economic aspects of biotechnology as central while tending to treat risks and ethical issues as secondary concerns" (Hilgartner 2007), and for representing complex socio-political problems as technological challenges solvable through technological innovation alone (Goven and Pavone 2015).

The Bioeconomy as a Biomass Economy

In this coinage, the essence of the bioeconomy is the increased and novel use of biomass, and especially the substitution of biomass for fossil sources of energy and industrial feedstock. This bioeconomy is presented as an inherently more environmentally sustainable economy. The "bio-" in this coinage stems from the matter (biomass, bioresources) to be utilized, rather than the technology applied to it.

The substitution of biomass for fossil fuels and other non-renewable inorganic resources in the production of energy, chemicals, industrial feedstock, and manufactured goods was the focus of one of the first policy documents utilizing the notion of a bioeconomy, "Fostering the Bioeconomic Revolution in Biobased Products and Bioenergy: an Environmental Approach" (2001), produced by the U.S. Biomass Research and Development Board. However, the biomass conception is particularly prevalent in Europe. The White Paper "The European Bioeconomy in 2030" (2011, 4) defines the bioeconomy as "the sustainable production and conversion of biomass into a range of food, health, fibre and industrial products and energy," while according to the European Commission working document on the bioeconomy ("Innovating for Sustainable Growth: A Bioeconomy for Europe") (2012, 16), "[t]he bioeconomy encompasses the production of renewable biological resources and their conversion into food, feed, bio-based products and bioenergy". The Finnish Bioeconomy Strategy defines the bioeconomy as "an economy that relies on renewable natural resources to produce food, energy, products and services" (Bosman and Rotmans 2016, 1). Japan's bioeconomy efforts have also been focused on biomass, for example, through the National Biomass Utilization Plan (Japan Government 2012).

The linking of the biomass economy with sustainability is more explicit in some European national strategies. For example, the German bioeconomy strategy defines the bioeconomy as "the knowledge-based production and utilization of biological resources to provide products, processes and services in all sectors of trade and industry within the framework of a sustainable economic system" (Federal Ministry of

Education and Research 2011). The Swedish strategy states that "[t]he conversion to a bio-based economy means a transition from an economy that to a large extent has been based on fossil fuels to a more resource-efficient economy based on renewable raw materials that are produced through the sustainable use of ecosystem services from land and water" (FORMAS 2012, 9). More specifically, a bioeconomy is based on the "sustainable production of biomass to enable increased use [of biomass]" with the object of "reduc[ing] climate effects and the use of fossil-based raw materials"; and on "an increased added value for biomass materials, concomitant with a reduction in energy consumption and recovery of nutrients and energy as additional end products" (FORMAS 2012, 9).

This conceptualization is also reflected in policy-related scholarly analyses, particularly those focusing on Europe. For example, in their study of the bioeconomy in Europe, de Besi and McCormick (2015, 10462) define the bioeconomy as "an economy in which the basic components of materials, chemicals and energy are derived from renewable biological resources." Similarly, for Pfau et al. (2014, 1223) the bioeconomy is taken to mean "an economy in which all (or most) fossil sources used for various forms of consumption and production are replaced by biomass resources."

This version of the bioeconomy has been criticized for equating the use of renewable biomass with sustainability, with critics pointing out that the increased use of biomass can instead have environmentally destructive (and socially unjust) effects (Asveld et al. 2011; Frow et al. 2009; Smolker 2008; de Besi and McCormick 2015; Sheppard et al. 2011). Interestingly, by adopting the bioeconomy-as-biomass concept, with its implicit or explicit claim to sustainability, the EU has opened itself to the criticism that its strategies for developing a bioeconomy favor economic growth and competitiveness at the expense of sustainability (Birch et al. 2010; Ramcilovic-Suominen and Pülzl 2016).

This has also played out as a struggle for control over the bioeconomy vision in Europe. For example, Levidow et al. (2012) and Schmidt et al. (2012) describe the mobilization of support for an "agro-ecological" approach to sustainability, emphasizing the role of farmers' knowledge and relationships in understanding and facilitating ecological processes to achieve ecologically sound increases in productivity, as opposed to an industry-focused and biotechnology-led approach that develops biomass in ways that are appropriable by large corporate actors. In Finland, the public agency SITRA (the Finnish Innovation Fund) has argued for a "distributed" bioeconomy emphasizing community-based enterprises as well as sustainable business and consumer practices (Luoma et al. 2011).

These approaches recall a much earlier use of the cognate term "bio-economics": in the 1970s, Nicholas Georgescu-Roegen coined the term "bioeconomics" to "make us bear in mind continuously the biological origin of the economic process and thus spotlight the problem of mankind's existence with a limited store of accessible resources, unevenly located and unequally appropriated" (Georgescu-Roegen 2011 [1978], 103). Georgescu-Roegen, considered a progenitor of both ecological economics and de-growth economics, argued that continual growth depends on the continual depletion of non-renewable resources and the continual production of waste. This calls for a new approach to economics, because "[t]he price mechanism cannot prevent bioeconomic catastrophes from happening", and neither can it bring about "the distribution of natural resources in a fairly satisfactory way among successive generations, [nor] even among contemporaneous ones" (Georgescu-Roegen 2011 [1978], 113). Agro-ecology approaches (as well as eco-economy approaches; see below) challenge the notion that simply substituting biomass for fossil resources will address the wider issues raised by Georgescu-Roegen. As a result of such challenges, Bugge et al. (2016) have argued for the existence of a bio-ecology (as distinct from bio-resource) vision of the bioeconomy.

*

While different bioeconomy strategies have assigned different degrees of emphasis to biotechnological innovation and biomass in their definitions of the bioeconomy, they are strikingly similar in the policy prescriptions they produce. These include: (1) increased public investment in science and its infrastructure, and in such a way that its commercialization is encouraged; (2) public support for private commercial actors, including a heavy emphasis on public–private partnerships; (3) ensuring regulation meets the needs of innovators and commercializers; (4) the development of new environmental management metrics and encouragement of private environmental regulation, especially through consumer purchasing choices; (5) government intervention in markets, particularly the creation of markets for new bioproducts; and (6) active government cultivation of public acceptance of bioeconomic activities (EU Commission 2012, 14–21; OECD 2009, 35–94; The White House 2012, 17–41).

Bioeconomy strategies have been a major driver of the rising interest in, and research on, the bioeconomy and of the increasing use of the term. Thus it is important to appreciate what may be embedded in the concept from its origins in bioeconomy strategies. Here we outline some of the key assumptions and implications associated with the bioeconomy as an object of governmental strategy.

Whether defined in terms of biomass or in terms of novel biotechnologies, bioeconomy strategies are strongly characterized by a discourse of present urgency and great future promise (Goven and Pavone 2015). This "high stakes" framing underpins a claimed necessity to re-make expectations, norms, and practices with regard to the roles of government, citizens, science, and the private sector in order to bring the bioeconomy to fruition. For this reason, as we have previously argued (Goven and Pavone 2015), the bioeconomy should be understood as a political project, in which political institutions, rights, and responsibilities are subordinated to the "needs" of a particular innovation regime.

Bioeconomy strategies position the bioeconomy as key to "global competitiveness", while the need for "global competitiveness" is taken for granted. As Fougner notes, international competitiveness has, with the neoliberal turn, been constituted "as a (if not *the*) central means to the resolution of more or less all other problems that the state is confronted with" (2006, 165; emphasis in original). This competitiveness framing then predetermines the preferred shape of, and pathway to, the bioeconomy; success in the bioeconomy race for global competitiveness requires accommodating the interests of large, globally competitive, corporate actors. This includes, but is not limited to, the removal of regulatory "barriers" to innovation. Not coincidentally, the public, too, is present in these strategies primarily as an obstacle to be overcome, through active cultivation of public acceptance of bioeconomy-strategy prescriptions (Goven and Pavone 2015).

By promoting the bioeconomy in such a way that it displaces other approaches that seek to address the actual (structural) causes of the urgent problems, bioeconomy strategies function to protect fundamental and problematic attributes of the political-economic status quo from the environmental and social crises they have generated (Goven and Pavone 2015). The promissory claims made around the bioeconomy's ability to address social and environmental problems can act to deflect or weaken opposition to the underlying causes of those problems (Marsden 2016; Goven and Pavone 2015).

An emerging critical literature on one such alternative, the eco-economy, highlights some of the unstated assumptions of bioeconomy strategies when it comes to the challenge of sustainability (Horlings et al. 2010; Horlings and Marsden 2011, 2014; Kitchen and Marsden 2011; Marsden 2016; Marsden and Farioli 2015). As a response to climate change and resource depletion, the bioeconomy "tries to ameliorate some of the most wasteful aspects of traditional industrial and agri-food processes whilst at the same time continuing to intensify production and supply" (Kitchen and Marsden 2011, 767, note 1). This translates into new product development, typically through the application of a universalizing, reductionist science to re-engineer living organisms, in an attempt to reduce dangerous externalities and minimize resource depletion while simultaneously boosting economic growth and corporate profits. The bioeconomy approach, pre-committed to economic growth and global competitiveness, is divorced from the local; its focus on "place" is largely limited to innovation "clusters". Its growth and competitiveness orientation makes always-local ecosystems largely irrelevant. While the bioeconomy approach prioritizes the technical over the social, the eco-economy takes a decentralized, co-productionist approach to innovation (Luoma et al. 2011; Schmidt et al. 2012). In contrast to the bioeconomy, the eco-economy "actively problematizes and attempts to integrate" the social and the natural (Marsden and Farioli 2015, 339), developing innovations to systems of production and consumption that would re-embed the social in the ecological. While an eco-economy faces significant obstacles, eco-economic approaches are already being enacted, for example: participatory plant breeding, in which "[b]reeding material is tested in farmers' fields at an earlier stage and with participation of groups of farmers; selection is done in each location, and it is location-based in terms of which material, plants and seeds to carry further and re-produce" (Marsden and Farioli 2015, 339); and demonstration eco-villages in Finland, which are "based upon integrated value networks, connecting horticulture, heating, waste treatment and energy production" and which demonstrate "how a more eco-economic, distributed networked system can be put in place, such that local communities can benefit and capture value from the new interconnected and co-produced systems" (Marsden and Farioli 2015, 340).

The Bioeconomy as a Novel Form of Capitalism

The third coinage arises out of the significant recent academic literature that explores the complex articulations between the life sciences and capitalism in areas as diverse as biobanking (Fortun 2008; Martin et al. 2008; Mitchell and Waldby 2010; Parry 2004; Waldby 2009; Waldby and Mitchell 2006); bioprospecting (Hayden 2003); the assisted reproduction platform of technologies (Franklin 2006, 2013; Waldby and Mitchell 2006; Waldby 2008; Waldby and Cooper 2010; Waldby and Cooper 2014); GMOs (Cooper 2008; Delvenne and Hendrickx 2013; Birch et al. 2010); pharmaceuticals (Rajan 2006; Cooper 2008; Cooper 2015); stem cells and regenerative medicine (Salter 2008, 2013; Salter and Salter 2010; Salter et al. 2014; Webster and Faulkner 2015; Hauskeller and Beltrame 2016); venture capital (Rajan 2006; Fortun 2008); and the life sciences in general (Birch 2006, 2012, 2016; Birch and Tyfield 2013).

While typically focusing on the local and global economies being built around biotechnological manipulation of tissues, organisms and bodies, much of this work until recently used neologisms such as biocapital and biovalue, rather than bioeconomy. More recently, the term bioeconomy has been more widely used, and earlier work is now more likely to be described as a theorization of the bioeconomy (Birch and Tyfield 2013; Tutton 2015). While this scholarly literature is too large and diverse to review here, particularly instructive for our purposes are two of its foci: the generation and appropriation of value in this (perhaps) new "biocapitalism"; and the bringing into being of new subjectivities.

Value in the Bioeconomy

The theoretical literature on what is now termed the bioeconomy can be seen to locate value generation and appropriation at one or more of four sites: "life itself" (bodies, tissues); human labor; speculation; and enclosure.

Life Itself

A number of scholars have conceptualized "life itself" as the source of value in the bioeconomy, arguing that the bioeconomy represents a new approach to capitalizing the productive and reproductive capacities of "life". Waldby (2002) coined the term "biovalue" to describe "the surplus of in vitro vitality produced by the biotechnical reformulation of living processes" (Waldby and Mitchell 2006, 32). Rose (2007b, 32) writes that "vitality itself has become a potential source of value" and defines biovalue as "the value to be extracted from the vital properties of living processes". This vitality is "stabilized, frozen, banked, stored, accumulated, exchanged, traded across time, across space, across organs and species, across diverse contexts and enterprises, in the service of bioeconomic objectives" (2007b, 37). Cooper (2008) writes of "life as surplus", Franklin of "biocapital" (2003), and Brown et al. (2011) of the "economization of life". Biological life, it is argued, is being incorporated in new ways into the capitalist accumulation regime, generating a surplus that can be captured and monetized through the technologically mediated commodification of tissues, bodies and biological processes (Cooper 2008; Franklin 2006; Waldby and Mitchell 2006; Waldby 2006). As economic production comes to be situated at the "genetic, microbial, and cellular level… life becomes, literally, annexed within capitalist processes of accumulation" (Cooper 2008, 19).

Labor

Others argue that a focus on technologically mediated life obscures the degree to which human labor is key to the production of value in the bioeconomy. For example, according to Birch and Tyfield (2013, 315) "the idea of biological or natural productivity is problematic since value results from the application of knowledge to nature, and the subjection of that knowledge to intellectual property rights." That is, what they call knowledge labor (together with enclosure, discussed below), rather than "life itself", produces value in the bioeconomy. Birch (2016, 4) seems to be tracing value to labor when he writes that value in the bioeconomy "is constituted primarily by the social practices of the political-economic actors who configure the financial value and valuation of firms." Mitchell and Waldby (2010) and Waldby and Cooper (2014) also trace value to labor, but a different kind of labor. In their work on national (population) biobanks, Mitchell and Waldby (2010, 339) argue that those providing samples and data engage in clinical labor—that is, "embodied biomedical work that produces economic value". While the clinical labor of biobank participants is quite light (but distributed across many people), the clinical labor of those involved in, for example, pharmaceutical trials, oocyte production for sale, and gestational surrogacy can be onerous and risky (Waldby and Cooper 2014). The organization of this labor is distinctly post-Fordist: risk and responsibility are devolved to individual contractors at the vulnerable end of economic inequality, social hierarchies, deindustrialization, and de-unionization. It often goes unacknowledged as labor at all; figured as altruistic donors, those providing reproductive or regenerative tissue are considered neither workers nor owners of that tissue, and therefore have no claim to the profits it may be used to generate.

Speculation

Speculation refers here to two intertwined phenomena: the role of speculative finance in the bioeconomy; and the widespread promissory practices—i.e., speculative claims of likely achievements—in the life sciences. A number of scholars have documented the importance of speculative finance to a wide range of bioeconomic activities (Cooper 2008; Martin 2015; Martin et al. 2008; Petersen and Krisjansen 2015; Rajan 2006). The promissory character of life-sciences research (Brown 2003; Glasner 2007; Hedgecoe 2004) itself is key to this relationship. In the bioeconomy, speculative finance meets promissory bioscience: according to Rajan (2006) it is this relationship that generates "biocapital" and gives rise to a novel form or phase of capitalism. Martin's (2015) analysis of the emerging "neurotechnology" industry has shown how "promissory organizations" such as trade associations and market-research firms are actively involved in constructing expectations of assets, which start-up companies use to secure venture capital in the present. Novas points to the role of patients motivated by these expectations: he posits a new "political economy of hope" where "patients' associations sponsor the value of transforming blood, tissue or DNA into resources for the generation of biovalue" (Novas 2006, 289).

Martin et al. (2008) discuss the "promissory bioeconomy" of private cord-blood banking, arguing that "hope itself is being capitalized as the basis of commodity value": it was only with the creation of the highly promissory field of regenerative medicine, holding out the prospect of a wide range of sometime-in-the-future therapies, that a wave of investment was directed toward commercial cord-blood banking. Speculative projections of advances in stem-cell research producing future cures (and future profits) not only mobilized investors; they also affected parents' feelings of obligation and calculations of prudential investment in private cord-blood accounts for their children. In the field of regenerative medicine, Salter et al. (2014) argue that the promissory practices of stem-cell science have catalyzed demand from health consumers that "has stimulated a global supply of what the Western model regards as 'illicit' therapies" and that has also produced political pressures for changes to the regulation of medical innovation. Thus, speculative claims of likely future research achievements (that can in some way be privately appropriated) can have effects in the present: on investors' willingness to invest, consumers' willingness to purchase, and governments' willingness or ability to regulate.

Enclosure and Assets
The importance of enclosure—that is, the establishment of private property rights—is widely recognized by theorists of the bioeconomy. Enclosure through intellectual property rights is seen as fundamental to processes of appropriating the surplus value generated by "life itself". Enclosure also works alongside labor and speculation in the generation and appropriation of value. Birch and Tyfield (2013) argue that it is enclosure (combined with knowledge labor), rather than either the generative and regenerative properties of living material or speculative claims, that generates value in the bioeconomy. According to Birch and Tyfield, surplus value can come only from assets (in this case, intellectual property): through the revenue streams (rents) generated by assets or the proceeds of trading the assets themselves.

*

There is considerable talking-past-each-other in these debates. "Value" is understood in a variety of ways and often goes undefined: economic value is interpreted variously, and non-economic forms of value are also incorporated by some but not others. For example, Rose and Novas (2005) posit three dimensions of biovalue: economic value, the value accorded to the enhancement of health, and ethical values. In Rajan's account, economic value is co-produced with moral and political value (Rajan 2006, 2012).

Birch and Tyfield (2013), however, object to what they see as the conflation of economic *value* and social *values* by Rose, Cooper, Rajan, and Waldby and instead focus on economic value as conceptualized within a Marxian labor theory of value. With the growing influence of valuation studies, others refuse that distinction, treating value more as a verb than a noun and seeing value, in whatever form (market, ethical, academic, epistemic, medical, etc.) as emerging from the practices and communities making up different bioeconomic fields (Frow 2015).

Differences also arise from focusing on different bioeconomic sectors. For example, Birch (2016) claims that "there is an overall dearth of products and services in the life sciences sector, with most firms never developing a marketed product" and bases his analysis on such firms, while others (Martin 2015; Martin et al. 2008; Hauskeller and Beltrame 2016; Petrini 2014) examine sectors where well-established companies are already offering a service (such as fertility clinics or cord-blood banks). Hauskeller and Beltrame's (2016) study of cord-blood banking highlighted the importance of non-enclosure (i.e., the failure to patent umbilical cord blood stem cells) to the industry's ability to generate profits through the provision of cord-blood banking services. In their study of global public-private pharmaceutical consortia, Lezaun and Montgomery (2015) show how intellectual property rights can be used, not as an asset that generates an income stream, but as a "ticket of admission" (for the owner) or an "incentive for entry" (for non-owners) to a network of cooperation that may or may not produce profits down the road; but any profits produced will not be based on rents from that patent.

Still other differences may arise from emphasizing different types of firms. Contra Birch and Tyfield's (2013) emphasis on assets as a determinant of value, Martin (2015, 427) argues that "for a key group of firms… assets are neither readily tradable nor can they be valued objectively. Instead, expectations of the future worth of particular innovations provide the foundation for the creation of 'promissory assets'".

Subjectivities and the Bioeconomy

Many scholars view the bioeconomy as productive of new subjectivities. Concepts such as "genetic citizenship" (Heath et al. 2007), "biosociality" (Rabinow 1999), and "biological citizenship" have been used to highlight the ways in which new predictive and diagnostic biomedical technologies are reshaping notions of self, kin, and community. In part, these concepts highlight the ways in which those suffering from or at risk of particular disorders, or their caregivers, form new communities of self-help and political and medical activism. In part, they highlight changes to conceptions of self and kin as a result of new types of knowledge, such as that of alleles indicating "genetic susceptibility". In part, they highlight the incorporation of new biomedical knowledge and technology into a neoliberal "regime of the self", in which the responsible individual prudently calculates and manages individual risk and "actively shap[es] his or her life course through acts of choice" (Rose and Novas 2005, 443, see also Novas and Rose 2000).

Importantly, while genetic knowledge becomes a way of individualizing responsibility, it also, it is argued, forms the basis of new forms of community and new discourses of rights (Heath et al. 2007; Novas 2006; Rose and Novas 2005; Rose 2007a, b). New patient associations and advocacy groups have organized around this knowledge, which constitutes communities (and identities) not along more traditional divisions of race or ethnicity, class, or nation, but rather on commonalities at the molecular level (a shared genetic identity and destiny) as well as common bodily experience of disease. Such organizations, sometimes constituted transnationally, have enabled members to pool knowledge and self-help strategies, challenge medical protocols, and lobby for extensions of rights and benefits from the state (see Heinemann 2016 for an overview of this literature).

Often utilizing a neo-Marxist perspective and drawing also from feminism, gender studies, and cultural studies, other authors link the emergence of new subjectivities with lived experiences of new biomedical treatments and practices, but emphasize the role of clinical labor. For example, Waldby and Cooper (2014) highlight the mutual constitution of the clinical labor performed by donors, surrogates, and clinical-trial participants and their understanding of their own identity and role in society. Sara Franklin (2013) also points to processes of mutual constitution. Asking what, beside babies, is (re)produced through ARTs, she argues that assisted reproductive technologies are reproducing existing family models and imaginaries, social categories, cultural patterns, and economic arrangements, while at the same enabling the emergence of new ones (Franklin 2013).

*

We have outlined what we've called different coinages of the term bioeconomy in an effort to mitigate confusion around the meaning of the term: it has multiple meanings, and they come with different assumptions and implications. The first two, both strongly associated with bioeconomy strategies, are much more closely related than either is to the third. However, given the challenges that the "sustainable biomass" framing has made possible, and the resulting explorations of the bioeconomy concept, there is reason to maintain a conceptual distinction between the two. Work encompassed by the third coinage investigates the kinds of activities that bioeconomy strategies seek to promote, developing a critical and conceptual apparatus almost entirely missing from the first two.

In this volume, we present a set of empirical studies of how different elements of the bioeconomy project play out in particular settings. They have been selected to highlight both key aspects of the bioeconomy project (institutional transformation; conflict, struggle, and resistance) and key disagreements in the academic literature (production and appropriation of value; subjectivities). Some chapters engage with the kinds of sites that have been central to the theoretical discussion (e.g., assisted reproduction, biobanking), while others draw our attention to relatively neglected corners of the bioeconomic landscape, such as immigration law and agricultural finance.

The first section of the book focuses on institutional transformation, a key element of the bioeconomy as political project. Here we ask: In what ways are bioeconomic promises and expectations performative? How can the bioeconomy as promissory construct bring about institutional transformations in the present? What do these transformations look like? What challenges do they face? One of the areas of practice most immediately, and potentially most significantly, affected is research. In their chapter, John Gardner, Andrew Webster, and James Mittra use Mazzucatto's concept of the Entrepreneurial State to elucidate how state initiatives in the UK are remaking research practices and institutions in the field of regenerative medicine in the name of "health and wealth". The European Innovative Medicines Initiative, also premised on the promise of "health and wealth", is shown by Luca Marelli and Giuseppe Testa to be co-productive with the making and legitimating of "Europe" itself. Beyond research, the bioeconomy project also remakes governance: through the lens of the AquaAdvantage GM salmon case, Tess Doezema and Ben Hurlbut show how the assumptions and imperatives of the bioeconomy project are reshaping the roles and responsibilities of citizens, scientists, and the state.

The second section of the book examines questions of value, specifically: how is value generated, appropriated, attributed, and distributed in the bioeconomies? Luisa Reis Castro shows how a project that aims to extract information and new forms of value from sewage raises questions about what it means to reconstitute a community in terms of its "collective microbiome". In his study of the Spanish reprogenetics sector, in which clinical labor and financialization are both relatively insignificant, Vincenzo Pavone uses the conceptual apparatus of bio-identification to show how value is produced and appropriated, placing emphasis on the sociotechnical processes constituting the value chain of the reprogenetics economy. Erik Aarden develops the metaphor of a social contract to explore the social and political dimensions of biobanks in three different national settings, demonstrating that they both generate value and impose values.

The third section of the book asks: What kinds of conflicts, struggles, and resistance are intertwined with or generated by bioeconomies? How are bioeconomies implicated in the constitution of geopolitical, legal, and political-economic power and marginalization? In his study of European population biobanks, Aaro Tupasela highlights the ways in which scientists resist the expectations placed upon them by the promissory bioeconomy as they struggle to negotiate the contradictory logics embedded in biobanks. In Sonja van Wichelen's comparative study of international surrogacy and migration regimes, we see how bioeconomies are growing up around, and enabling, inequitable formations of legal kinship and legitimate families. Pierre Delvenne's study of the development of the Argentinean soy bioeconomy shows how what appears to be a technoscientific and political-economic transformation is simultaneously a consolidation of political power through the displacement and denial of conflict.

The fourth section of the book examines the complex interplay of structural inequalities with altruistic acts and modes of being—such as donation and the often invisible practices of care—within a range of bioeconomies: assisted reproduction, stem cells, pharmaceutical development, and cultured red blood cells. More specifically, they address the following questions: how do the ordinary, the invisible, and the banal underpin the promises and practices of bioeconomic innovation, and how does this link bioeconomies to inequality? How are kin and care relationships framed, experienced, or exploited in globalized bioeconomies? Do these new relations change our understanding of these categories? In her study of Spanish fertility clinics, Sara Lafuente Funes shows how practices of egg donation depend upon and mobilize social inequalities and constructions of gender. Pablo Santoro and Carmen Romero-Bachiller explore, through case studies of private cord-blood banking and mastitis, the ambivalent positioning of parents with regard to bioeconomies of parenting, childbirth, and child health and the ways in which these bioeconomies rely on the activation of care. Rounding out this section, Emma King presents an analysis of attitudes toward the replacement of existing blood-donation systems with commercially produced cultured red blood cells (cRBCs), in which the acceptability of the cRBC product is shown to be inextricable from the social relations in which it is embedded.

Note

1. For other, overlapping, attempts to trace distinct bioeconomy concepts, see Bugge et al. (2016) and Staffas et al. (2013).

References

Asveld, L., Van Est, R., & Stemerding, D. (2011). Getting to the core of the bio-economy. A perspective on the sustainable promise of biomass.

Biomass Research and Development Board. (2001). *Fostering the bioeconomic revolution in biobased products and bioenergy*. Golden, CO: National Renewable Energy Laboratory.

Birch, K. (2006). The neoliberal underpinnings of the bioeconomy: The ideological discourses and practices of economic competitiveness. *Genomics, Society and Policy, 2,* 1–15.

Birch, K. (2012). Knowledge, place, and power: Geographies of value in the bioeconomy. *New Genetics and Society, 31,* 183–201.

Birch, K. (2016). Rethinking value in the bio-economy finance, assetization, and the management of value. *Science, Technology & Human Values*. 0162243916661633.

Birch, K., Levidow, L., & Papaioannou, T. (2010). Sustainable capital? The neoliberalization of nature and knowledge in the European "knowledge-based bio-economy". *Sustainability, 2,* 2898–2918.

Birch, K., & Tyfield, D. (2013). Theorizing the bioeconomy: Biovalue, biocapital, bioeconomics or… what? *Science, Technology and Human Values, 38,* 299–327.

Bosman, R., & Rotmans, J. (2016). Transition governance towards a bioeconomy: A comparison of Finland and the Netherlands. *Sustainability, 8,* 1017.

Brown, N. (2003). Hope against hype-accountability in biopasts, presents and futures. *Science & Technology Studies, 28,* 3–21.

Brown, N., Machin, L., & McLeod, D. (2011). Immunitary bioeconomy: The economisation of life in the international cord blood market. *Social Science and Medicine, 72,* 1115–1122.

Bugge, M. M., Hansen, T., & Klitkou, A. (2016). What is the bioeconomy? *A Review of the Literature. Sustainability, 8,* 691.

Cooper, M. (2008). *Life as surplus: Biotechnology and capitalism in the neoliberal era*. Seattle: University of Washington Press.

Cooper, M. (2015). In vivo economies: Temporalities of life and value. *Journal of cultural Economy, 8*(3), 257–259.

de Besi, M., & McCormick, K. (2015). Towards a bioeconomy in Europe: National, regional and industrial strategies. *Sustainability, 7,* 10461–10478.

Delvenne, P., & Hendrickx, K. (2013). The multifaceted struggle for power in the bioeconomy: Introduction to the special issue. *Technology in Society, 35,* 75–78.

EU Commission. (2011). *The European bioeconomy in 2030-Delivering sustainable growth by addressing the grand societal challenges*. Brussels: EU Press.

EU Commission. (2012). *Innovating for sustainable growth: A bioeconomy for Europe*. Brussels: EU press.
Federal Ministry of Education and Research. (2011). *National research strategy bioeconomy 2030: Our route towards a biobased economy*. Berlin: Bundesministerium für Bildung und Forschung.
Formas. (2012). *Swedish research and innovation strategy for a bio-based economy*. Stockholm: Edita AB.
Fortun, M. (2008). *Promising genomics: Iceland and deCODE genetics in a world of speculation*. Berkeley: University of California Press.
Fougner, T. (2006). The state, international competitiveness and neoliberal globalisation: Is there a future beyond 'the competition state'? *Review of International Studies, 32*, 165–185.
Franklin, S. (2003). Ethical biocapital: New strategies of cell culture. In S. Franklin & M. Lock (Eds.), *Remaking life and death: Towards an anthropology of biomedicine*. Santa Fe, NM: School of American Research Press.
Franklin, S. (2006). Embryonic economies: The double reproductive value of stem cells. *BioSocieties, 1*, 71–90.
Franklin, S. B. (2013). *Biological relatives: IVF, stem cells, and the future of kinship*. Durham: Duke University Press.
Frow, E. (2015). 12 Rhetorics and practices of democratization in synthetic biology. *Knowing New Biotechnologies: Social Aspects of Technological Convergence, 174.*
Frow, E., Ingram, D., Powell, W., Steer, D., Vogel, J., & Yearley, S. (2009). The politics of plants. *Food Security, 1*, 17–23.
Georgescu-Roegen, N. (2011). *From bioeconomics to degrowth: Georgescu-Roegen's 'new economics' in eight essays*. New York: Routledge.
German Bioeconomy Council. (2015a). Bioeconomy policy: Synopsis and analysis of strategies in the G7. Berlin: Office of the Bioeconomy Council.
German Bioeconomy Council. (2015b). *Bioeconomy policy: Synopsis of national strategies around the world*. Berlin: Office of the Bioeconomy Council.
Glasner, P. (2007). Cowboy cloners, mavericks and kings: A cautionary tale of a promissory science. *Twenty-First Century Society, 2*, 265–274.
Goven, J., & Pavone, V. (2015). The bioeconomy as political project a polanyian analysis. *Science, Technology and Human Values, 40*, 302–337.
Hauskeller, C., & Beltrame, L. (2016). The hybrid bioeconomy of umbilical cord blood banking: Re-examining the narrative of opposition between public and private services. *BioSocieties, 11*(4), (415–434).
Hayden, C. (2003). *When Nature Goes Public: The Making and Unmaking of Bioprospecting in Mexico*. Princeton: Princeton University Press.
Heath, D., Rapp, R., & Taussig, K.-S. (2007). Genetic citizenship. *A Companion to the Anthropology of Politics, 2*, 152.
Hedgecoe, A. (2004). *The politics of personalised medicine: Pharmacogenetics in the clinic*. Cambridge: Cambridge University Press.
Heinemann, T. (2016). Biological citizenship. In Henk ten Have (Ed.), *Encyclopedia of Global Bioethics*, (pp. 347–353). Cham: Springer International Publishing.

Hilgartner, S. (2007). Making the bioeconomy measurable: Politics of an emerging anticipatory machinery. *BioSocieties, 2*, 382–386.

Horlings, L. G., Kitchen, L. C., Marsden, T. K., & Bristow, G. I. (2010). *Exploring the potential contributions of the bio-economy and the eco-economy to agri-food and rural regional development.* BRASS: Centre for Business Relationships, Accountability, Sustainability and Society. Working Paper Series. No. 60. Cardiff University.

Horlings, L. G., & Marsden, T. K. (2011). Towards the real green revolution? Exploring the conceptual dimensions of a new ecological modernisation of agriculture that could 'feed the world'. *Global Environmental Change, 21*, 441–452.

Horlings, L. G., & Marsden, T. K. (2014). Exploring the 'new rural paradigm' in Europe: Eco-economic strategies as a counterforce to the global competitiveness agenda. *European Urban and Regional Studies, 21*, 4–20.

Japan Government. (2012). *Biomass industrialization strategy.* Tokyo: Governmental Press.

Kitchen, L., & Marsden, T. (2011). Constructing sustainable communities: A theoretical exploration of the bio-economy and eco-economy paradigms. *Local Environment, 16*, 753–769.

Levidow, L., Birch, K., & Papaioannou, T. (2012). Divergent paradigms of European agro-food innovation: The Knowledge-Based Bio-Economy (KBBE) as an R&D agenda. *Science, Technology & Human Values, 38*(1), 94–125.

Lezaun, J., & Montgomery, C. M. (2015). The pharmaceutical commons: Sharing and exclusion in global health drug development. *Science, Technology and Human Values, 40*, 3–29.

Luoma, P., Vanhanen, J., & Tommila, P. (2011). *Distributed bio-based economy—driving sustainable growth* (p. 24). Helsinki: Finnish Innovation Fund (SITRA).

Marsden, T. (2016). Exploring the rural eco-economy: Beyond neoliberalism. *Sociologia Ruralis, 56*, 597–615.

Marsden, T., & Farioli, F. (2015). Natural powers: From the bio-economy to the eco-economy and sustainable place-making. *Sustainability Science, 10*, 331–344.

Martin, P. (2015). Commercialising neurofutures: Promissory economies, value creation and the making of a new industry. *BioSocieties, 10*, 422–443.

Martin, P., Brown, N., & Turner, A. (2008). Capitalizing hope: The commercial development of umbilical cord blood stem cell banking. *New Genetics and Society, 27*, 127–143.

Mitchell, R., & Waldby, C. (2010). National biobanks: Clinical labor, risk production, and the creation of biovalue. *Science, Technology and Human Values, 35*, 330–355.

Novas, C. (2006). The political economy of hope: Patients' organizations, science and biovalue. *BioSocieties, 1*, 289–305.

Novas, C., & Rose, N. (2000). Genetic risk and the birth of the somatic individual. *Economy and Society, 29,* 485–513.
OECD. (2006). *Scoping document: The bioeconomy to 2030: Designing a policy agenda.* Paris: OECD.
OECD. (2009). *The bioeconomy to 2030: Designing a policy agenda.* Paris: OECD.
Parry, B. (2004). *Trading the genome: Investigating the commodification of bio-information.* New York: Columbia University Press.
Parry, B. (2007). Cornering the futures market in 'Bio-epistemology'. *Biosocieties, 2*(3), 386.
Petersen, A., & Krisjansen, I. (2015). Assembling 'the bioeconomy': Exploiting the power of the promissory life sciences. *Journal of Sociology, 51,* 28–46.
Petrini, C. (2014). Umbilical cord blood banking: From personal donation to international public registries to global bioeconomy. *Journal of Blood Medicine, 5,* 87–97.
Pfau, S. F., Hagens, J. E., Dankbaar, B., & Smits, A. J. (2014). Visions of sustainability in bioeconomy research. *Sustainability, 6,* 1222–1249.
Rabinow, P. (1999). From sociobiology to biosociality. In M. Biagioli (Ed.), *The science studies reader.* New York: Routledge.
Rajan, K. S. (2006). *Biocapital: The constitution of postgenomic life.* Durham: Duke University Press.
Rajan, K. S. (2012). *Lively Capital: Biotechnologies, Ethics, and Governance in Global Markets.*Durham: Duke University Press.
Ramcilovic-Suominen, S., & Pülzl, H. (2016). Sustainable development–A 'selling point' of the emerging EU bioeconomy policy framework? *Journal of Cleaner Production.* Retrieved December 30, 2016 from http://www.sciencedirect.com/science/article/pii/S095965261632220X
Rose, N., & Novas, C. (2005). Biological citizenship. In A. Ong & S. J. Collier (Eds.), *Global assemblages: Technology, politics, and ethics as anthropological problems* (pp. 439–463). Malden, MA: Wiley-Blackwell.
Rose, N. (2007a). Molecular biopolitics, somatic ethics and the spirit of biocapital. *Social Theory and Health, 5,* 3–29.
Rose, N. (2007b). *The politics of life itself: Biomedicine, power and subjectivity in the 21st century.* Princeton, NJ: Princeton University Press.
Salter, B. (2008). Governing stem cell science in China and India: Emerging economies and the global politics of innovation. *New Genetics and Society, 27,* 145–159.
Salter, B. (2013). Governing innovation paths in regenerative medicine: The European and global struggle for political advantage. *The Global Dynamics of Regenerative Medicine: A Social Science Critique,* 194.
Salter, B., & Salter, C. (2010). Governing innovation in the biomedicine knowledge economy: Stem cell science in the USA. *Science and Public Policy, 37,* 87–100.

Salter, B., Zhou, Y., & Datta, S. (2014). Health consumers and stem cell therapy innovation: Markets, models and regulation. *Regenerative Medicine, 9*, 353–366.

Schmidt, O., Padel, S., & Levidow, L. (2012). The bio-economy concept and knowledge base in a public goods and farmer perspective. *Bio-based and Applied Economics, 1*, 47–63.

Sheppard, A. W., Gillespie, I., Hirsch, M., & Begley, C. (2011). Biosecurity and sustainability within the growing global bioeconomy. *Current Opinion in Environmental Sustainability, 3*, 4–10.

Smolker, R. (2008). The new bioeconomy and the future of agriculture. *Development, 51*, 519–526.

Staffas, L., Gustavsson, M., & McCormick, K. (2013). Strategies and policies for the bioeconomy and bio-based economy: An analysis of official national approaches. *Sustainability, 5*, 2751–2769.

Styhre, A., & Sundgren, M. (2011). *Venturing into the bioeconomy: Professions, innovation, identity*. Basingstoke: Palgrave Macmillan.

The Republic of South Africa. (2013). Bio-economy strategy. In D. O. S. A. (Ed.), *Technology*. South Africa: Department of Science and Technology (DST).

The White House (2012, April). National bioeconomy blueprint. *Industrial Biotechnology, 8*, 97–102.

Tutton, R. (2015). *Book review: Melinda Cooper and Catherine Waldby, clinical labor: Tissue donors and research subjects in the global bioeconomy*. London: Sage.

Waldby, C. (2002). Stem cells, tissue cultures and the production of biovalue. *Health, 6*, 305.

Waldby, C. (2006). Umbilical cord blood: From social gift to venture capital. *BioSocieties, 1*, 55–70.

Waldby, C. (2008). Oocyte markets: Women's reproductive work in embryonic stem cell research. *New Genetics and Society, 27*, 19–31.

Waldby, C. (2009). Biobanking in Singapore: Post-developmental state, experimental population. *New Genetics and Society, 28*, 253–265.

Waldby, C., & Cooper, M. (2010). From reproductive work to regenerative labour. *Feminist Theory, 11*, 3–22.

Waldby, C., & Cooper, M. (2014). *Clinical labor: Tissue donors and research subjects in the global bioeconomy*. Durham: Duke University Press.

Waldby, C., & Mitchell, R. (2006). *Tissue economies: Blood, organs, and cell lines in late capitalism*. Durham: Duke University Press.

Webster, A., & Faulkner, A. (2015). Governing regenerative medicine. *The palgrave international handbook of healthcare policy and governance*. Basingstoke: Palgrave Macmillan.

Authors' Biography

Vincenzo Pavone is tenured research fellow the Institute of Public Goods and Policies (IPP) of the Spanish National Research Council (CSIC). His research activities address the relationship between neoliberalism and emerging bioeconomies, with a special focus on reproductive bioeconomies.

Joanna Goven is currently the Director of Kukupa Research, Deputy Director of the Centre for Integrated Research in Biosafety, Evaluation Director for the Engagement Programme of the Deep South National Science Challenge, and a researcher with the Centre for Integrated Biowaste Research (New Zealand). She has published on gender politics, antifeminism, transdisciplinarity, and the politics and political economy of science and technology.

PART I

Promissory and Performative: Remaking Institutions for the Bioeconomy

CHAPTER 2

The "Entrepreneurial State" and the Leveraging of Life in the Field of Regenerative Medicine

John Gardner, Andrew Webster, and James Mittra

Introduction: The Entrepreneurial State and Translational Medicine

Innovation has become the linchpin of industrial policy in advanced economies. It is seen as an important generator of economic growth; a mechanism for enhancing national strength through job and wealth creation. This imperative to innovate is particularly prominent in the biomedical and healthcare sectors. Biomedical innovation is being

J. Gardner (✉)
School of Social Sciences, Monash University, Melbourne, Australia

A. Webster
Department of Sociology, University of York, York, UK

J. Mittra
Science, Technology and Innovation Studies, University of Edinburgh, Edinburgh, UK

V. Pavone and J. Goven (eds.), *Bioeconomies*,
DOI 10.1007/978-3-319-55651-2_2

championed as both a source of novel treatments to address unmet clinical need and a new source of economic growth. In particular, advances in biology, such as stem cell biology, tissue engineering, gene-editing and synthetic biology, are being lauded for their potential to generate both health and wealth.

In many countries the State is seen as having an important role in facilitating biomedical innovation. However, rather than sponsoring specific innovation projects or commercial endeavours (that is, "picking winners"), state involvement today entails a more general, facilitative approach towards innovation. The economist Marianna Mazzucato (2015) has suggested that this is indicative of what she describes as the *Entrepreneurial State*, one that supports wealth-generating innovation by funding knowledge-production in labs and universities, mobilising the resources that allow knowledge and innovations to diffuse broadly, and developing strategies for technological advances in priority areas (2015, 39–40). Mazzucato calls for a reconceptualization of the roles of the State and the private sector in innovation: far from being a bureaucratic drag on innovation, she argues, the State has a track record of investing in key, innovation-underpinning research, particularly at the high-risk early stages before the private sector is willing to invest. The failure to recognize the State's entrepreneurial activities facilitates current inequitable practices whereby risk is socialized but rewards are privatized.

In the biomedical and healthcare sectors, this Entrepreneurial State activity is reflected in the Translational Medicine (TM) policy agendas of the US, Canada, Europe and the UK. TM policy aims to tackle a perceived set of challenges or barriers to the translation of basic science research into clinical therapies at the bedside (Mittra and Milne 2013). Promising new developments in biology (e.g. stem cell biology, tissue engineering, synthetic biology) are seen to have additional "translational" demands precisely because of their scientific, technical, regulatory and clinical complexities, which are far removed from traditional pharmaceutical drug innovation that has predominated in the healthcare ecosystem. TM policy has been designed to overcome the so-called "broken middle" or "valley of death" in drug development, which refers to the gap between basic science and initial invention, and downstream commercialisation and clinical utility (Mittra 2016). Various Government-supported initiatives have been launched, including: interdisciplinary research programmes that aim to foster collaborations between clinicians, researchers, and industry; the formation of research infrastructures to facilitate the gathering and exchange of data; mechanisms for securing intellectual property rights; and adjustments to governing systems such as regulatory frameworks. Hence, driven by promissory visions of health and wealth, the TM policy agenda has become an organising principle for reconfiguring the biomedical research landscape and healthcare sectors. New alliances are being forged, new collectives are emerging, and existing ones are continually being adapted. These changes characterize what has been called the "new health bioeconomy" (Mittra 2016).

In this chapter, we critically explore the reconfiguring of the biomedical research landscape and the healthcare sector in relation to the emerging field of regenerative medicine (RM) within the United Kingdom. RM—a field that is emerging from new developments in biology—has generated high expectations about its future "health and wealth" potential, and many governments have incorporated the field within their industrial policies. It thus represents a rich case study for examining the "Entrepreneurial State" and its role in shaping an emerging health bioeconomy. We examine some of the major, state-supported supply-side measures that are being implemented within the field of RM. New organizational forms are taking shape and existing infrastructures are being adapted within a collective initiative to establish a RM industry for "health and wealth". We illustrate that this entails: the establishment of "business-focused" innovation accelerator agencies; the promotion of interdisciplinarity; the reconfiguring of governance mechanisms, and the repurposing of the healthcare system as an innovation asset. Building on recent social science studies of value practices in the life sciences and medicine (e.g. Dussauge et al. 2015), we examine the values that are invoked within this collective initiative. In so doing, we argue that the power of the Entrepreneurial State in reconfiguring the bioeconomy derives from its capacity to appeal to diverse values, and consequently, mobilise and orientate actors into a common political project aimed at creating both health and wealth. We also highlight some of the tensions and countervailing processes at play in this process.

Methods

This chapter draws on both primary and secondary data collected as part of the UK ESRC-funded "REGenableMED" project, which is exploring the social dynamics of innovation in RM. Data include over 80 interviews with a range of stakeholders and practitioners in the field of RM, including stem cell scientists, clinicians, regulators, consultants, public officials, and representatives of commercial organizations, patient associations, and charities, many of whom are directly involved in the initiatives that form the focus of this chapter. Secondary data include publicly available government reports into RM, public agency meeting minutes, clinical trials databases, and company reports.

Our conceptual framework draws on recent work on value-practices in the life sciences (Dussauge et al. 2015) to explore the values that are appealed to and enacted within specific initiatives aimed at facilitating RM. Historically, the distinction has been made between value as it is understood economically (reflected in, for instance, the price of goods), and value(s) as they are understood sociologically (that is, shared standards or assumptions about what is important). However, we wish to avoid making an a priori distinction between economic value and other types or regimes of value, and instead focus on how actors articulate, both implicitly and explicitly, what they count as desirable or worth caring for or "knowing". As Dussauge et al. rightly note (2015, 10), a multiplicity of values may be enacted—both implicitly and explicitly—within a common project; these values may align or may be in constant tension or flux. This problematizes any attempt to strictly delineate the economic and the social components of value. In our analysis, we demonstrate that certain values relating to commercialization and global competitiveness feature heavily in many government-supported initiatives, but also that other values are invoked alongside these, reflecting the diverse communities of expertise and different practices that are being mobilised.

Regenerative Medicine and the Leveraging of Life

RM represents a heterogeneous collection of emerging technologies, techniques and practices. What counts as RM has varied somewhat over the last decade (Webster 2013), but usually it is defined as that which "replaces or regenerates human cells, tissues and organs, to restore or establish normal function" (Mason and Dunnill 2007, 4). In general terms it refers to the use of tissues, cells (embryonic stem cells, induced pluripotent stem cells, and adult stem cells) or genes to treat or manage illness and disease, and for this reason it is considered to be distinct from more conventional drug or device-based therapies. Advocates proclaim that the regenerative capacity of cells, tissues and genes means that the field has the potential to produce curative treatments for a range of conditions for which there is currently unmet clinical need, including cancers as well as cardiovascular, neurological, and autoimmune diseases. In addition to the promissory expectations about its clinical value, RM has become the focus of considerable optimism about its economic potential. Policymakers and industrialists, in particular, believe that RM may precipitate the creation of a new wealth-generating industry. In several countries, RM has become entwined with state initiatives aimed at generating a high-wealth, knowledge-based economy. Japan, for example, is aiming to take advantage of its strengths in stem cell science and has positioned RM as one of the pillars of its economic growth strategy (Ogawa 2015). Canada and several US states—particularly California—have made similar moves, and in the UK, RM has been named as one of the "Eight Great Technologies", which will propel the UK to future growth (Willetts 2013).

Currently, there are hundreds of clinical trials worldwide for RM therapies, mostly at clinical phases I and II. Although a few RM products have received regulatory approval in some jurisdictions, none have been widely and routinely adopted within healthcare systems. The field, then, is still in its infancy. Indeed, the complexity of cell, tissue and gene-based therapies presents a range of scientific, technical, regulatory and reimbursement challenges for investigators and manufacturers working within the field, which raises concerns that the translation of promising scientific developments into useful health- and wealth-generating RM therapies will be laboriously slow (Gardner et al. 2015). Producing RM therapies is often labour-intensive, and as yet there is little in the way of automated manufacturing platforms or established standards (Tait and Banda 2016) that would enable cost-effective scale-up of production and successful delivery to the clinic. The cost of maintaining appropriately licensed manufacturing facilities is high, and investigators are having to develop new assays for quality, safety and potency in order to meet regulatory requirements. Many promising RM therapies are likely to be, at least initially, high-cost, and this may create challenges for securing reimbursement, particularly for novel therapies for which there is little associated data on their cost-effectiveness and long-term clinical benefits (Malik 2014, 2016). RM therapies may also require specific skill-sets and supporting infrastructure that may be difficult to integrate with workflows in existing clinical settings (RMEG 2014).

It is in response to these challenges that several governments have launched strategic initiatives to facilitate the translation of RM therapies. These generally entail the establishment of innovation "accelerator" agencies (such as the Canadian-government-supported Centre for the Commercialization of Regenerative Medicine and the Californian Institute for Regenerative Medicine), funding initiatives to support research and to assess and adapt existing governance structures, notably regulatory frameworks and health technology assessment methodologies. In the UK, which serves as useful exemplar for this type of activity, the Government established the Cell and Gene Therapy Catapult innovation agency, funded the formation of an interdisciplinary collaboration—the UK Regenerative Medicine Platform—to tackle particular scientific and technical challenges, and launched a review of current governance structures, with the aim of assessing and potentially introducing "accelerated access" to innovative clinical treatments. Additionally, the National Health Service (NHS) is being encouraged to become a "more innovative" adopter of these new technologies, and a variety of activities have been instigated in an attempt to bring this about. Each of these can be seen as mobilising particular communities of expertise: groups of experts with skill-sets and professional backgrounds deemed to be "innovation-facilitative".

More broadly, these developments reflect a supply-side, neo-liberal strategy, as opposed to the demand-side Keynesian approach, that underpins the development of new products to the market—in effect the socialization of economic costs and risk. In this context, the State essentially intervenes in funding the early stage development of expensive new therapies (when the risk of failure is high), and then must purchase these very expensive therapies, if they make it to the clinic. State intervention, therefore, is aimed at facilitating the development of privately-owned products. In effect, public investment is converted to private profit.

In the remainder of this chapter we examine these four key state-sponsored activities. We describe the rationales that have been used to justify them and their socio-technical networks. In particular, we consider how the Entrepreneurial State is implicated in the emergence of the "new health bioeconomy", and discuss the invoking of values that relate to profitability, innovativeness, cost-effectiveness, public good, and patient-centredness, which serve to legitimate this process and drive the commercial and policy value of RM. However, such legitimacy depends on the interplay of different actors whose localised practices of valuation work in different ways to prepare RM products for the market.

The Commercialisation Imperative: The Cell and Gene Therapy Catapult

Over the last 5 years consecutive inquiries have sought to comprehensively assess the UK's capacity to support a RM industry, and this has driven the launch of particular innovation initiatives (RMEG 2014; HoL 2013; MRC 2012; BIS 2011). Generally, these inquiries emphasize the UK's significant potential to become a "world leader" in RM due to its world-class universities and hospitals, and its ability to attract and retain highly-skilled scientists and engineers. Realising this potential, the reports argue, requires a coherent, government-supported strategy to align existing infrastructures and create new collaborations to solve key scientific and technical challenges.

The inquiries reaffirm the importance of biomedical innovation in generating health and wealth, and they tend to advocate a particular *mode of innovation* that explicitly entails *commercialization*. That is, through the development of products and securing of intellectual property, it is expected that publicly funded research will eventually attract sufficient private investment, ideally from large multinational companies, to take the product to market. This commercialization imperative is most obviously reflected in the establishment of the Cell and Gene Therapy Catapult (CGTC).

The Cell and Gene Therapy Catapult was established in 2012 by a grant from the Technology Strategy Board (now Innovate UK), a public agency that reports to the Department for Business, Innovation and Skills (now part of the new Department for Business, Energy and Industrial Strategy). It is based in London, currently has approximately 100 staff, and its stated purpose is to: "Lead the UK cell therapy industry to create health and wealth from the UK's outstanding science foundation" (CGTC 2015). It is expected that long-term funding will be maintained by continued public investment, competitively-won contracts with businesses, and through its collaboration in both private and publicly-funded research projects. As with equivalent agencies in Canada (CCRM) and California (CIRM), the CGTC is an autonomous body with its own decision-making capacity. Its board of directors and management and advisory teams include extensive experience in industry, reflecting a strong emphasis on commercialization. The members of the management team, for example, have professional backgrounds in the life sciences industries, and the advisory group includes representatives from large companies including Johnson & Johnson, GlaxoSmithKline, AstraZeneca and GE Healthcare. In effect, a particular orientation to innovation (i.e., commercialization) has been institutionalized within the management structure of CGTC (Gardner and Webster 2017).

The commercialization priority is also reflected in the operations of the CGTC. The catapult functions like a consultancy. It provides assistance to manufacturers, clinicians and scientists on a range of innovation challenges: IP and legal assistance, detailed technical assistance on manufacturing processes, assistance with clinical trial design and adherence to regulatory frameworks, and assistance with reimbursement and commissioning challenges. CGTC staff also regularly visit UK universities in order to identify promising new developments, and facilitate collaborations between academic investigators, industry and regulators. Currently it is involved in over 40 RM projects. Its role involves the assessment of IP potential, delineating the "value proposition" of prospective RM products, how they can be developed and their potential markets. Through these activities, the CGTC coordinates what might otherwise be diverse research and clinical interests within RM into particular innovation pathways where commercialization is a main objective.

According to our interviewees, the commercial expertise of the CGTC provides it with considerable authoritative weight, and having CGTC involvement in their projects lends credibility that can lead to further investment from other funders, including other public funders. Here the CGTC is generative of important reputational value in addition to providing technical and organizational assistance. As one interviewee (a CEO of a small RM company) stated:

> The Catapult were important to us… they gave [our company] a vote of confidence by putting their support behind us… So that was a huge boost for us and a tick in the credibility box. It enabled us to get the [public agency] grant at that time.

The CGTC appears to have been endowed by stakeholders with the authority and legitimacy to judge and improve the commercial viability of RM projects, enabling it to "de-risk" these projects for other investors, an important role for the Entrepreneurial State in fostering new markets. It is worth noting in the above example that the reputational value of the CGTC led to further public funding to facilitate commercialization

(Gardner and Webster 2017). This is indicative of a context that equates public interest with innovation, and innovation with commercialization.

The ultimate realization of market value within RM requires socio-technical systems for producing, circulating, and exchanging RM products as commodities. The CGTC is enabling of these systems by providing assistance to manufacturers on logistics for cell and tissue transportation and, more significantly, with the construction of the large cell and gene therapy manufacturing centre (in Stevenage), at an initial cost of £55 million. This facility is intended to provide the capacity required to produce the quantity of RM products needed for phase III trials and commercial market supply. Companies will be able to rent space within the facility, take advantage of CGTC expertise, and therefore reduce their own costs by not having to build their own research infrastructure.

We also see other values, beyond the envisaged realisation of commercial value, reflected in official CGTC accounts of their activities, particularly in their annual reports. *Clinical value*, for example, is reflected in the CGTC rationale for supporting specific projects, which makes reference to considerable clinical need in particular disease areas (CGTC 2014). These reports also reflect what might be called *pathfinding value*, that is, a value that derives from creating new pathways to the clinic that can subsequently be used by other actors (CGTC 2014). In the life sciences, unlike many other technological fields, there is no advantage in being the first to build a route to market, given the high costs and risk of failure, so the CGTC is seen as having an important role in this context. We also observe what could be called *global competitiveness value*, which is most obviously illustrated by the official vision of the CGTC: "For the UK to be a global leader in the development, delivery and commercialisation of cell and gene therapies" (CGTC 2016). These qualities—satisfying clinical need, pathfinding, and global competitiveness—are framed within official accounts as being *valuable* in themselves and therefore provide legitimacy and motivation for CGTC activities.

The CGTC is an example of a community of expertise that has been mobilised to facilitate RM, and has, according to our respondents, been relatively successful so far. It positions itself with reference to a broad range of values, but it is the value of commercialization that is prominent in its interactions within the field of RM. We see priority given to this in its structure, its consulting-like work with stakeholders, and in its establishment of a new manufacturing centre. It represents an important means by which the State is attempting to facilitate RM with the creation of a publicly funded but autonomous agency with its own industry-oriented and commercially responsive decision-making capacity.

The Promotion of Interdisciplinarity

Interdisciplinarity broadly refers to mechanisms of knowledge production that traverse conventional disciplinary boundaries, and which have often been established to tackle specific scientific, technical or societal problems (Gibbons 1994). It has been championed as an important facilitator of R&D and as a crucial driver of innovation for at least the past two decades, particularly within biomedicine as part of the broader TM policy agenda. According to Barry et al. (2008), its emergence reflects a perception that the problems encountered in innovation are too complex to be managed within one disciplinary framework, alongside the view that science and innovation need to be accountable to the public. This perception is particularly apparent in the European Commission's Responsible Research and Innovation framework, which proclaims that accountability is best achieved if science and innovation are directed towards user or societal need, which is more likely to be seen if multiple perspectives, including potential end-users, are actively informing R&D and innovation processes (Von Schomberg 2013).

Interdisciplinarity can broadly be seen, then, as an attempt to incorporate a wider range of values into innovation activities. It has gained further traction in regard to the "innovation challenges" that characterise RM—expanding and directing the growth of cell lines, creating cost-effective manufacturing processes, navigating regulatory and reimbursement hurdles, etc. These are all complex problems that have been deemed to be in need of interdisciplinary problem-solving (MRC 2012). Specific government-supported initiatives have been launched to encourage and support interdisciplinarity within the emerging field, and as a result, interdisciplinarity (and new communities of expertise) has become institutionalized. It is, as Mittra (2016) notes, becoming increasingly consolidated in "bricks & mortar" infrastructural forms (indeed within the CGTC itself).

Academic networks are also being redrawn in accordance within this emphasis on interdisciplinarity. One example is the establishment of the UK Regenerative Medicine Platform (UKRMP) which, like the CGTC, was launched in response to the government inquiries into RM. The

UKRMP was established in 2013 by several of the UK research councils with £25 million in funding. The official, stated aim of the platform is to address "key translational challenges in regenerative medicine" via:

> ...five interdisciplinary and complementary research Hubs that collectively provide a national resource through the generation of new tools, protocols, and resources that can be utilised by other UK research groups in both academia and industry... [The UKRMP is] a cornerstone of the broader and integrated UK research strategy...which is seeking to support high quality UK research activity and translational activity that will help deliver the great promise of regenerative medicine to benefit both patients and future economic growth. (UKRMP 2012)

Here we see that the official rationale for establishing UKRMP makes reference to several values: *clinical value* and of course, *the value of commercialization itself*—also reflected in statements such as "help[ing] to de-risk future commercial investment" (UKRMP 2015, 4). It represents a community of expertise, then, which is oriented towards a particular, commercialization mode of innovation.

Other values are also articulated. These include the value of safety, and also epistemic/scientific value, manifested in practices aimed at creating shared infrastructures and standards that transcend disciplinary boundaries. This is exemplified in the following rationale of a leading scientist within the platform:

> Safety is obviously paramount, and 'safety' is about good science. It will come from the science... 'Safety' requires a holistic approach – bringing together pharmacology, pathology, physiology... (Interview field notes)

These values are reflected in the structure of the platform, which consists of five hubs, each being a multi-institutional, interdisciplinary collaboration involving academic scientists from various disciplinary backgrounds. Two of the hubs focus on clinically-oriented biological problems; two are exploring technical challenges "further" along the translational pathway; and one is focused on identifying safety challenges and methodologies to assess risk. This last hub brings together a particularly diverse array of specialisms (stem cell biology, biostatistics, nanochemistry, multimodal imaging, and clinician disciplines such as nephrology and hepatology), and it reflects an emphasis within the UKRMP on what can be called *clinical safety* value.

The UKRMP, we suggest, is an exemplar of the new life sciences academic networks within advanced economies. These are networks of expertise that are increasingly interdisciplinary, appear to reflect a range of values, and are emerging from often government-supported translational medicine initiatives aimed at facilitating a wealth-generating industry. A similar example in the US is the National Institutes of Health (NIH) Road Map Initiatives launched in 2004 with the aim of "fostering collaboration" for "high-risk/high reward research" (NIH 2011). Such initiatives are disrupting conventional disciplinary boundaries and conventional institutional and professional norms (Mittra 2016). The rise of the "clinician-scientist", perceived as being particularly capable of moving between conventional disciplinary spaces and the clinic, is an example of this (Wilson-Kovacs and Hauskeller 2012).

Reconfiguring Governance Mechanisms: Accelerating Access to Regenerative Medicines

For much of the second half of the twentieth century, promising therapeutic medical technologies were regulated under either pharmaceutical or medical-device frameworks. Both emerged as politically-supported governance mechanisms in response to much-publicized medical scandals: the thalidomide scandal in respect to pharmaceuticals, and a spate of defective cardiac pacemakers in respect to medical devices. Regulatory agencies such as the US's Food and Drug Administration (FDA) and the European Medicines Agency (EMA) were thus mandated to protect citizens, and the resulting regulatory frameworks, which have been subject to ongoing incremental adjustment, impose complex requirements on manufacturers to demonstrate the safety and efficacy of their products before they can be placed on the market. The consequence of these requirements is that bringing promising therapeutic technologies to the market is a costly process that takes considerable time. In response to pressure from industry and patient groups, regulatory provisions have been launched by both the FDA and the EMA to provide expedited access to promising therapies in certain circumstances. These include the "orphan drug" legislation and "fast-track" drug approval processes, which provide incentives for manufacturers to innovate in areas of considerable unmet clinical need (Milne and Tait 2009). With the adoption of such measures, the role of regulators is recast not only as protectors of citizen well-being, but also as facilitators of innovation. The potential conflict between these two positions has not been acknowledged.

Developments within RM have tested the limits of conventional regimes. In both the US and Europe, it has been decided that RM technologies are sufficiently incommensurable with conventional frameworks to warrant new frameworks that both protect patients and facilitate innovation (Omidvar et al. 2014). Hence, after considerable debate and consultation, EMA established the Advanced Therapies Medicinal Products (ATMP) framework, and the FDA has established its Cellular and Gene Therapy Products framework, both of which have been tailored to mitigate the particular safety and quality-related risks presented by cell and gene-based medicines, while also providing what is intended to be an innovation-facilitative degree of stability and harmony to the emerging field. Currently, however, few RM therapies have been approved within these frameworks (eight ATMPs have received market authorisation by the EMA). This has fuelled calls to explore additional measures to accelerate translational medicine (RMEG 2014). A more permissive regulatory regime, it is argued, will encourage the investment required to tackle technical and scientific challenges related to the complex biology of the field (HoL 2013).

The "reimbursement hurdle" has also been identified as a problematic governance mechanism (RMEG 2014; HoL 2013). Of those ATMP products that have received market approval, none have been widely adopted, as manufacturers have struggled to secure commissioning arrangements from payers. In most jurisdictions, commissioning decision-making is guided by a formal Health Technology Assessment of the candidate therapy, in which cost-effectiveness and clinical benefit are determined. In the UK, one influential report (RMEG 2014) suggested that such mechanisms could unfairly disadvantage novel regenerative medicine therapies due to their potentially high upfront costs and lack of longer term data.

Hence, there has been a concerted, government-supported movement in several countries towards additional mechanisms that would "accelerate access" to RM therapies. The most radical of these is Japan's RM legislation that allows "conditional marketing authorization". In effect, it enables a product to be placed on the market after completion of phase II clinical trials once safety has been demonstrated and efficacy has been predicted. The product is then subject to close post-market surveillance to monitor safety and obtain efficacy data, and the manufacturer must then reapply for "full" marketing authorization within 7 years (Sipp 2015). During this time, patients pay a significant proportion of the costs (Nature Editorial 2015). Similarly, the EMA is piloting an adaptive pathways approach that is intended to allow for "early and progressive patient access to a medicine" (EMA 2016) in the case of new medicines for which there is unmet clinical need and which are intended to treat chronically debilitating or life-threatening conditions. This would represent a significant change compared to already in-place "fast-track" provisions (Mittra 2016).

Similarly, the UK government has launched the Accelerated Access Review with the intention to "speed up access to innovative drugs, devices and diagnostics for NHS patients" (GOV.UK 2016). It has been tasked with examining three areas of reform: existing regulation and the potential for quicker methods of assessing safety and efficacy; reimbursement and possible adaptation of health economic systems to better accommodate recent technology advances; and clinical adoption and better ways to support and drive innovation within the NHS (DoH 2015). The initiative has an external advisory group whose composition reflects an apparent commitment to industry-orientated innovation, but also a commitment to including a wider range of values and perspectives. It includes representatives from industry and from patient-advocacy and charity groups, senior academic clinicians, and regulatory experts.

The review group's interim report (Accelerated Access 2015) is based on a number of propositions. The first is that "patients should be given a stronger voice at every stage of the innovation pathway" (2015, 11). This can be achieved by "directing innovation towards the outcomes that matter most to patients" and taking into account the "patient appetite for risk". As a next step, the review will thus "explore the scope for identifying and codifying patient-led outcome measures" (2015, 12). Here we see a clear attempt to align an early marketing of products with patient-centred values defined primarily in terms of their preparedness for risk, which assumes that patients are less risk-averse than the regulatory system has previously been on their behalf. Other values are invoked in the remaining propositions. Proposition two is "getting ahead of the curve", and includes a set of potential measures aimed at ensuring that the UK remains on the leading edge of innovation and remains a "go to" place for industry (2015, 13). Here we have the invoking of *global competitiveness value*, and *commercialization value*. Proposition three, "Supporting all Innovators", invokes "innovation" as a value in its own right that requires constant support, via, for example, creating a more flexible and streamlined health technology appraisal process. Proposition four, "galvanising the NHS", suggests ways in which the NHS can be improved to better support innovators and provide more space for patient-involvement. (This will be more broadly discussed in the following section.)

The accelerated access review, then, represents a significant government-mandated initiative to adjust governance mechanisms to make them more facilitative of innovation. As Mazzucato has noted in regard to earlier orphan drug "fast-track" regulatory provisions, such initiatives are an important means by which the Entrepreneurial State can attempt to "de-risk" innovation and support the formation of new markets in biomedicine (Mazzucato 2015, 81). What we see within this UK initiative is that a range of values are invoked to justify this "de-risking" activity, several of which—such as commercialization value and global competitiveness value—are actively championed in relation to other TM initiatives such as the CGTC.

Reframing the Healthcare System as an Innovation Asset

Healthcare providers such as hospitals have always played a key role in innovation. However, over the last decade TM initiatives have been launched in an attempt to enhance and make more visible the hospital's role in innovation. In 2006 in the UK, for example, the National Institute for Health Research (NIHR), a government body, was launched to coordinate research across the NHS. The NIHR describes itself as "improving the health and wealth of the nation through research". A major NIHR project was the establishment of 11 Biomedical Research Centres (BRCs) across the UK. Each represents a partnership between leading NHS research hospitals and universities, and they receive substantial levels of funding to create and consolidate innovation-facilitating dynamics, such as interdisciplinary collaboration. The BRCs are intended to "drive innovation", "translate advances in biomedical research into benefits for patients" and "provide a key component of the NHS contribution to our nation's international competitiveness" (NIHR 2016). There is, then, a clear articulation of values: *innovation as a value* in its own right; *clinical value;* and *global competitiveness value.*

Despite such initiatives, there is an ongoing discourse in UK policy circles in which the hospital, and the wider healthcare system, is represented as being a hindrance to innovation. For example, recent government-supported reports into the UK's readiness for RM (RMEG 2014; HoL 2013; BIS 2011) have stated that existing hospital workflows present a significant hurdle to the uptake of clinical developments in RM, which require new skill sets and supporting infrastructures. There is, then, ongoing pressure to make the NHS more conducive to, and supportive of, innovation.

This is exemplified by the fourth proposition of the Accelerated Access Review, "galvanising the NHS". The review notes that the NHS is widely perceived to be too slow in adopting new technologies and is resistant to change. In light of this, various initiatives have been proposed, such as: providing clear incentives to induce innovation leadership among particular NHS organisations; an earmarked fund or specialized supplementary funding to facilitate the implementation of promising technologies; and making use of clinical leaders by encouraging them to act as "champions of change" (Accelerated Access 2015, 20–21).

These proposed changes would align with current NHS strategic objectives. In 2011, a specific innovation strategy for the NHS was published, titled *Innovation: Health and Wealth* (2011). One outcome of this was the formation of the Government-supported regional Academic Health Science Networks in an attempt to create better linkages between the NHS, academia and the private sector, and ensure that healthcare delivery is better aligned with clinical research (NHSE 2016a). Making the NHS more supportive of innovation has also been identified as a priority in NHS England's *5 Year Forward View*, which outlines the official strategic trajectory of NHS England (NHS 2014). This led to the establishment of seven NHS England "test-bed" sites, which are being used to test new clinical delivery models based on novel combinations of interconnected technologies (NHSE 2016b).

The NHS, then, is being reimagined in official discourse as an important source of innovation generating "health and wealth", as well as being a provider of universal healthcare. Consequently, various government-supported initiatives have been launched to reconfigure elements of the healthcare system, particularly hospitals, in ways that are perceived to be more facilitative of innovation. New research infrastructures are being established, interdisciplinary collaborations are being actively encouraged and strengthened, new expectations are being placed on health professionals, and new roles, such as the "research champion", are being created. Within this activity is a clear articulation of clinical value, the value of innovation, and of course, global competitiveness value. In effect, the NHS, as with healthcare providers in other countries, is being *instrumentalized as an innovation system.*

Discussion and Conclusions

Government-supported TM endeavours are instrumentalizing biomedical R&D and the healthcare sector in the name of "health and wealth". The formation of accelerator agencies such as the CGTC, the institutionalization of interdisciplinarity, the adjustment of governance frameworks, and the repositioning of healthcare providers as innovators, constitute a significant reconfiguring of the biomedical landscape. Mittra (2016) has described this as the new health bioeconomy, in which novel therapies emerge from new innovation pathways involving heterogeneous actors and interests. Who these actors are and how they are aligned will foster forms of clinical innovation that might reflect traditional distinctions of public vs private (health) goods, but increasingly are seen to reflect a hybridised relation favouring a wealth-generative bioeconomy.

The Entrepreneurial State is actively championing and enacting a particular mode of biomedical innovation that entails commercialization. This is reflected in the establishment of commercialization-oriented accelerator agencies such as the CGTC that draw on considerable expertise from industry and commercial enterprise. There is, as illustrated above, a clear prioritisation of commercialization value in these translational medicine endeavours, and this is closely tied to the value of global-competitiveness—actively invoked in all the initiatives explored above. In effect, this Entrepreneurial State involvement in driving the new health bioeconomy is a form of nation-building driven by promissory visions of health and wealth: state resources are being redirected into innovation-facilitating activities in order to create future wealth-generating markets. However, it remains open to doubt that the capital generated by such activity is in some way bounded within a UK-specific bioeconomy, given the mobility of capital (and its returns) in a global economy. It is likely that an RM industry would entail a variety of markets, characterised by different technological platforms, business models and regulatory pathways. However, in broad terms, the envisaged RM markets will align with what Aspers (2007) refers to as fixed-role, standard-based markets: buyers are expected to be national health services and health insurers, while sellers are expected to be large pharma companies with the resources necessary to obtain marketing authorisation, and which have swallowed up smaller, specialist companies that drove the initial development of the project. The status of the RM therapies themselves (especially their status as "safe" and "clinically effective") will derive from their adherence to standards entrenched in regulatory provisions.

This envisaged market raises some important issues relating to the State's role in attempting to serve as a broker between public and private good. In effect, large companies stand to profit significantly from products whose original development was, to a considerable extent, funded with public resources. National health systems such as the NHS will be expected to buy what are likely to be high-cost products whose development they had been repurposed to facilitate. There is, then, a socialization of risk and a privatization of rewards. Mazzucato notes that such a situation exacerbates inequality, and that current taxation would fail to sufficiently compensate the state for its investment. She suggests several measures for addressing this injustice: the establishment of a national innovation fund in which royalties from state-supported technological breakthroughs could be paid; the State could grant income-contingent loans and equity; and the establishment of state development banks. Currently, none of these options have been mooted in the UK for RM.

The pursuit of "health and wealth" in RM involves the delegation of innovation-facilitating, "de-risking" work to particular communities of expertise, and a range of actors are therefore involved in the creation of new innovation pathways. Mittra suggests that this complex ecology of the new health bioeconomy provides openings for the enactment of a multiplicity of values that, in the past, may have been excluded from innovation processes, values that may inform and shape innovation processes. To what extent such values will actually be enacted as the RM field unfolds is uncertain, but our assessment of the current situation is that this delegation to communities certainly involves the invoking and articulation of various values, in addition to those relating to commercialization and global competitiveness. These include, of course, values relating to safety, innovation as a value in its own right, and also patient-centeredness as a value. We suggest that the power and influence of the translational medicine agenda derives from its capacity to appeal to, and ultimately legitimate, such diverse interests, and consequently mobilise a range of actors (industry, scientists, clinicians and patients) into a common project aimed at generating "health and wealth".

Indeed, the example of RM provides valuable insights into the mechanisms by which an essentially neo-liberal, market-creating (and sustaining) political project gains a sense of legitimacy among various groups. In light of this, it is necessary for social scientists to examine which groups are being excluded from the rewards of such a project, and whose aims and interests are elided as governments engage in nation-building via the deployment of supply-side policy. It is worth noting that in England, this translational medicine agenda is being deployed at a time when public funding is being withdrawn from social services, and when health providers are grossly underfunded. (Again, it is important for social scientists to be attentive to the ways in which hospital managers and clinicians respond to the innovation imperative in such a climate.) As the field unfolds, it is also necessary to be attentive to how particular values are enacted in specific settings. "Patient-centredness" and the emphasis on "patient empowerment", for example, can be mobilised as a means of disciplining patients into particular neo-liberal technological projects (Gardner 2016). More generally, if the societal goal is to maximize health and well-being of citizens, it is necessary to ask whether public funds might be more usefully deployed elsewhere, rather than towards the instrumentalization of biomedical R&D and the health sector in the name of commercialization. Finally, it is important to note that the four state strategies described above, toward which significant public funds have been directed, do not guarantee "wealth", for success in the market-place requires customers prepared to buy the products, either through reimbursement (via insurance or public-sector procurement) or through private sale. Underpinning by the neo-liberal state only goes so far.

It should also be noted that the government-supported activities outlined above have encountered some resistance from particular groups, which themselves reflect various, potentially conflicting values. Clinical groups, for example, have shown some resistance to the commercialization mode of innovation that is embodied in the CGTC. More generally, it can be questioned whether such a mode is likely to render public health gains that could be secured through less commercial routes. Relatedly, another government-funded institution, the UK Stem Cell Bank, has received criticism that its policy of only providing non-exclusive licences for the use of deposited stem cell lines—a policy which is intended to support open science (and which thus enacts scientific value)—will discourage commercial investment in the field. Furthermore, despite the optimistic discourse on interdisciplinarity, interdisciplinary organizational forms have been difficult to operationalize. A recent study has noted, for example, that interdisciplinary grant proposals are much less likely to receive funding (Bromham et al. 2016). Proposed governance changes to accelerate access to innovative medicines have also received criticism. A group of scientists, for example, have voiced concern about the assumptions underlying the EMA's Adaptive Pathways pilot project, particularly the apparent assumption that existing frameworks stifle innovation and are "bad for all parties" (Epha 2016). Similarly, a recent *Nature* editorial suggests that such initiatives will create a plethora of safe but clinically ineffective therapies (Editorial 2016). There are, then, tensions and countervailing processes at work within the emerging new health bioeconomy. Its emergence may be driven by powerful government-supported initiatives aimed at generating health and wealth, but it will inevitably entail a set of entangled and contested processes in which a multiplicity of conflicting values will be enacted so that assembling and leveraging such values in order to drive the translational process in RM is, and will remain, a challenge for the Entrepreneurial State.

References

Accelerated Access. (2015). *Accelerated access review: Interim report.* London: The Crown.

Aspers, P. (2007). Theory, reality, and performativity in markets. *American Journal of Economics and Sociology, 66*(2), 379–398. doi:10.1111/j.1536-7150.2007.00515.x.

Barry, A., Born, G., & Weszkalnys, G. (2008). Logics of interdisciplinarity. *Economy and Society, 37*(1), 20–49. doi:10.1080/03085140701760841.

BIS. (2011). *Taking stock of regenerative medicine in the United Kingdom.* London: Department for Business Innovation & Sills.

Bromham, L., Dinnage, R., & Hua, X. (2016). Interdisciplinary research has consistently lower funding success. *Nature, 534*(7609), 684–687. doi:10.1038/nature18315.

CGTC. (2014). *The cell therapy catapult: First review to March 2014.* London: Cell and Gene Therapy Catapult.

CGTC. (2015). *The cell therapy catapult: Growing a UK cell therapy industry, delivering health and wealth.* Cell and Gene Therapy Catapult. Retrieved September 23, 2015, from https://ct.catapult.org.uk/.

CGTC. (2016). *Our purpose.* Cell and Gene Therapy Catapult. Retrieved June 20, 2016, from https://ct.catapult.org.uk/about-us/purpose-vision-and-mission/.

DoH. (2015). *Terms of reference: Innovative medicines and medtech review*. London: Department of Health.

Dussauge, I., Helgesson, C. F., & Lee, F. (2015). *Value practices in the life sciences and medicine*. Oxford: Oxford University Press.

Editorial. (2016). FDA should stand firm on stem-cell treatments. *Nature, 535*, 7–8. doi:10.1038/535007b.

EMA. (2016). *Adaptive pathways*. London: European Medicines Agency. Retrieved June 28, 2016, from http://www.ema.europa.eu/ema/index.jsp?curl=pages/regulation/general/general_content_000601.jsp.

Epha. (2016). *Scientists voice concerns about adaptive pathways*. Epha. Retrieved July 1, 2016, from http://epha.org/.

Gardner, J. (2016). Patient-centred medicine and the broad clinical gaze: Measuring outcomes in paediatric deep brain stimulation. *BioSocieties*. doi:10.1057/biosoc.2016.6.

Gardner, J., Faulkner, A., Mahalatchimy, A., & Webster, A. (2015). Are there specific translational challenges in regenerative medicine? Lessons from other fields. *Regenerative Medicine, 10*(7), 885–895. doi:10.2217/rme.15.50.

Gardner, J., & Webster, A. (2017). Accelerating innovation in the creation of biovalue. *Science, Technology, & Human Values*. Early access online. doi: 10.1177/0162243917702720.

Gibbons, M. (1994). *The new production of knowledge: The dynamics of science and research in contemporary societies*. London: Sage.

GOV.UK. (2016). *Accelerated access review*. The Crown. Retrieved June 29, 2016, from https://www.gov.uk/government/organisations/accelerated-access-review.

HoL. (2013). *Regenerative medicine report*. London: House of Lords Science and Technology Committee.

Malik, N. N. (2014). Reimbursement and adoption of advanced therapies: The 5-C framework. *Regenerative Medicine, 9*(5), 573–578. doi:10.2217/rme.14.51.

Malik, N. N. (2016). Pay-for-performance pricing for a breakthrough heart drug: Learnings for cell and gene therapies. *Regenerative Medicine, 11*(3), 225–227. doi:10.2217/rme-2016-0014.

Mason, C., & Dunnill, P. (2007). A brief definition of regenerative medicine. *Regenerative Medicine, 3*(1), 1–5. doi:10.2217/17460751.3.1.1.

Mazzucato, M. (2015). *The entrepreneurial state: Debunking public vs. private sector myths*. London: Anthem Press.

Milne, C.-P., & Tait, J. (2009). Evolution along the government-governance continuum: Impacts of regulations on medicines innovation in the United States. In C. Lyall, J. Smith, & T. Papaioannou (Eds.), *The limits of governance: The challenge of policy-making for the new life sciences* (pp. 107–132). London: Ashgate.

Mittra, J. (2016). *The new health bioeconomy: R&D policy and innovation for the twenty-first century*. Basingstoke UK: Palgrave Macmillan.

Mittra, J., & Milne, C. P. (2013). *Translational medicine: The future of therapy?* Singapore: Pan Stanford.

MRC. (2012). *A strategy for UK regenerative medicine*. London: Medical Research Council.

Nature Editorial. (2015). Stem the tide. *Nature*, 163–164.

NHS. (2011). *Innovation: Health and wealth*. London: NHS.

NHS. (2014). *Five year forward view*. London: NHS.

NHSE. (2016a). *Academic health science networks*. England: NHS. Retrieved June 30, 2016, from https://www.england.nhs.uk/ourwork/part-rel/ahsn/.

NHSE. (2016b). *Test beds*. England: NHS. Retrieved June 3, 2016, from https://www.england.nhs.uk/ourwork/innovation/test-beds/.

NIH. (2011). *NIH roadmap and roadmap-affiliated initiatives*. National Institute of Environmental Health Sciences. Retrieved June 27, 2011, from http://www.niehs.nih.gov/funding/grants/announcements/roadmap/.

NIHR. (2016). *Biomedical research centres*. NHS. Retrieved June 30, 2016, from http://www.nihr.ac.uk/about/biomedical-research-centres.htm.

Ogawa, N. (2015). *Japan working hard on regenerative medicine*. Tokyo.

Omidvar, O., De Grijs, M., Castle, D., Mittra, J., Rosiello, A., & Tait, J. (2014). *Regenerative medicine: Business models, venture capital and the funding gap*. Edinburgh: Innogen Institute.

RMEG. (2014). *Building on our own potential: A UK pathway for regenerative medicine*. London: Regenerative Medicine Expert Group.

Sipp, D. (2015). Conditional approval: Japan lowers the bar for regenerative medicine products. *Cell Stem Cell, 16*(4), 353–356. doi:10.1016/j.stem.2015.03.013.

Tait, G., & Banda, J. (2016). *Proportionate and adaptive governance of innovative technologies: The role of regulations, guidelines and standards*. Edinburgh: Innogen.

UKRMP. (2012). *UK regenerative medicine platform*. UK Regenerative Medicine Platform. Retrieved June 22, 2012, from http://www.ukrmp.org.uk/.

UKRMP. (2015). *Annual report*. London: United Kingdom Regenerative Medicine Platform.

Von Schomberg, R. (2013). A vision of responsible research and innovation. In R. Owen, J. Bessant, & M. Heintz (Eds.), *Responsible innovation: Managing the responsible emergence of science and innovation in society* (pp. 51–74). London: Wiley.

Webster, A. (2013). *The global dynamics of regenerative medicine: A social science critique*. Basingstoke: Palgrave Macmillan.

Willetts, D. (2013). *Eight great technologies*. London: Department of Business, Innovation and Skills.
Wilson-Kovacs, D. M., & Hauskeller, C. (2012). The clinician-scientist: Professional dynamics in clinical stem cell research. *Sociology of Health & Illness, 34*(4), 497–512. doi:10.1111/j.1467-9566.2011.01389.x.

Authors' Biography

John Gardner is a research fellow in sociology, School of Social Sciences, at Monash University in Melbourne, Australia. His research examines the social and political dimensions of medical innovation, with a focus on regenerative medicine and neurotechnologies.

Andrew Webster is Professor of Sociology of Science and Technology and Director of the Science and Technology Studies Unit, Department of Sociology, University of York. He has published extensively in the sociology of the biosciences.

James Mittra is Deputy Director of the Innogen Institute, and a Senior Lecturer in Science, Technology and Innovation Studies, based within Science Technology and Innovation Studies at the University of Edinburgh, UK. He is an interdisciplinary researcher who has published widely on the bioeconomy.

CHAPTER 3

Technologies of Governance: Science, State and Citizen in Visions of the Bioeconomy

Tess Doezema and J. Benjamin Hurlbut

Science is the highest personification of the nation because that nation will remain the first which carries the furthest the works of thought and intelligence.
—Louis Pasteur

Introduction

In 2010 when J. Craig Venter announced that he had created "synthetic life," the Guardian ran an article with the headline "synthetic life breakthrough could be worth over a trillion dollars." The breakthrough technology was a bacterium with a synthetic genome whose purpose was to demonstrate the feasibility of converting a simple organism into a flexible industrial platform by installing new genetic "software." It was the product of "Venter's ambition to create organisms that are not only new, but lucrative" (Sample 2010).

T. Doezema (✉) · J. Benjamin Hurlbut
School for the Future of Innovation in Society & School of Life Sciences, Arizona State University, Tempe, USA

V. Pavone and J. Goven (eds.), *Bioeconomies*,
DOI 10.1007/978-3-319-55651-2_3

Venter's project reflects a vision of biotechnology wherein novel biological constructions become platforms for transforming agricultural and industrial production, ushering in a new bioeconomy. The "bioeconomy," as its proponents imagine it, would achieve a radical realignment—and redemption—of both life and market. Modifications of life at the molecular level would generate an unprecedented harmony between economic expansion, industrial innovation, environmental responsibility, and human wellbeing.

This chapter examines some visions of the bioeconomy in order to surface their tacit constructions of rightly ordered relationships between science, state, and citizen. We argue that dominant visions of the bioeconomy reflect, most fundamentally, an imaginary of governance: they construct technoscience as the agent capable of enhancing social wellbeing, and outline the corollary political commitments that are prerequisite for desirable technological futures. As such, they are grounded in presumptions about the nature of sociotechnical change, the intrinsic potential of biotechnology, and the ways of knowing that are necessary to achieve a better world.

Here we focus on two vision statements about the bioeconomy (the US Bioeconomy Blueprint and the OECD's Bioeconomy to 2030) and a recent, high-profile example of US biotechnology regulation that has been seen as a test-case for the next generation of agricultural biotechnology products: the AquaAdvantage (AA) Salmon. Our analysis relies on data from interviews and meetings with approximately 35 U.S. regulators and biotechnology industry insiders, as well as a variety of publicly available documents such as news media, biotechnology industry public relations materials, and public documents from U.S. federal regulatory agencies.

The visions of the bioeconomy we analyze are grand and aspirational "vanguard visions" (Hilgartner 2015). At the same time, they draw upon and crystalize widely shared notions of the rights, roles, and responsibilities of political subjects in relation to both science and the state. In this respect they are exemplary of these shared notions and offer an entry-point for elucidating them. We argue that the perceived authority and legitimacy of these visions are underwritten by particular sociotechnical imaginaries—collectively held imaginations of right knowledge, legitimate authority, and progress that are expressed in scientific and technological visions and projects (Jasanoff and Kim 2015). With these vanguard visions, elites claim a privileged vantage point from which to foresee potential for sociotechnical futures that the average citizen cannot, and thus claim the authority to declare imperatives of policy action in the present in the name of realizing those futures. They do so by drawing upon imaginations of the relationship between science and social progress that are increasingly part of a cultural commonsense. This commonsense lends legitimacy to elites' demands for democratic deference. The project of this chapter is to explore the forms that such legitimation takes, and the symmetries between how they play out both in grand, aspirational visions of the bioeconomy and in the particular practices of governance that are implicated in the bioeconomy's realization.

The imaginaries that inform the bioeconomy visions we examine below inform the discourses and practices of a multitude of actors and institutions, finding subtle but consequential expression in the politics of certain regulatory moments. As we demonstrate using the case of AquaAdvantage Salmon, such moments are sites of enactment, contestation, and formation of precedent in which broad visions of progress inform—and are ratified through—practices of governance. The controversy over the salmon illustrates how the imaginaries that underwrite the bioeconomy vision shape what questions get asked and what questions are left unasked in US regulatory politics as routines are established for determining what new life forms—and associated forms of social life—may enter our world. In imagined technological futures, elite visions, shared imaginaries, and particular practices of governance come together to inscribe the arenas in which ideas of potential, progress, and right political authority crucially take shape, with real and concrete consequences for life in the present.

We make three primary arguments. First, imaginations of the potential of biotechnology to transform society are simultaneously imaginations of governance, what we call an imaginary of innovation-as-governance. We use the term "governance" advisedly and in the broadest sense to evoke the roles, long associated with the state, that have come to be more widely distributed and conceptualized in functionalist terms (e.g. network governance [Slaughter 2005]), such that allocations of authority and responsibility are often left tacit—both in theoretical conceptualization and in practice. Governance in its etymological origins was nautical—"steering the ship"—and this is the sense in which we use it. Insofar as political authority has come to be grounded in the capacity to navigate society toward desirable futures, technoscience, with its putative ability to envision and enact such futures, has acquired a central and powerful role in governance. The project of this chapter is, in part, to elaborate the imaginary that underwrites this political transformation. The visions of the bioeconomy that we examine below offer accounts of the future that are simultaneously accounts of the necessary and right relationships between science, the state, and its citizens, because they allocate authority for imagining and enacting desirable futures. Importantly, the market figures in this arrangement more as a feature of an aspirational future order than as an agent for enacting and governing that future. The state's responsibility for ushering in the bioeconomy gets focused on creating a set of conditions for generating (intrinsically valuable) innovations—transformations of life that are good for human lives, with the corollary presumption that markets will thereby inevitably be brought into being.

Below, we first consider the various ways in which state-market relationships in the bioeconomy have been theorized, arguing that accounting for state-science relations is key to understanding the bioeconomy project. Next we outline how these visions establish the bioeconomy as simultaneously necessary, existentially urgent, potentially threatened, and yet inevitable by examining how important elements of this discourse are leveraged in the controversy over the AA Salmon. In the latter part of the chapter we examine how the imaginary of innovation-as-governance undergirds subtle reconfiguration of the relationships between government, technoscience, and citizen, illustrating the power of this imaginary with a discussion of how responsibilities are contested and allocated in the case of the AA Salmon.

Theorizing the Bioeconomy

Some have argued that the political economy of biotechnology is not unique (Birch and Tyfield 2013), and for the circulation of existing bio-based goods, this may be so. Yet imaginations of life's technological potential do invite forms of speculation and promising wherein authorized imaginations of future novelty (and lucrativeness) attract capital and move markets in unique ways (Sunder Rajan 2012). The same imaginations inform the underlying legal, political, and normative commitments that are the conditions of possibility for making life into "lively capital" (Jasanoff 2012; Sunder Rajan 2012; Hurlbut 2016). These dynamics are biopolitical before they are "bioeconomic." Allocations of authority and responsibility—between science and citizen, state and market—get delimited and codified through the making of biotechnology (Jasanoff 2005), not least through processes of imagining biotechnological futures. The programmatic visions of the bioeconomy of the sort we analyze here are exemplary of these dynamics. As Goven and Pavone observe, "'the bioeconomy' is a political project, not simply or primarily a technoscientific and economic one" (2015). We also adopt a theoretical stance that attends explicitly to how dominant modes and practices of imagining technological futures underwrite and sustain particular (re)configurations of power. However, rather than bringing to light the particular concerns that may be passed over by the bioeconomy project, the injustices caused in the name of progress, or the manner in which the interests of some are privileged over others, we seek to illuminate the construction and stabilization of the conditions of possibility for this panoply of effects such that they are easily explained as the products of natural practices of prudent governance. This process is crucially one in which technoscience and political authority are coproduced, while—equally crucially—in bioeconomic visions and arenas of governance, their separateness is taken for granted.

The bioeconomy vision is a neoliberal one insofar as markets—and policy that makes way for markets— are valorized. Yet this diagnosis offers insufficient analytic purchase for making sense of its construction of desirable state-*science* relations. As we argue below, state-science relations inform corollary ("neoliberal") constructions of the right relationship between state and market. Neoliberal discourse has become "the common sense frame for conducting and evaluating policy" in the U.S. (Cahill 2011, 487) and offers a language that can be strategically mobilized to achieve divergent political ends (Kennedy 2016). As such, "neoliberalism" is a plural category with varied practical instantiations (Mirowski and Plehwe 2009; Bakker 2010), and neoliberalisms have thus become easy to find wherever one goes in search of them. Thus, while "the bioeconomy" may be articulated in a neoliberal idiom, it mobilizes an imaginary of innovation-as-governance that figures science and state as cooperative prime movers in the making of desirable futures, and the market as *consequence*, not cause, of progress.

Potential and Urgency

The proliferation of bioeconomy strategies, blueprints, and plans that have emerged in the last decade offer a vision of a world transformed by biotechnology, together with prescriptions for pathways to this future. Each of these statements merits analysis in its own right, and although they share important similarities, they are too heterogeneous to be lumped together (Staffas et al. 2013). Here we focus on two documents, the US Bioeconomy Blueprint (BEB), which was released by the Obama

Administration in 2012 and reflects many of its priorities vis-à-vis innovation and US economic policy, and the OECD's earlier document, "The Bioeconomy to 2030" (BE2030). There are important consonances between the US and the OECD documents (Goven and Pavone 2015), which we trace to a shared imaginary wherein normative constructions of science-state-citizen relations are mediated by an imagination of the role of biotechnology innovation in governance.

The central claim of both the BEB and BE2030 is that biology holds untold potential for benefiting society by generating technologies that will drive profound, cross-sectorial economic transformations, and that realizing this potential is an imperative of governance. Both documents cite a litany of technologies with the potential to "revolutionize" society and economy. "Tomorrow's bioeconomy relies on the expansion of emerging technologies such as synthetic biology…proteomics…and bioinformatics… as well as new technologies as yet unimagined" (White House 2012, 1–2). The benefits of growing such a "bioeconomy," they promise, go beyond simple economic growth or incremental enhancements of quality of life. Biotechnology's potential is emancipatory, producing fundamental infrastructures for human security, health, and wellbeing. The bioeconomy will "allow Americans to live longer, healthier lives, reduce our dependence on oil, address key environmental challenges, transform manufacturing processes, and increase the productivity and scope of the agricultural sector while growing new jobs and industries" (White House 2012, 1).

The corollary to potential is urgency. Both reports construct biotechnology innovation as a necessary response to urgent health, security, and geopolitical challenges of food scarcity, climate change, environmental collapse, and resource demands that outstrip natural reserves. It is an essential pathway to human survival and security. The urgency of growing the bioeconomy is premised on the notion that that basic human needs will go unmet in the absence of it: "A growing U.S. population requires increased health services and more resources such as food, animal feed, fiber for clothing and housing, and sources of energy and chemicals for manufacturing. In short, it *needs a new and more potent bioeconomy fueled by innovative ideas and practices* that can help address these needs in new, more powerful ways" (White House 2012, 1, emphasis added). In this respect, the figure of "the bioeconomy" stands for the realization of life's intrinsic technological potential and as delivering on the state's duty to ensure the security and welfare of the lives in its care.

There is an important temporal disjunction between the project of catalyzing (ostensibly) valuable innovation and the realization of a market in which it is value-able (i.e. is valued by economic agents via market exchange). The story of potential is told in teleological terms: life's potencies, harnessed by science, will generate valuable technologies and, thus, the *bio*economy. The imaginary is linear but the logic of justification is circular: the documents invoke the market to validate the presumption that innovation is intrinsically valuable (if there is a market, innovation is by definition responding to people's needs), even as they assert the imperative to first bring about that innovation in order to engender that market.

Achieving a secure and beneficent future is a matter of actualizing biotechnology's "promises [for] a vibrant bioeconomy with vast societal benefit" (White House 2012, 8). Yet narrow is the way, and thus the dependency between state and science is mutual: biotechnology's potential relies on the state to make straight the path by removing "barriers" to innovation such that biotechnology's potential can be actualized. In this imagination, the technological future is at once inevitable and contingent, sui generis and vulnerable. This is evident in the way the bioeconomy vision naturalizes the technological future while marking the social and political conditions that might shape or constrain it as contingent and in need of reformation—"barriers" to the bioeconomy's otherwise inevitable realization.[1]

Both documents construct biotechnology as autochthonous, and thus independent of—even if subject to the constraints of—political and economic order: "Additional scientific and technological revolutions are needed to fundamentally improve the approaches needed to confront the complex societal challenges of the future" (White House 2012, 15). At the same time, they suggest that the conditions of possibility for realizing "the full potential of the U.S. bioeconomy" depend on policies that prepare the way for technoscience's putatively self-driven unfolding. Unfettering technoscience from externally imposed constraints thus becomes an imperative of governance. Social and economic arrangements cannot themselves engender progress, but only stand out of (or in) its way: "Many of these technologies will not reach their potential unless they can overcome economic and social barriers to their development" (OECD 2009, 7). Formulations like "barriers to innovation" appear throughout both the BEB and the BE2030. Policymakers are tasked with removing "barriers" that stand in the way of realizing the "full potential of the bioeconomy," and what policy reforms will liberate science to "expand and develop *essential* bioeconomy technologies" (White House 2012, 3). Governing science and technology is limited to "preserving safety" (White House 2012, 29).

While it is not clear that these visions translate linearly into policy action (much to the frustration of their proponents), the discourses and imaginaries that they codify are evident in the mundane practices of regulation whose consequences are at once more defined and less equivocal. We now turn to one exemplary moment.

Building a Better Fish

In 1995 a company called A/F Protein established an Investigational New Animal Drug application with the U.S. Food and Drug Administration's (FDA) Center for Veterinary Medicine to seek regulatory approval for what would eventually be branded the AquAdvantage (AA) salmon. The product that the FDA approved for human consumption in November of 2015 is a genetically engineered Atlantic salmon with genetic material inserted from the Chinook Salmon and the Ocean Pout. The company boasts that the fish grows to market size twice as fast as its non-engineered counterpart.

The fish, which is widely referred to as a "game changer," is seen as paving the way for products already in the early stages of development and for a host of "new technologies as yet unimagined" (White House 2012, 2). As such, the fish is widely considered a test case for the (in)adequacy of biotechnology regulation and the U.S. government's commitment to providing a path for the bioeconomy to reach its "full potential." As a large group of scientists and biotech executives put it in a letter to President Obama, "there is much more at stake here than a fish" (Letter to President Obama 2014).

The AA salmon has been cited by both proponents and critics as the sort of technology whose novelty poses challenges for existing regulatory processes (Pollack 2015; Miller 2015). Proponents argue the approval process revealed existing regulations to be obsolete and "barriers to innovation" that demand reform. According to a Harvard expert on biotechnology regulation, the fish has "forced government regulation to play catch-up with science" (Juma 2016). Food and environmental protection groups likewise asserted the inadequacy of FDA's assessment. Some filed a lawsuit, charging that the agency rendered an "unprecedented decision" without "considering or fully disclosing the environmental and other risks" (Institute For Fisheries Resources et al. and V. Sylvia Mathews Burwell et al. 2016).

There is indeed more at stake here than a fish. The AA salmon is widely characterized as an exemplar of bioeconomic promise and thus as an important precedent, not only for regulatory approaches, but for the norms and commitments that will shape the future of the bioeconomy. Mirroring the discourse of the bioeconomy vision discussed above, biotechnology advocates, academic scientists, and news media have framed the salmon as a solution to urgent environmental, food security, and human health challenges (see e.g. GMO Answers; Zohar 2010; Pollack 2010). Aquabounty too describes the AA salmon as environmentally beneficial: it is "Locally Raised" and "Climate-Smart" (Aquabounty 2017). Thus public discourse figures the fish as precisely the sort of sustainable, innovative biotechnology that is "needed to feed the world." A portent of the bioeconomy to come, it promises abundance without cost: "consumers pay less for healthier food, nature is conserved—everyone wins" (Campbell 2016).

Although the FDA ultimately approved the AA Salmon, some of the strongest proponents of the bioeconomy vision have characterized the approval as a failure. One biotechnology company executive explained in an interview that although the process reduced uncertainty by clarifying regulatory requirements, it replaced it with the certainty that the bar is untenably high. "Nobody is going to go down that path again, right? It doesn't matter whether it's successful or not…why would you ever [use the techniques used to develop the AA salmon] again?" The Salmon thus represents a setback on the path to achieving a better world through biotechnology, and the results, in his telling, are tragic: "We've lost a generation."

Ordering Institutions

The BEB offers its prescriptions in the name of living happily ever after. It calls for disciplining regulation, markets, and social attitudes to recognize not only technoscientific assessments of risk, but corollary assessments of opportunities lost—and the resulting human, economic, and environmental costs—because of frictions and "barriers" to innovation. The effect is not merely to limit regulation to narrow forms of risk assessment, but the delegation of responsibility for rendering judgments about what does (not) need regulating to technoscientific authorities who are in a position to see such future costs and benefits.

The prescriptions these documents offer explicitly mark the market as a potential barrier, as its shortsightedness may fail to encourage the most valuable technological advances. OECD warns, "Biotechnology offers technological solutions for many of the health and resource-based challenges facing the world…yet [it] is unlikely to fulfill its potential without appropriate regional, national and, in some cases, global policies to support its development and application." Thus they call for regulatory restraint on the part of governments, intervening only where "preserving safety" warrants it (White House 2012, 29). In practice, this means limiting regulation to a narrowly construed form of risk assessment, grounded in technical authority (Jasanoff 1999; Thompson 2003; Winickoff et al. 2005). What the BEB calls for does not amount to a displacement of regulatory authority from state to market. Rather, it reflects a subtle—but consequential—reallocation of responsibility to imagine and enact the public good from state and society–and even market–to technoscience.

For both the BEB and BE2030, the value of the imagined technologies for society are presumed and treated as beyond question. Although they both affirm the need to ensure safety, beyond this there is no room for ambivalence about, let alone rejection of, their imagined technological future. While specific technologies may pose risks to safety that must be contained (through technocratic means), the bioeconomy as such is presumptively good. Precisely the opposite applies to socio-political arrangements that affect innovation. Those that are seen as interfering with the autonomy of technoscience are presumed to be "barriers to innovation." On this account, good governance requires a partnership between science and the state, one in which science assumes the authority to envision and realize technological futures and the state makes straight the way. This political partnership has the effect of guaranteeing science something more than political autonomy; it confers upon it a privileged agenda-setting role in governance because science alone has the ability to foresee what futures are possible, what benefits they offer, and what stands in their way.

Importantly, the bioeconomy vision entails a particular imagination of the nature of "science" as well as state, citizen, and market. The "science" of the BEB and BE2030 is the "New Biology": an evolutionary moment in the history of science marked by the "convergence" of biology with the physical sciences, information sciences and engineering (Committee on a New Biology 2009). The social roles and responsibilities attributed to the New Biology, including by these documents, are underwritten by an imagination of what technical powers this convergence portends. According to a report published by the U.S. National Academy of Sciences (which the BEB cites in describing how the biosciences will realize the bioeconomy), the New Biology will "enunciate and address broad and challenging societal problems" (Committee on a New Biology 2009, 3). This remarkable expansion of biology's remit is rooted in the notion that emerging powers over life promise to address society's needs in a way that policy and economy alone cannot.

Thus the "science" we refer to above is part and parcel of the imaginary of innovation-as-governance. It is the endless frontier, the republic of science, that most special institution of modernity that generates (through its own autonomy) knowledge, technique, and progress. While an enormous body of critical work in STS has demonstrated the untenability of this construction of science, our concern is instead with how this construction of science (no matter how problematic) figures in imaginations of what allocations of authority and responsibility in governance are legitimate (cf. Hurlbut 2017). This construction of science as autonomous institution worthy of protection is essential to accounts of how the bioeconomy will be achieved and what kinds of threats to its realization must be addressed.

Failing the Future

Both the BEB and the BE2030 note that public preferences, expressed through markets, are potential barriers to the bioeconomy, and as such must be managed. "Social attitudes to biotechnology… influence market opportunities," with significant consequences for whether the "full potential" of biotechnology is realized, for instance by "driv[ing] firms to alter the *type* of biotechnology that they use" (OECD 2009, 11). In such circumstances, markets "fail" as agents of progress, and the state must step into ensure particular avenues of innovation are not hampered by the failure of existing capital investment and markets to nurture them. These documents position technoscience as best able to declare what forms of innovation are possible, desirable, and good, and thereby to diagnose market failures that are the consequence of disjunctions between potential future benefits and present consumer preferences—preferences that will presumably be corrected once the intrinsic benefits of a technology are immediate. That is, once the technology is available as a new, better or, cheaper product, its intrinsic value will (presumably) overcome social resistance. Such failures of imagination on the part of consumers reflect "misplaced economic/marketplace concerns and reactionary fears" (Letter to President Obama 2014). Containing fear and engendering the right sort of citizen-consumer subject is thus presented as a task of governance—one that is legitimate because it is undertaken *in the name of* those self-same citizen-consumers who are not yet in a position to understand why the technological future whose realization they are inhibiting is in fact in their interest.

Thus OECD notes, "Social attitudes to biotechnology will continue to influence market opportunities, but public opinion can change, for instance when biotechnology products provide significant benefits for consumers or the environment" (OECD 2009, 11). The role of the state is to mitigate the effects of these attitudes. The BEB argues, "a major justification for government investments in science and technology is to overcome market failures; these occur when private investors invest less in technology than the *socially optimal* level" (White House, 2, emphasis added). What is "socially optimal," then, is not "known" by the market, but by technoscience—that is, by those who are in a position to see ahead into a future where particular technological potentials are realized and specific urgent needs are thereby met. The BEB continues, "In this context, scientific discovery is a public good that benefits all" (White House 2012, 2). Thus the market is only implicated in driving innovation as a post hoc validation of the intrinsic value of the technological products. A story is told about a market that will inevitably come to be by virtue of future demand for "as yet unimagined" goods. The (future) fact of the market legitimates the presumption that biotechnology innovation "is a public good that benefits all." It signals the right direction for innovation, even though it is itself an imagined future. In this way, the bioeconomy blueprint is articulated in the language of marketization and market governance, even as it subverts the premise of market as best able to "know" and govern progress, dislocating that construction of agency to science and technology.

Spawning the Bioeconomy

AA salmon offers an instructive example of how notions of what must be known and allocations of responsibility for knowing shape not only what biotechnologies are brought to market, but what kinds of markets they are brought to. One of the reasons the FDA took a long time to review the AA Salmon was because it repeatedly sought public input in the name of "transparency." The FDA confronted a deluge of public comments from environmental protection, anti-GMO advocacy groups, and many concerned citizens, including 38,000 comments on its draft environmental assessment, These ranged from concerns about the AA Salmon escaping into existing salmon populations (FDA 2015b) to criticisms of the scientific method of the regulatory assessment itself (Hansen 2010). The FDA dismissed as inapplicable to their review all but 90 comments (FDA 2015b). Nevertheless, AA salmon advocates alleged that the process was sullied. "Protests by anti-technology interests are, we believe, the root cause of the unconscionable delays in the FDA approval. These groups' criticisms are based on misplaced economic/marketplace concerns and reactionary fears of people who either don't understand or choose not to understand the science behind the AquAdvantage Salmon" (Letter to President Obama 2014). However, these critics located blame not primarily with publics who fail "to understand the science," but with a regulatory process that creates space for such public interventions: "American people, and indeed all people everywhere, are best served by a trusted objective regulatory process truly based on sound science, a system which can be counted upon to evaluate and act on the applications it receives *without fear of political interference*" (Letter to President Obama 2014, emphasis added). This repeated call to limit decision-making to considerations deemed "scientific" not only privileges scientific authority in in governance over other forms of knowledge (cf. Felt and Wynne 2007; Wynne 2007), but it defines the terms and political parameters of legitimate deliberation in the politics of imagining the future. The problem is not primarily epistemic (i.e. that the public gets it wrong), but political (i.e. that the process allows for such interventions in the first place). Insofar as the science of the bioeconomy "is a public good that benefits all," its integrity requires autonomy and deference from everyday democratic politics. If governance requires innovation, the conditions of possibility for innovation define the parameters of appropriate politics.

Criticism notwithstanding, FDA did in fact take a consonant approach in evaluating public concerns. The agency sorted the comments into three categories—those who opposed the approval of the salmon because they appeared to have an incorrect understanding of the relevant facts and the regulatory process, and therefore, the FDA surmised, "had not carefully read" the documents; second, those who opposed the approval based on suggestions that are not "reasonable" or "feasible" because they "may not be within the FDA's jurisdiction" (FDA 2015b); and third, those offering "a favorable opinion about the AquAdvantage Salmon application noting the thoroughness of the scientific, risk-based review and the benefits of AquAdvantage Salmon for sustainable aquaculture" (FDA 2015b). Expressions of doubt are subject to explanation, whereas attestations to the positive potential of the fish are affirmed because they align with the regulatory agency's understanding of itself as midwife of bioeconomic progress.

Rather than reading opposition to the AA salmon as an expression of democratic will, or even of as evidence of the need for deliberating issues of a sort that the FDA is ill-suited to engage, social concerns were dismissed as irrelevant. They were dismissed because they were not offered in the authorized idiom, but also because they were seen as a temporary impediment to be overcome. As John Entine of the Genetic Literacy Project predicted, "ten years from now the educated public is going to look back with a great sense of embarrassment at how fiercely groups that call themselves 'progressive' whipped up emotional concerns about whether GMOs are safe or sustainable. This opposition is based on ideology trumping science, at least in the short term" (Sentenac 2014). Whereas public concerns are insubstantial (even if inconvenient), the forward march of the bioeconomy is presumed to be natural and inevitable, with the public eventually falling into line.

Thus, the argument goes, once the technology brings a market into being, all will become clear, and ambivalence will give way to enthusiasm even as the market validates experts' prior judgments about the intrinsic value of the technology. But for this to happen, it must go to market, and market conditions must be arranged in advance for the technology to be viable. Policies that do not adequately sweep away barriers, including by insulating innovation from the public's "reactionary fears" and failures to "understand" the benefits of biotechnology, lead to "market failures." Thus the process that gives rise to the products of technology, including the exercise of technocratic authority to silence public dissent, is presumed to be virtuous, even if specific products are subjected to scrutiny and potential rejection. Thus the imaginary of innovation-as-governance that underwrites the vision of the bioeconomy defines the parameters of legitimacy such that the notions of what questions must be asked and answered, in what terms, drawing upon (and excluding) what forms of knowledge, is heavily insulated from democratic critique. This is not merely an epistemic posture. It is also rooted in an imagination of right governance: of the roles and responsibilities of science, state, and public in achieving security and well-being.

In this imaginary, the market is invoked to legitimate these arrangements: the intrinsic value of biotechnology will be demonstrated in (and validated by) the future bioeconomy. The fact that there will be a market even for "technologies as yet unimagined" (and the BEB presents this as an indubitable fact) is invoked to justify the (value-laden) judgment that technologies like the salmon are good for the world. To illustrate this point, let us unpack a sentence from the BEB. "*If it is to be successful and thrive, the bioeconomy will be based on a steady flow of new products and services that address American needs. To ensure this flow, policies must be developed and taxpayer dollars must be used responsibly to foster an ecosystem that supports discovery, innovation, and commercialization*" (FDA 2012, 3). The first sentence is relational—it constructs different actors in different roles which taken together constitute an economy. Thus the bioeconomy is composed of "products and services" flowing into a market in response to consumer demand, "address[ing] American needs." Producers' respond to consumers' demands. The second sentence, however, posits a temporal disjunction between present and future. The relationships of the first sentence are, therefore, aspirational—they exist in a future that can only be realized by policy action in the present. To ensure future supply, steps must be taken to create conditions amenable to biotechnology innovation even where a market for those innovations does not yet exist—e.g. where potential consumers are ambivalent and/or capital refuses to invest. In the context of the second sentence, "address[ing] American needs" is a task of governance undertaken in partnership with technoscience. Far from fading into the background, the state mitigates "market failures" in the name of enacting a desirable technological future.

Governing Future Consumers

Thus visions of the bioeconomy are (also) visions of governance: of the responsibilities of the state to manage the conditions of possibility for achieving a better future, where "better" is imagined in terms of the fruits of innovation. This task of diagnosing barriers and mitigating market failures is grounded in the notion that, together, state, and science are in a position to *know* the future in a way that citizens and markets are not. The science that the bioeconomy vision valorizes will "enunciate and address broad and challenging societal problems" (Committee on a New Biology 2009, 3). Seeing the "societal problems" to which the bioeconomy offers "solutions" requires the capacity to "enunciate" those problems, to discern what potential "solutions" are latent in living systems, and thus to determine what policy reforms are necessary to realize those solutions. Limiting questions of benefit and harm to narrowly construed notions of risk to public health and safety excludes other ways of knowing and rejects ambivalence about technological change that is not grounded in authorized constructions of risk (Levidow and Carr 1997; Wynne 2001). Importantly, this is not merely an effect of the epistemic parameters of risk assessment, but of a normative, political imagination of what forms of authority—epistemic and political—must be authorized to govern if progress is to be achieved.

Expert knowledge thus becomes a locus of political authority, even as public knowledge becomes an object of regulation. The following statement from the US BEB reflects this dual move: "In addition, because externalities and social attitudes influence market opportunities in the bioeconomy as they do in other economic arenas, a clear understanding of the benefits and risks of bioproducts is critical to the future bioeconomy" (White House 2012, 29). "Clear understanding" in the quote above can be read in two ways—(1) by experts, so that "sound science" governs not only innovation, but also governance itself such that it is placed in the service of scientific progress and thereby ensures that benefits will flow, and (2) by publics who, absent a clear understanding, might jeopardize an otherwise guaranteed technological future. Either way, "clear understanding" is figured as accountable only to itself. It is meant to govern, rather than derive from, the politics of democratic judgment about what forms of understanding are in the service of a good and just social order. Thus, good governance is construed as achieving right understanding—that is, of authorizing and empowering those ways of knowing that open the way to "socially optimal" technological transformation—and with the corollary that wrong ways of understanding must be mitigated and contained. But that which demands "clear understanding" is an imagined future world, a figment of imagination that is treated as more solid and real than the social world it is expected to mold and transform.

The AA salmon case reveals on a micro-level how such molding of social order is undertaken in the name of the intrinsic value of biotechnology. The FDA's evaluation of the acceptability of the salmon for the consumer was simultaneously an evaluation of what kind of consumer is appropriate to the salmon. The salmon was regulated in two stages. First, it was evaluated as a new veterinary drug for safety. As a technology, it was seen as radically novel, with all the hallmarks of bioeconomic promise. The FDA understood its task to be to "preserve safety" while opening the way to realization of these promises. The second stage regulated product labeling, i.e. the information that sellers would be required to provide to consumers. At this stage, the salmon (now deemed safe) shifted in FDA's process from radically novel technology to normal foodstuff, naturalizing a host of ethical, environmental, and biological particularities in a manner that is familiar from prior regulation of GM plants (Jasanoff 2005; Levidow and Carr 2000). For purposes of labeling, Aquabounty maintained that consumers need not (and ought not) be informed that their salmon steaks were a biotechnology product. All consumers needed to know, in FDA's estimation, was that the fish looked the same, tasted the same, and had the same nutritional value as regular salmon.[2]

The FDA is, most fundamentally, a regulator of information (Carpenter 2010). The statutory language that governs labeling specifies that consumers should receive any information that is "material." This leaves the FDA broad discretion. For AA Salmon, the FDA determined that the only information material for consumers was that it was Atlantic salmon. To justify this judgment, the FDA engaged in a remarkable exercise in ontological surgery (Jasanoff 2011). Whereas the fish was a novel drug at stage one, at stage two the fish was a fish. This was established by conducting a mitochondrial DNA analysis to establish that AA Salmon was, indeed, descended from Atlantic salmon and thus *is* Atlantic salmon.

The FDA's judgments about the label reflect a particular imagination of the agency and role of the consumer in the bioeconomy. The conditions of possibility for members of the public to exercise genuinely informed and rational choice obtain only when the benefits are immediate and therefore clearly understood: when salmon is cheaper and more available. Therefore, an organism-turned-technology can be—and should be—straightforwardly assimilated into a preexisting ontological commonsense. "Food from AquAdvantage Salmon is as safe to eat and as nutritious as food from other non-GE Atlantic salmon and…there are no biologically relevant differences" (FDA 2015a). Salmon is salmon.

This ontological move—assimilating the radically novel to the familiar—is not an innocent one. Although it appears to hide something from consumers that, on some accounts, they "have a right to know," it is justified by the notion that it is something they need not know, since knowing can only inhibit and not facilitate the unfolding of a world that is being made in their name. The notion that experts have a monopoly on judging the intrinsic value of technology (as opposed to whether it is value-able for consumers) is underscored by producers' right to remain silent: "a developer's First Amendment rights…protects it from forced prejudicial speech" where making the biotechnology (i.e. the engineered living entity) visible, as opposed to the market value of its product (i.e. its flesh) is "prejudicial speech" (Entine 2015).

The promise of the bioeconomy is that global society's most basic needs will be met and desires satisfied without cost or sacrifice: a world reengineered to make overconsumption into "sustainable" consumption while maintaining or increasing consumption. For such a world, the living inputs to systems of production are radically transformed, and habits of consumption persist, untroubled and unaware of the radically altered orderings of life undertaken in their name. The bioeconomy is assimilated into the economy.

Remaining so untroubled depends on the dual affirmation of technological transformation and the ontological stability of its products. Salmon is a novel technology. Salmon is salmon. This dual affirmation is foundational for the project of governance that we have described above, and it is to its political dimensions, not its ontological sleight of hand, to which we want to draw attention. The definitive deployment of the copula in each of these formulations, admitting of no ambiguity or ambivalence, grounded in the unequivocality of epistemic authority (sound science demonstrates…), reflects an imaginary of the right mode of governing the world-to-be-transformed. Responsibilities of governance lodge in the duty to engender rapid technological progress, undertaken in the name of addressing urgent societal needs. Given life's potential and society's urgent needs, there is no time for the sorts of frictions that democratic politics invites. Ambivalence is instead parsed as a temporary failure of politics and market. Dissent can be brushed aside as an expression of ignorance–a product of misunderstanding of what is valuable and good—only because such normative judgments get recast as epistemic ones. The *political* move of dismissing dissent is legitimated by the notion that defining (future) public goods—what society will need, want and value—is, in the present, an *epistemic* problem to be delegated to experts.

Conclusion

We have argued that the bioeconomic vision of innovation is a vision of governance. Indeed, the "concerned scientists and executives" who maintained that "more than a fish is at stake" declare that at stake are the very foundations for—and legitimacy of—government: "Your Administration's oft-stated commitment to sound science and the rule of law will only ring true if it is followed by prompt regulatory action, including actions that may not be politically popular in some circles." (Letter to President Obama 2014) For, the good of "the American people–and indeed all people everywhere–are best served by a trusted and objective regulatory process…which can be counted upon to evaluate and act on the applications it receives without fear of political interference." In this vision, political unpopularity—or, rather, the dissent of the governed—is a problem to be overcome by elevating the *political* authority of scientific expertise in the name of fulfilling responsibilities of government.

Planning and policy intervention by the state are underwritten by scientific expertise, legitimated in turn by the notion that since it is experts who are in the position to invent the future, they too are in the best position to know how it should be governed—to "*enunciate* and address broad and challenging societal problems." Imagining technoscience as a wellspring of capacity to satisfy the state's future responsibilities of governance, the state places itself in the service of technoscience in the present.

Thus the BEB outlines policy imperatives that must be undertaken in the here and now in the name of satisfying future demands and desires—i.e. bringing about future markets. Insofar as existing markets (as expressions of aggregate preferences and desires) fail to push innovation in the "right" direction, these are marked simultaneously as "market failures" and as political problems of recalcitrant publics—of retrograde "social attitudes." It thus becomes a policy imperative to produce the conditions in the present necessary to encourage innovation and to reshape markets: governments are responsible for supporting the "innovation ecosystem" while simultaneously mitigating social "barriers to innovation."

In this imaginary of innovation-as-governance, imagined technological futures are treated as more real—and thus are made more politically consequential—than the political expressions of ambivalence and dissent that stand in their way. The imagined potential of life as a technological cornucopia and the anticipation of urgent societal needs to which political institutions alone are incapable of ministering underwrite the notion that those who know how to innovate also know best how to govern—and that the impediments to progress that they (and only they) can foresee defines the agenda for political reform in the present. The choice we are presented with is between abundance and disaster. We are told that it is a choice on which we cannot delay: the fate of the future must be decided now. It is a choice that can tolerate no ambivalence, for friction and contestation, virtues of democracy's checks and balances, are the poison pill of true progress. Yet this vision of progress is an imaginary of governance. Lest we be too quickly swept up in its flight from ambivalence, we must ask not only what powers are expressed in the authority to know, but what imagination of progress stands behind the certitude of its truths.

Notes

1. Put differently, this is a constitutional relationship (Jasanoff 2011; Hurlbut 2017).
2. Although FDA determined that no label was necessary, a U.S. Senator from Alaska added a provision in a Congressional spending bill that directed the agency to develop a label. This process is ongoing at the time of writing.

References

Aquabounty. (2017). Retrieved May 2, 2017, from http://www.aquabounty.com.

Bakker, K. (2010). The limits of 'neoliberal natures': Debating green neoliberalism. *Progress in Human Geography, 34*(6), 715–735.

Birch, K., & Tyfield, D. (2013). Theorizing the bioeconomy biovalue, biocapital, bioeconomics or… What? *Science, Technology & Human Values, 38*(3), 299–327.

Cahill, D. (2011). Beyond neoliberalism? Crisis and the prospects for progressive alternatives. *New Political Science, 33*(4), 479–492.

Campbell, H. (2016). GMO salmon needed to feed the world. *San Diego Union-Tribune*. Retrieved August 7, 2016, from http://www.sandiegouniontribune.com/news/2016/jun/22/agriculture-gmo-salmon/.

Carpenter, D. (2010). Reputation and power: Organizational image and pharmaceutical regulation at the FDA. *Princeton studies in American politics: Historical, international, and comparative perspectives* (1st ed.). Princeton: Princeton University Press.

Committee on a New Biology for the 21st Century: Ensuring the United States Leads the Coming Biology Revolution; National Research Council. (2009).

A new biology for the 21st Century. Washington, DC: The National Academies Press.

Entine, J. (2015). Why we oppose GMO labeling: Science and the law. *Huffington Post*. Retrieved August 7, 2016, from http://www.huffingtonpost.com/jon-entine/gmo-labeling-science-and-_b_8871680.html.

Felt, U., & Wynne, B. (2007). *Taking European knowledge society seriously: Report of the expertgroup on science and governance to the science, economy and society directorate, directorate-general for research, European commission*. Luxembourg: Office for Official Publications of the European Communities.

Goven, J., & Pavone, V. (2015). The bioeconomy as political project: A polanyian analysis. *Science, Technology and Human Values, 40*(3), 302–337.

Hansen, M. (2010). Comments of consumers union on genetically engineered salmon, food and drug administration docket no. FDA-201034-N-0001, Veterinary Medicine Advisory Committee Meeting. Consumers Union. Retrieved August 7, 2016, from http://consumersunion.org/pdf/CU-comments-GE-salmon-0910.pdf.

Hilgartner, S. (2015). *Capturing the imaginary*. Science and Democracy: Making Knowledge and Making Power in the Biosciences and Beyond. Abingdon: Routledge

Hurlbut, J. B. (2016). Promising waste: Biobanking, embryo research, and infrastructures of ethical efficiency. *Monash Bioethics Review, 33*(4), 301–324.

Hurlbut, J. B. (2017). *Experiments in democracy: Human embryo research and the politics of bioethics*. New York: Columbia University Press.

Institute For Fisheries Resources et al., & V. Sylvia Mathews Burwell et al. (2016). Case no. 16-Cv-01574-Vc, United States District Court for the Northern District of California.

Jasanoff, S. (1999). The songlines of risk. *Environmental Values, 8*, 135–152.

Jasanoff, S. (2005). *Designs on nature: Science and democracy in Europe and the United States*. Princeton: Princeton University Press.

Jasanoff, S. (2011). Making the facts of life. In S. Jasanoff (Ed.), *Reframing rights: Bioconstitutionalism in the genetic age* (pp. 59–84). Cambridge: MIT Press.

Jasanoff, S. (2012). Taking life: Private rights in public nature. In K. Sunder Rajan (Ed.), *Lively capital: Biotechnologies, ethics, and governance in global markets* (pp. 155–183). Durham: Duke University Press Books.

Jasanoff, S., & Kim, S. H. (Eds.). (2015). *Dreamscapes of modernity: Sociotechnical imaginaries and the fabrication of power*. Chicago: University of Chicago Press.

Juma, C. (2016). *Innovation and its enemies: Why people resist new technologies*. New York: Oxford University Press.

Kennedy, D. (2016). *A world of struggle: How power, law, and expertise shape global political economy*. Princeton: Princeton University Press.

Letter to President Obama. (2014). Retrieved August 7, 2016, from www.ftrw.org/scientist_executive_letter_to_president_obama_on_biotechnology.docx.
Levidow, L., & Carr, S. (1997). How biotechnology regulation sets a risk/ethics boundary. *Agriculture and Human Values, 14*(1), 29–43.
Levidow, L., & Carr, S. (2000). Normalizing novelty: Regulating biotechnological risk at the US EPA. *Risk, 11*, 9.
Miller, H. (2015). How the genetically engineered salmon floundered in regulatory limbo for 20 Years. Forbes. Retrieved September 28, 2016, from http://www.forbes.com/sites/henrymiller/2015/11/20/how-the-genetically-engineered-salmon-floundered-in-regulatory-limbo/#70d2fb6e3c59.
Mirowski, P., & Plehwe, D. (2009). *The road from Mont Pelerin*. Cambridge: Harvard University Press.
OECD. (2009). The bioeconomy to 2030: Designing a policy agenda.
Pollack, A. (2010). Panel leans in favor of engineered salmon. *New York Times*. Retrieved September 28, 2016, from http://www.nytimes.com/2010/09/21/business/energy-environment/21salmon.html.
Pollack, A. (2015). Genetically engineered salmon approved for consumption. *New York Times*. Retrieved September 28, 2016, from http://www.nytimes.com/2015/11/20/business/genetically-engineered-salmon-approved-for-consumption.html?_r=0.
Sunder Rajan, K. (2012). *Lively capital: Biotechnologies, ethics, and governance in global markets*. Durham: Duke University Press Books.
Sample, I. (2010, May 20). Synthetic life breakthrough could be worth over a trillion dollars. *The Guardian*. Retrieved August 7, 2016, from https://www.theguardian.com/science/2010/may/20/craig-venter-synthetic-life-genome.
Sentenac, H. (2014). GM salmon: Will supermarket wariness sabotoge rollout of first genetically engineered animal? *Generic Literacy Project*. Retrieved August 4, 2016, from https://www.geneticliteracyproject.org/2014/03/12/gm-salmon-how-much-can-one-product-overcome/.
Slaughter, A.-M. (2005). *A new world order*. Princeton: Princeton University Press.
Staffas, L., Gustavsson, M., & McCormick, K. (2013). Strategies and policies for the bioeconomy and bio-based economy: An analysis of official national approaches. *Sustainability, 5*(6), 2751–2769.
Thompson, P. B. (2003). Value judgments and risk comparisons. The case of genetically engineered crops. *Plant Physiology, 132*(1), 10–16.
United States Food and Drug Administration. (2010). FDA Response to Public Comments to the Veterinary Medicine Advisory Committee. Retrieved May 1, 2017, from https://www.fda.gov/AnimalVeterinary/DevelopmentApprovalProcess/GeneticEngineering/GeneticallyEngineeredAnimals/ucm466221.htm.

United States Food and Drug Administration. (2012). AquAdvantage Salmon Draft Environmental Assessment.

United States Food and Drug Administration. (2015a). FDA has determined that the aquadvantage salmon is as safe to eat as non-GE salmon. Retrieved August 7, 2016, from http://www.fda.gov/ForConsumers/ConsumerUpdates/ucm472487.htm.

United States Food and Drug Administration. (2015b). Aquadvantage salmon—Response to public comments on the environmental assessment. Retrieved August 7, 2016, from http://www.fda.gov/AnimalVeterinary/DevelopmentApprovalProcess/GeneticEngineering/GeneticallyEngineeredAnimals/ucm280853.htm.

White House. (2012). National bioeconomy blueprint.

Winickoff, D., Jasanoff, S., Busch, L., Grove-White, R., & Wynne, B. (2005). Adjudicating the GM food wars: Science, risk, and democracy in world trade law. *Yale Journal of International Law, 30*(81), 81–123.

Wynne, B. (2001). Creating public alienation: Expert cultures of risk and ethics on GMOs. *Science as Culture, 10*(4), 445–481.

Wynne, B. (2007). Public participation in science and technology: Performing and obscuring apolitical–conceptual category mistake. *East Asian Science, Technology and Society: An International Journal, 1*, 99–110.

Zohar, Y. (2010). Genetically modified salmon can feed the world. *CNN*. Retrieved September 28, 2016, from http://www.cnn.com/2010/OPINION/09/22/zohar.genetically.engineered.salmon/.

Authors' Biography

Tess Doezema is a doctoral student in the School for the Future of Innovation in Society at Arizona State University

J. Benjamin Hurlbut is assistant professor in the School of Life Sciences at Arizona State University

CHAPTER 4

"Having a Structuring Effect on Europe": The Innovative Medicines Initiative and the Construction of the European Health Bioeconomy

Luca Marelli and Giuseppe Testa

Will it resemble the face of some persona
whom we believe we know: Europe?
And if its non-resemblance bears the traits
of the future, will it escape monstrosity?
Jacques Derrida, The Other Heading

L. Marelli (✉) · G. Testa (✉)
Department of Experimental Oncology, European Institute of Oncology (IEO), Milan, Italy

G. Testa
Department of Oncology and Hemato-Oncology, University of Milan, Milan, Italy

V. Pavone and J. Goven (eds.), *Bioeconomies*,
DOI 10.1007/978-3-319-55651-2_4

Introduction

In recent years, the European Union (EU) has experienced unprecedented turmoil. Ongoing contestations about the lack of democratic credentials of its founding treaties and institutions; largely ineffective programs to prop up its ailing economy vis-à-vis an unrelenting and self-perpetuating crisis; and strategic geopolitical challenges, ranging from disputed trade agreements to the rise of intense political instability across its borders: all these elements define the contours of a critical existential struggle. Longstanding and as yet unresolved, issues about the EU's political identity keep resurfacing along with questions on the legitimacy of its trans-boundary exercise of power (Lucarelli et al. 2011; Habermas 2015; Streeck 2014a): what should "Europe" stand for, and what are the sources and the defining tenets of "Europeness"?

Building on arguments advanced within constructivist research in the social sciences (see e.g. Christiansen et al. 1999), this chapter moves beyond the formal institutional structures of the EU and accounts for the *mise en politique* of the technosciences as a strategic site for the formation and (attempted) legitimization of a distinctively European identity (Callon 2004). In particular, the chapter probes how the intertwinement of biotechnology and capital that goes under the rubric of the "bioeconomy" is increasingly being recruited to frame the basic elements of the European supranational order in the making. Unsurprisingly for a political entity that proclaims its embrace of a knowledge-driven economy, the promissory expectations revolving around the bioeconomy have played a significant role in performing political and institutional reconfigurations throughout the continent (Goven and Pavone 2015). How, and to what effect, is what we aim to lay bare in the following pages, by specifically focusing on the revealing case study of the Innovative Medicines Initiative (IMI), the world's largest public-private partnership (henceforth: PPP) in the life sciences, and a mainstay of the European health bioeconomy sector.

The chapter is composed of three parts. First, we tease out different relevant strands within European public discourse that have propelled the construction of supranational imaginaries, and highlight the important role played therein by EU engagement with bioscientific research and innovation. Next, we introduce the European discourse on the bioeconomy, with specific regard to the biopharmaceutical sector, and expound its key tenets. Finally, we probe the latter through a case study-based analysis of IMI. Our findings reveal how the vision underpinning the governance set up of IMI has been geared towards a redefinition of the public-private boundary within the European biopharmaceutical research, and a skewed redistribution of agency among the academic and industrial actors involved in the PPP. Most notably, we show how the initiative programmatically endows "*European* pharmaceutical corporations"—itself a legitimacy framing that we problematize in the chapter—with the capacity to set the research agenda and the epistemic standards of this newly crafted, integrated public-private research and innovation space at EU level. Accordingly, we trace how the striving for competitiveness that is enshrined in the construction of the European health bioeconomy entails tensions and trade-offs with ideals of public accountability, fair distribution of resources, and science as a public good—thus raising salient issues for the making of the European identity that will hopefully spur sustained scholarly scrutiny.

Analytic Perspective and Methodology

Against the backdrop of public debates and scholarly efforts tackling issues of identity and legitimacy in the EU, we propose here an empirically-based account of a prominent science policy tool to probe the normativity that underpins the European project itself. Specifically, we draw from the notion of "sociotechnical imaginaries" (Jasanoff and Kim 2009, 2013, 2015), advanced within the field of Science and Technology Studies (STS), to capture technoscientific systems as powerful vehicles through which political visions and normative commitments gain material traction within the fabric of the EU polity.

We thus tackle the thorny issues of EU identity and legitimacy by taking socio-technical practices as windows onto the broader political dynamics that are ordering our continent. Such an approach is based on a twofold consideration. First, at a general level, we contend that it is impossible to make sense of current EU political developments—from the single currency project to the launch of the EU agenda for economic growth—without taking into account their structuring technoscientific elements. In turn, we maintain that focusing on the materiality of socio–technical *practice* represents a productive heuristics to expose the normative and political-economic ideal of society (the so-called "idea of Europe") encoded in and performed by EU policies; besides overcoming the pitfalls of idealistic approaches devoid of attention to real–world dynamics,[1] such heuristics also represents a revealing litmus test for uncovering, among the competing imagined futures, which one are elevated, and how, to a dominant position in the unfolding European order.

This study took its inception from a PhD research project on the governance of translational research based on the induced Pluripotent Stem Cell (iPSC) disease-modeling paradigm completed in 2016 at the European School of Molecular Medicine (SEMM), the European Institute of Oncology (IEO) and University of Milan, and relies on various sources of primary and secondary data. The latter comprise scientific, policy, and regulatory documents, journal articles, and videos. Documents have been accessed through official EU, IMI, and European Federation of Pharmaceutical Industries and Associations (EFPIA) websites, and further retrieved through PubMed search of "Innovative Medicines Initiative" (which yielded 75 results in title/abstract, out of 125 overall). Furthermore, from September 2014 to December 2015 participant observation through attendance at two meetings in one IMI project helped to solidify our understanding of the topic. In addition, fifteen semi-structured interviews were conducted from September 2014 to October 2016 with relevant actors, such as European Commission officials (identified in the text with the code "EC official"), IMI representatives ("IMI Rep"), Small–Medium Enterprises (SMEs) representatives ('SME Rep'), and academic scientists ("IMI Scientist") and partners ("IMI Academic Partner") involved in five different IMI-sponsored projects. Relevantly, we were refused interviews with one solicited representative of the pharmaceutical industry. The interviews explored respondents' perspectives on IMI, and all but two, for which interviewees did not give their consent, were recorded. Recordings were later transcribed and manually coded to look for emerging themes, and analyzed by means of critical discourse analysis methodologies.

Discursive Genealogies of European Construction

As argued by constructivist scholars in the social sciences, answers to questions of EU identity cannot be provided by reference to institutional dynamics alone (Rumford and Delanty 2005). Against the perceived lack of democratic legitimacy of its technocratic institutions and the absence of an "imagined community" (Anderson 1991) of European citizens supporting efforts at integration, the consolidation of Europe as a unified political space has become increasingly dependent, other than on treaties and communitarian rules, upon ideational resources shaping up as political imaginaries that articulate common European discourses, publics, and values (Jasanoff 2005). The configuration of this ideational space entails three main domains (Cerutti 2008: 3–4).[2]

The first is represented by long standing philosophical efforts towards a *substantive*, albeit-ever shifting, definition of the very concept of "Europe". From Hegel (1975 [1837]) to Husserl (1970 [1936]), from Heidegger (2000 [1953]) to Derrida (1992), to name but a few, the notion of Europe has elicited multiple interpretations (for a review, see Ferraris 1991) that recurrently engage, albeit critically, with the idea of a historically—and culturally—situated instantiation of an allegedly universal reason. Oftentimes, such framings of "Europe" have been produced *ex negativo* in processes of identity-construction, in relation to an ever-shifting "otherness" from which to gain differentiation (Said 1978).

The second domain consists of normative efforts at conceptualizing the European *political* space. Interestingly, whereas the 1700s and 1800s saw conceptions of Europe that grounded its identity in its political diversity, the idea of a politically-unified Europe started gaining traction in the intellectual discourse only in the 1900s (Becchi 2015). Alongside von Coudenhove-Kalergi's influential manifesto *Pan-Europe* (Coudenhove-Kalergi 1926 [1923]), notable early contributions are the often cited *Manifesto di Ventotene* (Rossi and Spinelli 2006 [1944]), a pamphlet calling for the establishment of a federal European State, as well as Friedrich Von Hayek's essay *The economic conditions of interstate federalism* (Hayek 1939). The latter advocated the removal of impediments to the movement of people, goods, and capital, and the creation of common rules of law, a single market as well as a uniform monetary system—thereby laying down the foundational principles for the political developments that have animated the continent since the 1970s. More recently, a vast body of literature has emerged following the start of the integration process in the 1950s, with Habermasian ideas of a "supranational democracy" (Habermas 2013) proving to be particularly well-received, especially within progressive circles. Representing competing answers to the questions of European political identity, the relevance of these different philosophical and normative models lies in their orientation of public discourse, in providing the normative underpinnings of different regimes of justification (Boltanski and Thevenot 2006)—as well as contestation—for the European project in the making.

A third major resource for European identity-building has been the engagement with *scientific research and innovation*, which has progressively led to the elaboration of a European science and technology (S&T) policy. Since its inception with the Rome Treaty of 1957, the project of European integration has indeed gained traction through, among other things, the joint articulation of science and democracy "as matters of progress and perfectibility" (Laurent 2015: 2). At stake in EU research and innovation policies, consequently, has been nothing less than a "socio-technical imaginary" of a common Europe (Felt 2015), that is, a shared idea of a socio-political order to be attained, which is encoded and enacted in the design of technoscientific projects (Jasanoff and Kim 2015).

In the domain of the biosciences, the establishment of European research programs is rooted in efforts spearheaded within the European Commission in the mid-1970s, which materialized in the early 1980s with the launch of the "Biomolecular Engineering Programme (BEP)" (1982–1986) (Aguilar et al. 2013). Since then, and ever increasingly since the turn of the 2000s, such programs have become strategic sites for the formation of a uniquely European political agenda (Nordmann 2009) and the establishment of a common sense of belonging that could legitimize the EU as a political project (Badii 2008). Significantly, therefore, the focus on bioscientific innovation as a key area for EU policy intervention aspires to more than typical efforts by liberal democratic states to legitimize themselves through instrumental uses of science and technology (Ezrahi 1990). Instead, in the absence of a European polity built on common language and culture, it was meant to frame certain issues, and the urgency of their resolution, as eminently European matters of concern and intervention (Rabeharisoa and O'Donovan 2014), and thus to *make Europe ontologically legible*. For, as Sheila Jasanoff has poignantly observed (2005: 93), "like nineteenth-century nation-states, the EU has found it necessary to specify the problems it wants to solve in order to consolidate and legitimate its political existence".

From the Lisbon Agenda to the Health Bioeconomy

Since the launch of the Lisbon Agenda in 2000, the expectations underpinning European policies in the biosciences have been increasingly articulated in relation to the so-called "knowledge-based bio-economy" (KBBE) (Chichocka et al. 2011). Against the backdrop of its explicit neo-liberal undertones praising the "benefits" delivered by the single currency, "stability-oriented monetary policies", "sound fiscal policies", and "wage moderation", the Lisbon Agenda set as its overarching objective that of turning Europe into "the most competitive and dynamic knowledge-based economy in the world" (EU Council 2000: 2). Biotechnology in particular, framed in subsequent EU policy documents as "probably the most promising of the frontier technologies" (EC 2002: 8), was envisaged to play a prominent role in this strategy.

What is evidently at stake in the Lisbon Agenda (and captured under the very rubric of "knowledge-based economy") is, firstly, the policy materialization of the mutually constitutive relationship between knowledge and (economic) value (Jasanoff 2004; Rajan and Leonelli 2013): new articulations of knowledge are explicitly recognized as leading to new forms of value-creation and, at the same time, new forms of value-creation are seen as supportive of new configurations of knowledge. Furthermore, and of special relevance for our analysis here, the narrative of the Lisbon strategy conceives of the new knowledge-based (bio-) economy as the lynchpin around which to consolidate the process of European integration and re-orient the European polity along the lines of competitiveness and innovation. Indeed, whereas earlier EU research and innovation policies pursued the objectives of enacting S&T cooperation across member states (Gusmão 2001) and closing the "competitiveness gap" in research with the US and Japan (Menéndez and Borrás 2000), in the post-Lisbon age such policies take on the additional and in fact defining task "of helping to create an integrated European knowledge society, and thus *Europe itself*" (Nordmann 2009: 280, italics ours).

Rooted in the framework of the Lisbon Agenda, which discursively set the stage for subsequent policy action, the European discourse on the bioeconomy was then further advanced and refined in the following decade (McCormick and Kautto 2013). Major milestones were the adoption of the first comprehensive *Life Sciences and Biotechnology* (LSB) strategy by the European Commission in 2002 (EC 2002), and the organization of a major conference on the Knowledge-Based Bio-Economy in Brussels in 2005 (EC 2005a). Furthermore, the so-called *Cologne Paper*, tracing a 20-year vision for the development of the bioeconomy in Europe, was outlined in 2007 at a conference hosted by the German Presidency of the Council of the European Union (EU Council 2007). More recently, the European Commission presented the *Innovating for Sustainable Growth: A Bioeconomy for Europe* strategy (EC 2012b), which represents a pillar of the current economic growth strategy of the EU (the so-called "Europe 2020 strategy").

Throughout all these policy steps, the EU discourse on the bioeconomy has been articulated around some core concerns aimed at naturalizing framings that reinforce specific political economic assumptions and desired courses of action, thereby operating in the preemption of alternative futures and undesired critical responses (Birch et al. 2012). A number of these concerns are especially relevant for the articulation of the discourse on the health bioeconomy in the biopharmaceutical sector, and the development of IMI that we trace below.

The first concern revolves around the need to accelerate the clinical and commercial translation of scientific findings (Mittra 2015). Framed as a "translational lag" between discovery and application (Aarden et al. 2015), and already identified as a "European paradox" in the first *European Report on Science and Technology Indicators* in 1994 (Nordmann 2009), it points to the alleged failure of converting constant advancements in science-based findings into tangible wealth-generating innovations. The translational lag has been variously described as a "gap" in R&D spending with the United States and Japan (and prospectively China) (EC 2006: 13), a systemic "failure" due to "major bottlenecks" in R&D processes (IMI 2014: 8), as well as an "(in)flexible approach" in the process of medicines' approval (Eichler et al. 2012: 234). Efforts to address these concerns have resulted in a vast array of translational initiatives—most notably within the 7th Framework Program and the Horizon 2020 funding scheme, the latter designed with the explicit intent of "taking great ideas from the lab to the market" (EC 2016)—as well as recent regulatory reforms intended to facilitate patients' access to innovative therapies and create a marketplace for them (EMA 2016). In times of austerity and fiscal consolidation, such initiatives are seen as prominent (supply-side) levers to sustain the competitiveness of the European biotech sector, by driving "smart, sustainable, and inclusive" growth throughout the continent (EC 2010), so as to "innovate Europe out of the crisis" (EC 2012b).

The second concern, expressed through multiple variations of the metaphor of a global race, addresses the *pace* of the desired change (Felt 2015). A sense of urgency pervades the European science policy discourse: Europe is pictured as having to "*act now* to harness the full benefits of the opportunities presented" (EU Council 2000: 1, italics ours). Immediate adhesion to the bioeconomy agenda is further solicited by portraying the threat of dreaded alternatives: accelerating the pace of innovation is mandatory, for "to be dilatory is to lose position in the global competition for advantage" (Salter 2013: 198).

Third, in order to "improve the speed and spread of innovation", the bioeconomy agenda calls for a "better alignment of EU research and innovation funding with the established priorities of bioeconomy-related policies" (EC 2012c: 11), so as to ensure that the process of knowledge-production lends itself to the pursuit of, and is shaped from the outset by, such priorities. Specifically, by the mid-2000s, research in the life sciences becomes increasingly and instrumentally enrolled as a key *means* for solving "complex and inter-related" societal challenges (EC 2012c). In turn, the latter are defined through a powerful framing that links economic competitiveness to sustainability. Whereas in the discourse on the "green" bioeconomy this framing is geared toward reconciling economic growth with ecological and environmental sustainability (EC 2012a; Birch et al. 2010), in the discourse on the health bioeconomy it explicitly aligns health with wealth (Birch et al. 2012): setting up schemes meant to make "a crucial contribution to delivering better health and wellbeing for all" is also seen as being conducive to "positioning Europe as a leader in the rapidly expanding global markets for health and wellbeing innovations" (IMI 2014: 12).

Fourth, *integration* is advanced as the privileged means to achieve, synergistically, the aforementioned goals. The Lisbon Agenda set indeed as one of its priorities the establishment of a unified European area of research and innovation (ERA), perceived as functional to turning Europe into a leading knowledge society and knowledge-based economy. Therefore, paramount to the realization of the bio-economy agenda becomes the—yet to be achieved—integration of different sets of public policies, of public and private resources, as well as of the distinct research and innovation efforts at Member States' level (EC 2012c). The epistemic correlate of this objective is the (attempted) definition of common S&T standards at EU level, with the creation of a European "technological zone" (Barry 2001) as an integrated techno-scientific space, increasingly seen as the pre-requisite for the achievement of ERA, and the economic and political integration of Europe itself (Callon 2004).

Against this backdrop, in order to probe the modes through which integration is achieved and agency is distributed among the actors involved, we turn now our attention to a revealing case study that represents the mainstay of the European health bioeconomy: the Innovative Medicines Initiative (IMI).

The Origins of IMI in the European Technology Platforms (ETPs) and Joint Technology Initiatives (JTIs)

IMI, the world's largest public–private partnership in the life sciences, couples the European association of pharmaceutical industries (EFPIA) with the European Commission (EC), and originated in 2004 as one of the European Technology Platforms (ETPs), which were created within the

Lisbon Agenda for the achievement of ERA. The ETPs were established within FP6 (2003–2007) as industry-led stakeholder *fora*, sustained by public and private funding, aimed at shaping the developmental trajectories of new technologies while creating a marketplace in Europe for new R&D-intensive products (EC 2007). In EU policy documents, the establishment of ETPs is described as a "bottom-up process" initiated by industrial stakeholders, namely, "top executives from leading companies in the sector" reaching a consensus view on a "strategic vision document" setting out desired research and technological development priorities (EC 2005a: 5). As such, ETPs were explicitly conceived as a way to "better align EU research priorities with industry's needs" (EC 2007: 1), with the overarching aim of "boost[ing] competitiveness" and establishing "world leadership for the EU" in distinct (bio)economic sectors (such as food, agriculture, and biotechnology) (Aguilar et al. 2009: 381).

In order to cement a more formal partnership between the EC and industrial actors, FP7 (2007–2013) subsequently oversaw the launch of the Joint Technology Initiatives (JTIs). The JTIs were meant to implement the agenda of a limited number of ETPs (whose scale and scope of objectives required dedicated governance mechanisms, which the various ETPs were not able to provide (EC 2015)), with the intent to "make a unique contribution to Europe's industrial competitiveness in strategic technologies" (EC 2005b: 10). Among the few ETPs that evolved into JTIs was IMI, officially launched in 2008. As an IMI representative recalled:

> IMI was a bet, something absolutely innovative in the political landscape of the time, in the relationship between industry and the European Commission. Apparently, IMI had a great return, so much so that the European Commission wanted to invest more. There was a lot of interest towards IMI. IMI strived to obtain visibility and credibility, and apparently, today, that goal has been reached. Possibly, among the JTIs, IMI is the one that reached the highest visibility, and that was among the key objectives set forth by the governing board of the European Commission and EFPIA. (IMI Rep 1)

The development of IMI occurred in two distinct phases. In the first phase (IMI1), which ran from 2008 to 2013, IMI was endowed with 2 billion Euros provided jointly by EC and EFPIA, the latter through "in kind" provisions. Within the framework of IMI1 around forty projects were launched, in a wide variety of areas ranging from cancer to stem cell research (EC 2012c). IMI2 was established as a 6-year initiative in 2014 within Horizon 2020, and was endowed with an increased budget of 3.3 billion Euros. With respect to its progenitor, IMI2 maintains a more translation-oriented approach, covering areas of research that are closer to clinical application, and it involves a broader range of actors, such as SMEs, ICT companies, and non-profit organizations.

The "Strategic Research Agenda" and the Translational Goals of IMI

IMI's overarching objectives were outlined in a strategic vision document drafted by EFPIA in 2004 and later adopted, through a process of consultation with relevant actors, by IMI's *Scientific Research Agenda* (SRA) in 2008 (IMI 2011). The scope and goals of IMI2 were set in 2014, following a similar process, by the *Strategic Research Agenda* of IMI2 (IMI 2014). While reflecting a shifting scientific and historico-political *milieu*, with IMI2 unfolding amidst the lasting consequences of the economic crisis, the two agendas advance largely similar priorities. In particular, they re-produce and articulate a well-established discursive code that ties together the acceleration of clinical translation to industrial growth and capitalization of biotechnology, grounding the urgency for action upon a purported lag vis-à-vis the US and, as for the 2014 document, BRIC countries such as China, Brazil, and India (IMI 2011: 10, 2014: 13).

More specifically, IMI was conceived as "the most translational end of the EU research landscape" (IMI Scientist 1), designed to "revitalize the European biopharmaceutical research environment" (EFPIA 2004: 4) and to "create a truly unified space of public and private scientific and technological research" at EU level (IMI Scientist 2). In line with the "translational lag narrative" that frames the acceleration of translation as the *removal* of the impediments hindering the seamless transition from scientific discovery to clinical application (Aarden et al. 2015), as well as with the KBBE agenda that insists on *institutional* as much as technological change (Birch et al. 2012), IMI was introduced as an organizational innovation targeting one of the major translational "bottlenecks", namely the persistent productivity crisis in pharmaceutical R&D (Pammolli et al. 2011). As for EFPIA's original vision document of 2004, the problem is diagnosed in terms of the high attrition rate during the late stages of drug development, which entails significant costs for the industry. However, whilst being an issue perceived as pressing within the whole pharmaceutical sector worldwide (even though contested, see

Light and Warburton 2011), the problem is presented here as a *specific matter of concern for Europe*. Failures within the industry's own R&D model are framed as a lack of European public investments, resulting in a pattern of industrial disinvestment from the EU towards the US; conversely, initiatives to reduce development costs are seen as a pre-requisite for allowing Europe to "again become a place where industry chooses to invest" (EFPIA 2004: 5). In the words of a former representative of IMI's top management:

> In the early 2000s, the discussion for establishing IMI revolved around two things: the first and most important was that at the time there was a trend of pharmaceutical companies moving out of the EU, especially R&D moving out to the US and other parts of the world. Europe used to be called "the pharmacy of the world", and there were fears this could change because of decisions of the pharma industry. The other dimension was that the pharma industry said: if you want to change this, you have to develop another model to promote therapeutic innovation in Europe, and this model might be a public-private partnership. (IMI Rep 3)

Enacting the Public–Private Partnership: "Changing the Way We Do Research in Europe"

The public-private partnership (PPP) thus emerged, at the beginning of the 2000s, as the preferred solution to address the industry's shortcomings. Normatively, the implementation of PPP is based upon the goal to remove the adversarial character of the public-private interaction (Linder 1999), so as to foster co-operation in capital- and knowledge-intensive domains where the "complexities of the challenges" require "a more efficient sharing of expertise" among public and private actors (EC 2013: 3). In the domain of the life sciences, PPPs reflect the increasingly heterogeneous, multi-disciplinary, and cooperative nature of research practices (Penders et al. 2015). Specifically, as contingent alignments between different institutional logics within one organizational entity (Skelcher and Smith 2015), PPPs testify to the *socio-economic hybridity* that underpins large-scale "biomedical platforms" (Cambrosio et al. 2009), which are a defining feature of contemporary research infrastructures (Webster 2015).

Biomedical platforms pursue the objective of *integrating* a diverse array of resources, skills, and expertise (Keating and Cambrosio 2003). In doing so, however, they do not merely establish an interface between well-defined, self-contained organizations—each pursuing different aims according to distinct systems of incentives (Cambrosio et al. 2009). Rather, as they feed off the organizations' own agency to create a new institutional configuration, at the same time they redefine the very identity, agency, and goals of the organizations themselves (Latour 2005).

For IMI, the capacity to reconfigure the biomedical research landscape became a defining tenet of its mission. As explained by the CEO of Roslin Cell, leading partner in the IMI-funded EBiSC project, speaking at the IMI stakeholders forum in Brussels in 2014:

> "IMI projects are big. And we are here not just to deliver the research, but actually to have a structuring effect in Europe, and change the way we do research in Europe" (Courtney 2014).

This desired "structuring effect" that IMI intends to achieve becomes manifest at a variety of levels. In the first place, IMI is envisaged to perform the ***reconfiguration of the identities, roles and functions of industrial and academic actors*** in the process of biomedical knowledge-production at the European level. For one thing, the IMI framework is meant to push academic research to become more translation-oriented, and engage in the commercialization of the knowledge it generates. In the words of an academic partner of an IMI-sponsored project, IMI was conceived as an "emergency procedure" intended to make European academia a site of innovation in its own right:

> I think academia in Europe still does not see the whole of the translational chain. It is basic research, but it does not go all the way towards the clinic, towards commercial exploitation. Universities in Europe don't have functional Technology Transfer units. If a professor comes up with something worthy of commercial exploitation, it's complex, there is not the infrastructure in play, the translation still doesn't work very well. In the US they are much more professionalized. The IMI projects are supposed to be a way to stop this.... I think there was a feeling at the EU level that initiatives put in place in the past [to foster commercialization of academic research] were not done efficiently. Universities have the knowledge, but they don't exploit that efficiently. IMI-funded projects are then something like an emergency procedure. (IMI Academic Partner 1)

More specifically, our analysis reveals the rationale for turning academia into a generator of commercially actionable innovation to be threefold.

From the part of EU officials, the intention has been to gather support from the EU citizenry for continued, and in fact increased, funding of biomedical research and innovation in order to advance the bioeconomy and innovation agenda of the EU. As a discursive construct, "clinical translation" (the envisaged endpoint of IMI projects), is thus mobilized both as a *desired outcome* and a *justificatory tool* for the initiative, as attested by the following remarks from a prominent official at the Directorate-General for Research and Innovation of the European Commission:

> Ten, fifteen years ago, in the early days, the industrial-manufacturing landscape in the EU was quite different. The idea was, if academia comes up with good ideas, they would have been picked up by industry and developed. And there was almost a kind of antipathy between academia and the industry. But what we have seen is that at the EU level the resources going into biomedical research have increased in recent years through the different framework programs, largely because of the EU parliament and their influence, and they are representing the citizens of EU, who want to see something back for their investment. To demonstrate that money is justified we need to see something come out of it—and the upstream, more fundamental part of research is seen as a bit of a luxury. That's been the reason [for the emphasis on translation]. Now the whole debate about the EU project, referenda being in the balance in some cases… So the trend, the driver, has been to connect more with citizens, and try to get something out of it. That's my take on the change in emphasis… On the luxury side, this is the public perception of scientists in academic and public labs in today's cold, wintery world in Northern Europe, that they have got a nice lab, a permanent staff, they are doing something which they like, it's probably their hobby. In contrast to the precarious world outside… (EC official 1)

In addition to that, the increased engagement of academic scientists with translational research was also perceived as a crucial asset that pharmaceutical industries could leverage in order to address "gaps" and failures in their R&D model, thereby accelerating the development of novel and commercially exploitable products. As a SME representative observed, IMI has represented a prominent platform where the industry could promptly "reach out to academia and understand how to use novel technologies [pioneered] in academic research" (SME Rep 1). More broadly, as a prominent German scientist similarly noted, academic involvement in research alongside the industry has been part of an overall strategy, "of which IMI is a prominent instrument, but which is generalized to the whole of Horizon 2020", for "outsourcing R&D in the academic sector":

> Until a few years ago, the trend in the pharmaceutical industry was to have the whole of R&D in house. But this turned out, for them, to be problematic. It's inflexible, costly, rigid, the field moves fast and they have to keep pace with it, and adapt accordingly… And they don't have access to the top league of scientists, who are in academia. Industry has gaps, and for them is good to have academic scientists solving problems. (IMI Scientist 1)

From the part of academia and academic spin-offs, the rationale for partaking in IMI projects has instead been based on two distinct considerations. On the one hand, and in line with recent STS research findings about physicians' involvement with pharmaceutical companies as a way of overcoming the financial constraints of existing research infrastructures (Wadmann 2014), IMI projects are seen as "cash cows" (IMI Scientist 3) that, notwithstanding the provision of data access to the industry, represent a relatively uncumbersome way for "getting funding for one's own research" (IMI Scientist 2). On the other hand, the more translation-oriented, large-scale academic labs conceive of IMI as a translational "niche" (Schoot and Geels 2008) in which they can test and develop their products in a pre-competitive and protected setting ahead of commercialization. Similarly, academic spin-offs and SMEs see EU-funded IMI projects as a way to reduce financial risk while building capacity and thus establish an advantage over competitors:

> As we are a translational lab, we use IMI as a platform for our technology. With IMI, we are very close to our potential clients, the industry, and we can listen to their needs at a pre-competitive phase. (IMI Scientist 4)

> [IMI-sponsored project] is extremely important to us, because it's an EU grant-funded activity, so it's the best also if you are losing money. What it allows our company to do is to achieve scale, which is a key competitive advantage over other companies. (SME Rep 2)

Other than endeavoring to change the well-established *modus operandi* of academic actors, especially by turning academia into a site of commercially exploitable innovation, a further prominent reconfiguration is enacted by IMI *within* the boundaries of the pharmaceutical industry itself. EFPIA 2004's "strategic vision document" already placed emphasis upon pre-competitive collaboration as a way to address the deficiencies of industrial R&D (EFPIA 2004: 5; Birch et al. 2012). Since IMI1, and more prominently since the launch of IMI2, the removal of entrenched competitive barriers among industries becomes, accordingly, a key priority. Besides creating closer links between academic and industrial research (to the point of embedding them within each other), IMI was also primarily geared to foster *collaboration* between different pharmaceutical companies at a "pre-competitive stage" of research. The latter notion, ubiquitous in IMI documents, refers to what Webster and Eriksson (2008) conceptualize as the realm of "basic innovation", that is the early stage of knowledge-generation characterized by the goal of reducing epistemic uncertainty by collectively coming to an agreement on common standards. As an IMI representative noted:

> For EFPIA, one of the first priorities that led to the establishment of IMI was to optimize resources of all the companies conducting the same kind of research, to identify and address more rapidly and at a lower cost issues that everyone had (but everyone was working in its own corner, thus duplicating expenses), without competing in that specific domain. For instance, if we find out that a certain compound is toxic (or not)... this information is of interest for everyone. IMI was launched to ease the worries of the industry in sharing these kinds of data. (IMI Rep 2)

The Leading Role of the Pharmaceutical Industry

While other PPPs in the life sciences were launched in the 2000s in the US, such as the FDA's Critical Path Initiative and the NIH's Public–Private Partnership program,[3] IMI differs in being, programmatically, an industry-led initiative (Goldman 2012). In addition to what we have observed so far, two further elements should be outlined in this regard.

First, IMI projects stem primarily from pharmaceutical companies, which set the projects' agendas and define their topics. IMI projects are thus launched when a group of pharmaceutical industries find their agreement upon a line of (pre-competitive) research to be pursued collaboratively, elaborate a call and find suitable academic partners. Finally, once the consortium is assembled, the project receives IMI funding, and can thus start its operations (see Fig. 4.1). As an IMI project manager argued:

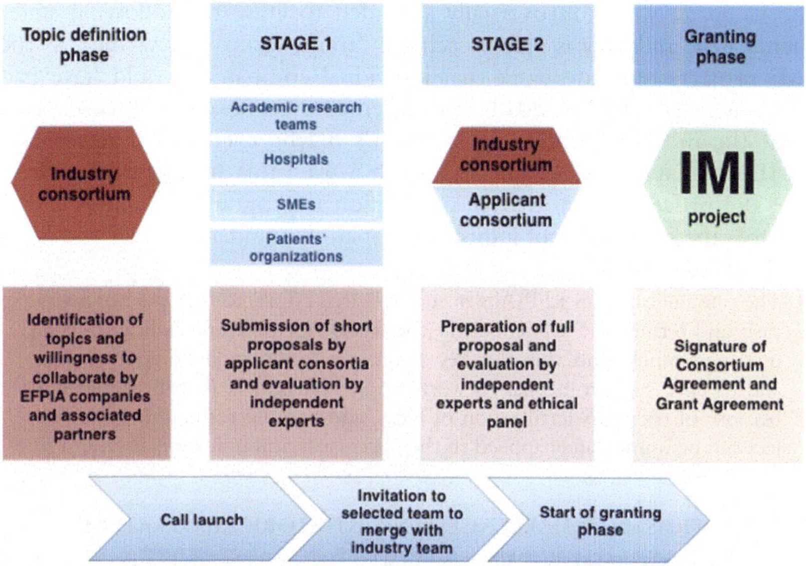

Fig. 4.1 Topic definition in IMI projects *Source*: adapted from IMI presentation (https://www.imi.europa.eu/)

> Under IMI and IMI2, *because* industry is giving half of the contribution, the topic idea is coming from industry... Generally, [the topic] is something on which the industry is already working, and the proposal they put forward relates to that. It would be very difficult for them to commit to a totally new idea, the ideas have to come internally. (IMI Rep 4)

Thus, if IMI projects are meant to create a unified European space of public and private technoscientific research, it is the industry that defines the boundaries of this integrated landscape. Conversely, public funding provided by the European Commission takes the form of an overt subsidy to the industry's R&D strategy, reinforcing, with few or no challenges, the demarcations traced by the industry within the political geography of this central domain of European biomedical research.

Second, the contributions provided by EFPIA are "in kind". That is, rather than committing their own financial resources, EFPIA contributes its 50% share to the IMI budget by providing facilities, equipment, and staff time. This way, as EFPIA itself acknowledged on its website (before the webpage was controversially amended to erase the following statement),[4] the industry is able to achieve "tremendous cost savings, as the IMI projects replicate work that individual companies would have had to do anyway" (EFPIA 2011) (one scientist even observed: "who knows what the in kind contributions are? It's maybe even less than 50%..." (IMI Scientist 1)). More importantly, however, this arrangement ensures that the process of knowledge-production is programmatically oriented towards the generation of industry-compatible standards:

> The one billion from EFPIA is in staff research effort, so they are not giving cash and letting the academics do the work, but they are dedicating their own personnel. And this is a key point: we fund a collaborative project, and to have staff from pharma working on that project is really important, because of the cross-fertilization of ideas, and because results from the project can be immediately applied to the pharmaceutical industry. (IMI Rep 5)

Criticisms, Challenges and Rhetorical Strategies of Legitimacy: "When You Dine with the Devil, You'd Better Have Long Forks"

In light of the lopsided governance arrangements shown above, IMI has also attracted a substantial amount of criticism. An investigative report published in 2015 by the German magazine *Spiegel Online*, Swiss public broadcaster *SRF* and the Belgian newspaper *De Standaard* pointed to a number of issues posing "key challenges for IMI's credibility" (Spiegel Online 2015). Under the heading of "the people pay, corporations cash in", the news outlets identify, among them, the enrolment of patient advocacies as lobbyists for the pharmaceutical industry, the focus on industrial profits rather than medical innovation, and the lack of transparency in projects and budget management. Similarly, in a featured article in the British Medical Journal, prominent Italian scientist Silvio Garattini and colleagues reported the insurmountable difficulties encountered, due to barriers posed by the industry, in accessing data within a collaborative IMI project from which they later withdrew (Garattini et al. 2013). These criticisms followed in the footsteps of concerns raised by the League of European Research Universities (LERU) and the European University Association (EUA) as to the opacity of IP arrangements, said to largely favor the industry, and the peripheral role of academia in the steering of research projects (LERU 2010; EUA 2010).

Other pressing challenges emerging from our analysis concern the diverging expectations of the actors involved, and the frictions generated by attempts at epistemic and regulatory standardization within IMI projects.

With regard to the former, albeit turning to IMI for research funding, the academic partners in IMI projects we interviewed were consistent in pointing to a mismatch of expectations, owing to both the different aims pursued by academic and industrial partners and their distinct ways of operating. For instance, whereas academic scientists pursue "innovation" in the form of epistemic *novelty*, pharmaceutical corporations frame "innovation" as the generation of tangible outputs, to be achieved through highly *standardized* "systems of production" laden with connotations such as robustness, efficiency, and quantitation of output (Rheinberger 1997).

With regard to the frictions generated within the PPP, relevant stumbling blocks have proven to be the differences in normative standards as well as "language, way of working, style of interacting with each other, and communicating" (IMI Academic Partner 1) between industrial and academic actors groups. As one prominent academic scientist evocatively summed up his own experience with IMI and the difficulties in navigating its research environment: "when you dine with the devil, you'd better have long forks" (IMI Scientist 1). Consider, in addition, the following interview excerpts:

> Pharma is very pushy with regard to the IP, and scientists believe that it's like in other EU programs, such as Horizon[2020], where the IP is balanced between the industry and academia. So... The scientists are astonished that pharma gets access to academic R&D, but that's the reality, and they'd better be prepared! Pharma is sometimes very pushy, and you have to be alert... (IMI Scientist 1)

> IMI projects are not about innovation, they are about standardization, which is a completely different thing. They are not meant to sustain the flow of innovation, but to freeze it. That's what the industry needs—but it's different from our aims. (IMI Scientist 3)

> They don't understand that [the informed consent form] should be like a contract between the donor and the facility. It is not marketing, it shouldn't have pretty pictures on it. But [the one proposed] it does. It is high gloss, 160-gram paper, happy children, sunshine, green landscape... There is a problem of translation, from commercial to academic language. And the reason for that is that all the documents were done by [an

> international law firm]...which spoke the language of commercial partners. (IMI Academic Partner 1)

Against such criticisms and concerns, IMI's rhetorical strategies for legitimacy have mostly revolved around two main framings. Firstly, IMI representatives have strived to ***depoliticize*** IMI's role, highlighting its independence and mediating role between EFPIA and the EC, the latter acting as the spokesperson (Latour 2005) for both European taxpayers and academic research institutions. In official documents and journal publications, IMI has thus been variously described as a "neutral and autonomous organization", an "honest broker", a "neutral platform", a "neutral ground for conflict resolution", a "neutral trusted organization", and a "neutral third" (see e.g. Goldman 2011, 2012). Acting as a "compensation chamber for conflict resolution" (interview with IMI Rep 5), IMI's role is thus rhetorically presented as that of a "boundary organization" (Guston 2000), meant—on behalf of its two "principals", EFPIA and the EC—to stabilize the public–private boundary through the generation of commonly agreed standards, and to facilitate interactions between academic scientists and industrial representatives. As remarked by an IMI project manager:

> We are the executive office that manages the process. The concepts, the ideas come from both our partners. Everything is set by them together, and we run the process according to the rules we have been given. (IMI Rep 4)

Secondly, as a discursive strategy in support of its funding scheme, IMI rhetorically bestows upon major pharmaceutical corporations partaking in IMI projects a ***European identity and citizenship***. This legitimacy framing becomes manifest in the requirement, within the IMI grant agreement, that only costs sustained within the EU and partner countries (such as Switzerland) are eligible for EU funding—implying the existence of something like a "*European* pharmaceutical industry". However, this provision seems to largely ignore the quintessentially global nature of multinational pharmaceutical corporations, which are oftentimes registered overseas, and operate on global markets on a trans-continental scale. As such—and even notwithstanding a later provision allowing the industry to sustain a large percentage of its operating costs outside of the EU (field notes 2015)—it comes perhaps as no surprise that this framing plays out more as a legitimacy strategy for justifying relevant public funding towards the industry, rather than an effective provision for enacting a profound reconfiguration in the industry's inner organizational structure.

Conclusions

Since the 2000s the discourse on the knowledge-based bioeconomy, conceived as a pillar of the growth strategy of the EU, has been playing an increasingly prominent role in defining the sociotechnical imaginary around which the EU polity was meant to coalesce. Heralded as nothing less than a "new Renaissance", as in the EC's final report on ERA (EC 2012a), the mandate to "innovate Europe out of the crisis" (*ibid.*) has become a *mantra* underpinning the promise of the European project, by conjoining the notions of innovation and crisis without, however, problematizing either.

First, the call for *more innovation*, mostly taken as self-explanatory, largely sidesteps the question of what kind of innovation should be pursued, and for whom, while failing to address some prominent causes of the crisis currently unfolding in the EU: namely, the very model of innovation that led financial models to increasingly produce, rather than describe, markets' dynamics (MacKenzie 2008), the relentless pursuit of austerity policies and, at an even deeper level, the macro-economic imbalances generated by the adoption of the single-currency (Bagnai 2012; Lapavitsas 2012). In doing so, it reproduces a political economic narrative that controversially identifies *supply-side* mechanisms as the privileged means to address a *demand-side* crisis—the latter attested, and indeed propelled, by declining investments, wage stagnation, and rising inequality and unemployment rates, as well as a persistent deflationary environment throughout the continent (Bagnai 2014).

Second, the pressing objectives of *accelerating translation* and *enhancing S&T integration* within the bioeconomy agenda run the risk of reinforcing the lopsided political geography, traced by the institutional developments following the Maastricht Treaty in 1992, between a European "center" and its "peripheries". On the one hand—as analysts have noted, for instance, with regard to unequal research capacity across Member States leading, through competitive Framework Programmes funding, to a concentration of European public resources in a small number of elite institutions (Naczinsky 2016)—these imperatives are conducive to the amplification of imbalances between Member States' innovation systems, thus weakening the very idea of European cohesion they were purported to support (Felt 2015). On the other hand, as paradigmatically attested by IMI's governance arrangements traced in this chapter, they also underpin the redefinition of the public–private boundary and a lopsided allocation of agency among institutions and actors groups within the biomedical research landscape

(Nowotny 2015). Indeed, failures in the industry's R&D model are framed in IMI as matters of concern for the whole EU, and thus identified as priorities to be addressed by leveraging public resources and academic research know-how. In addition, pharmaceutical corporations are then endowed in IMI projects with agenda-setting prerogatives, and are thus able to define epistemic standards from the outset of the process of knowledge production, with academic institutions running the risk of ancillary engagements. Accordingly, the bioeconomy agenda advanced through IMI appears as a prominent means through which contemporary biosciences are enrolled in a specific form of empire-making (Jasanoff 2006), one geared to increase the power of *institutional* (rather than geographical) "centers" vis-à-vis "peripheries" of science and technology. Furthermore, the ideal of economic competitiveness and the sense of urgency enshrined in such initiatives increasingly generates tensions and trade-offs with ideals of fair distribution and access to public resources and science as a public good.

For all of the above, the bioeconomy emerges as not only an economic development but indeed as a core axis of a full-fledged political project (Goven and Pavone 2015), that, also via initiatives such as IMI, lays down a relevant part of the material infrastructure through which the EU polity is envisaged to consolidate itself. As such, 17 years after the launch of the Lisbon Agenda, the units of meaning of the biosciences, defined through the molecular gaze in the guise of genes, cells, and sequencing reads (Nowotny and Testa 2011), seem to increasingly take up the role once assigned to coal and steel as building blocks for European economic and political integration.

In conclusion, our analysis brings to the fore the enabling role performed by technoscientific systems in materializing normative ideas about Europe and operationalizing them within the fabric of the European polity. In this light, prominent initiatives such as IMI are exposed here as something more than sites for the mere "rehearsal" or "continual exercise" (Felt 2015) needed to *maintain* a shared imagination of a common Europe. Rather, they represent prominent "experiments" (Nordmann 2009) through which normative ideas about the Europe-to-be acquire the performativity to *produce* effects and thereby sculpt into being the twenty-first century image of that *persona* of "Europe" that, with Derrida, is said to eschew any given once-and-for-all representation (Derrida 1992).

Notes

1. For a compelling critique of idealistic approaches to issues of EU identity and legitimacy, see Streeck's criticism of Habermas' defense of EU monetary integration (Streeck 2014b).
2. Methodologically we share here the perspective advocated by, among other scholars, Furio Cerutti, who proposes a phenomenological, rather than normative, approach based on a reflexive notion of identity, whereby identity can be defined as "what the citizens and the elites perceive as shared values and principles: a process of self-identification" (Cerutti 2008: 3). As we clarify in the chapter accordingly, philosophical and political ideas of Europe become relevant for issues of European identity insofar as they are *mobilized* and *re-elaborated* in the public discourse, and thus become powerful resources for the formation of performative imaginaries.
3. In 2004, following publication of its influential report on Stagnation/ Innovation, the FDA launched its PPP, the Critical Path Initiative, as a means to share data, expertise, and resources so as to drive innovation in the development of medical products. To this end, the C-Path Institute was created. An Arizona-based non-profit body, the institute was meant to support this initiative by fostering collaborations between industry, academia and regulators. Funding sources were varied and included grant funding from the FDA, fees from participating member organizations, donations from private and philanthropic organizations. Around the same time, as an outgrowth of the NIH Roadmap launched in 2003, the NIH initiated its Public–Private Partnership program, with the goal of developing an advisory support with various non-governmental organizations, such as industry, foundations, and advocacy organizations, in setting up complex, multi-sector arrangements oriented to the acceleration of translation of laboratory research findings (Goldman 2012).
4. Cf. Spiegel Online (2015).

Acknowledgments This work was supported by a PhD scholarship from the University of Milan and the Umberto Veronesi Foundation for the Advancement of Science grant (L.M.); the EPIGEN Flagship Project of the Italian National Research Council (G.T.); and the COST action 'Bio-objects and their Boundaries: Governing Matters at the Intersection of Society, Policy and Science' (G.T. and L.M.).

References

Aarden, E., Blasimme, A., Holloway, D., & Marelli, L. (2015, May 18–19). *Making sense of clinical translation: Ethical, regulatory and policy challenges for the EU and the US.* Position paper presented at the international workshop Making Sense of Clinical Translation. Hermance: Brocher Foundation.

Aguilar, A., Bochereau, L., & Matthiessen, L. (2009). Biotechnology as the engine for the knowledge-based bio-economy. *Biotechnology and Genetic Engineering Reviews, 26*(1), 371–388.

Aguilar, A., Magnien, E., & Thomas, D. (2013). Thirty years of European biotechnology programmes: From biomolecular engineering to the bioeconomy. *New Biotechnology, 30*(5), 410–425.

Anderson, B. (1991). *Imagined communities.* London: Verso.

Badii, R. (2008). EU red biotechnology regulation and European values. In *The search for a European identity: Values, policies and legitimacy of the European Union.* Abingdon: Routledge

Bagnai, A. (2012). *Il tramonto dell'Euro.* Reggio Emilia: Imprimatur.

Bagnai, A. (2014). *L'Italia può farcela: Equità, flessibilità, democrazia. Strategie per vivere nella globalizzazione.* Milano: Il Saggiatore.

Barry, A. (2001). *Political machines: Governing a technological society.* London: A&C Black.

Becchi, P. (2015). È la fine dell'Europa? *A/Simmetrie WP 2015/11.* Rome: Italian Association for the Study of Economic Asymmetries.

Birch, K., Levidow, L., & Papaioannou, T. (2010). Sustainable capital? The neoliberalization of nature and knowledge in the European "knowledge-based bio-economy". *Sustainability, 2*(9), 2898–2918.

Birch, K., Levidow, L., & Papaioannou, T. (2012). Self-fulfilling prophecies of the European knowledge-based bio-economy: The discursive shaping of institutional and policy frameworks in the bio-pharmaceuticals sector. *Journal of the Knowledge Economy, 5*(1), 1–18.

Boltanski, L., & Thévenot, L. (2006). *On justification: Economies of worth.* Princeton: Princeton University Press.

Callon, M. (2004). Europe wrestling with technology. *Economy and Society, 33*(1), 121–134.

Cambrosio, A., Keating, P., Bourret, P., Mustar, P., & Rogers, S. (2009). Genomic platforms and hybrid formations. In *Handbook of genetics and society* (pp. 502–520). London: Routledge.

Cerutti, F. (2008). Why political identity and legitimacy matter in the European Union. In *The search for a European identity: Values, policies and legitimacy of the European Union* (pp. 3–22). London: Routledge.

Christiansen, T., Jorgensen, K. E., & Wiener, A. (1999). The social construction of Europe. *Journal of European Public Policy, 6*(4), 528–544.

Cichocka, D., Claxton, J., Economidis, I., Högel, J., Venturi, P., & Aguilar, A. (2011). European Union research and innovation perspectives on biotechnology. *Journal of Biotechnology, 156*(4), 382–391.

Coudenhove-Kalergi, R. N. (1926 [1923]). *Pan-Europe*. New York: Alfred A. Knopf.

Courtney. (2014). EBiSC: The European Bank for induced pluripotent Stem Cells. Retrieved from https://www.youtube.com/watch?v=GQ_Om8OMZUo&index=3&list=PLvrEEDAAI_jFF0fSLckLaTAgXVww8AKEt.

Delanty, G., & Rumford, C. (2005). *Rethinking Europe*. Abingdon: Routledge.

Derrida, J. (1992). *The other heading*. Bloomington: Indiana University Press.

EC. (2002). Life sciences and biotechnology. A strategy for Europe. Retrieved from http://ec.europa.eu/biotechnology/pdf/com2002-27_en.pdf.

EC. (2005). New perspectives on the knowledge-based bioeconomy. Retrieved from https://ec.europa.eu/research/conferences/2005/kbb/pdf/kbbe_conferencereport.pdf.

EC. (2005). Report on European technology platforms and joint technology initiatives: Fostering public-private R&D partnerships to boost Europe's industrial competitiveness. Retrieved from https://ec.europa.eu/research/fp7/pdf/tp_report_council.pdf.

EC. (2006). Creating an innovative Europe. Report of the independent expert group on R&D and innovation. Retrieved from http://www.eua.be/Libraries/research/aho_report.pdf?sfvrsn=0.

EC. (2007). Third status report on Euroepan technology platforms. Retrieved from ftp://ftp.cordis.europa.eu/pub/technology-platforms/docs/etp3rdreport_en.pdf.

EC. (2010). Europe 2020. A strategy for smart, sustainable and inclusive growth. Retrieved from http://ec.europa.eu/eu2020/pdf/COMPLET%20EN%20BARROSO%20%20%20007%20-%20Europe%202020%20-%20EN%20version.pdf.

EC. (2012a). Innovating for sustainable growth: A bioeconomy for Europe. Retrieved from http://ec.europa.eu/research/bioeconomy/pdf/official-strategy_en.pdf.

EC. (2012b). The new Renaissance: Will it happen? Innovating Europe out of the crisis. Third and final report of the European research area board. Retrieved from https://ec.europa.eu/research/erab/pdf/3rd-erab-final-report_en.pdf.

EC. (2012c). Report of the independent expert panel accompanying the report assessment of European innovative medicines initative 2. Retrieved from https://ec.europa.eu/research/health/pdf/imi-ppp-expert-panel-report_en.pdf.

EC. (2013). Public-private partnerships in Horizon 2020: A powerful tool to deliver on innovation and growth in Europe. Retrieved from http://eur-lex.europa.eu/LexUriServ/LexUriServ.do?uri=COM:2013:0494:FIN:EN:PDF.

EC. (2015). DG research & innovation. Overview on JTIs. Retrieved from http://ec.europa.eu/research/jti/index_en.cfm?pg=home.

EC. (2016). Horizon 2020. The EU framework programme for research and innovation. Retrieved from https://ec.europa.eu/programmes/horizon2020/en/what-horizon-2020.

EFPIA. (2004). Creating biomedical R&D leadership for Europe to benefit patients and society. Retrieved from http://ec.europa.eu/research/fp6/p1/innovative-medicines/pdf/vision_en.pdf.

EFPIA. (2011). The innovative medicines initiative (IMI). Retrieved from https://web.archive.org/web/20111023095832, http://www.efpia.eu/Content/Default.asp?PageID=515.

Eichler, H. G., et al. (2012). Adaptive licensing: Taking the next step in the evolution of drug approval. *Clinical Pharmacology and Therapeutics, 91*(3), 426.

EMA. (2016). Human medicines: Regulatory informations. Retrieved from http://www.ema.europa.eu/ema/index.jsp?curl=pages/regulation/landing/human_medicines_regulatory.jsp&mid=.

EU Council. (2000). Presidency conclusions. Retrieved from http://www.europarl.europa.eu/summits/lis1_en.htm?textMode=on.

EU Council. (2007). En route to the knowledge-based bioeconomy. Retrieved from, https://dechema.de/dechema_media/Cologne_Paper-p-20000945.pdf.

EUA. (2010). Stakeholders request urgent revision of Innovative Medicines Joint Technology Initiative (IMI). Retrieved from http://www.eua.be/activities-services/news/newsitem/10-07-14/Stakeholders_request_urgent_revision_of_Innovative_Medicines_Joint_Technology_Initiative_IMI.aspx.

Ezrahi, Y. (1990). *The descent of icarus. Science and the transformation of modern democracy*. Cambridge: Harvard University Press.

Felt, U. (2015). Social science expertise in European innovation policy. In *Future directions for scientific advice in Europe*. Cambridge: Centre for Science and Policy.

Ferraris, M. (1991). Postfazione to *Oggi l'Europa*. Milano: Garzanti.

Garattini, S., Bertele, V., & Bertolini, G. (2013). A failed attempt at collaboration. *BMJ, 347*, f5354.

Goldman, M. (2011). Reflections on the innovative medicines initiative. *Nature Reviews Drug Discovery, 10*(5), 321–322.

Goldman, M. (2012). Public-private partnerships need honest brokering. *Nature Medicine, 18*(3), 341–342.

Goven, J., & Pavone, V. (2015). The bioeconomy as political project: A polanyian analysis. *Science, Technology and Human Values, 40*(3), 302–337.

Gusmão, R. (2001). Research networks as a means of European integration. *Technology in Society, 23*(3), 383–393.

Guston, D. (2000). *Between politics and science. Assuring the integrity and productivity of research.* Cambridge: Cambridge University Press.

Habermas, J. (2013). Democracy, solidarity and the European crisis. *Roadmap to a Social Europe.* Retrieved from http://www.pro-europa.eu/index.php/en/at-issue/european-identity/11-jürgen-habermas-democracy,-solidarity-and-the-european-crisis?path=/home/proeuro/public_html/.qidb.

Habermas, J. (2015). *The lure of technocracy.* Cambridge: Polity Press.

Hayek, F. A. (1939). The economic conditions of interstate federalism. *New Commonwealth Quarterly, 131,* 49.

Hegel, G. W. F. (1975 [1837]). *Lectures on the philosophy of world history.* Cambridge: Cambridge University Press.

Heidegger, M. (2000 [1953]). *Introduction to metaphysics.* New Haven: Yale University Press.

Husserl, E. (1970 [1936]). *The crisis of European sciences and transcendental phenomenology: An introduction to phenomenological philosophy.* Evanston: Northwestern University Press.

IMI. (2011). The Innovative Medicines Initiative (IMI) scientific research agenda. Revision 2011. Retrieved from https://www.kowi.de/Portaldata/2/Resources/horizon2020/coop/2014-IMI-SRA.pdf.

IMI. (2014). The right prevention and treatment for the right patient at the right time. Strategic research agenda for innovative medicines initiative 2. Retrieved from http://www.imi.europa.eu/sites/default/files/uploads/documents/IMI2_SRA_March2014.pdf.

Jasanoff, S. (2005). *Designs on nature: Science and democracy in Europe and the United States.* Princeton: Princeton University Press.

Jasanoff, S. (2006). Biotechnology and empire. *Osiris, 21*(1), 273–292.

Jasanoff, S. (Ed.). (2004). *States of knowledge: The co-production of science and the social order.* London: Routledge.

Jasanoff, S., & Kim, S. (2009). Containing the Atom: Sociotechnical imaginaries and nuclear power in the United States and South Korea. *Minerva, 47*(2), 119–146.

Jasanoff, S., & Kim, S. (2013). Sociotechnical imaginaries and national energy policies. *Science as Culture, 22*(2), 189–196.

Jasanoff, S., & Kim, S. H. (Eds.). (2015). *Dreamscapes of modernity: Sociotechnical imaginaries and the fabrication of power.* Chicago: University of Chicago Press.

Keating, P., & Cambrosio, A. (2003). *Biomedical platforms: Realigning the normal and the pathological in late-twentieth-century medicine.* Cambridge: MIT Press.

Lapavitsas, C. (2012). *Crisis in the Eurozone.* London: Verso Books.

Latour, B. (2005). *Reassembling the social*. Oxford: Oxford University Press.
Laurent, B. (2015). Perfecting European democracy. Science as a problem of technological and political progress. Retrieved from https://halshs.archives-ouvertes.fr/hal-01142379/document.
LERU. (2010). LERU letter on the Innovative Medicines Initiative (IMI). Retrieved from http://www.leru.org/index.php/public/news/imi/.
Light, D. W., & Warburton, R. (2011). Demythologizing the high costs of pharmaceutical research. *BioSocieties, 6*(1), 34–50.
Linder, S. H. (1999). Coming to terms with the public-private partnership: A grammar of multiple meanings. *American Behavioral Scientist, 43*(1), 35–51.
Lucarelli, S., Cerutti, F., & Schmidt, V. (2011). *Debating political identity and legitimacy in the European Union*. London: Routledge.
MacKenzie, D. (2008). *An engine, not a camera: How financial models shape markets*. Cambridge: MIT Press.
McCormick, K., & Kautto, N. (2013). The bioeconomy in Europe: An overview. *Sustainability, 5*(6), 2589–2608.
Menéndez, L. S., & Borrás, S. (2000). Explaining changes and continuity in EU technology policy, The politics of ideas. CSIC.
Mittra, J. (2015). *The new health bioeconomy: R&D policy and innovation for the twenty-first century*. Basingstoke: Palgrave Macmillan.
Naczinsky, C. (2016). A plea for a newly united Europe. ERA Austria. Retrieved from https://era.gv.at/object/document/2575/attach/0_20160419_Policy_Brief_EN.pdf.
Nordmann, A. (2009). European experiments. *Osiris, 24*(1), 278–302.
Nowotny, H. (2015). The radical openness of science and innovation. *EMBO reports*, e201541546.
Nowotny, H., & Testa, G. (2011). *Naked genes: Reinventing the human in the molecular age*. Cambridge: MIT Press.
Pammolli, F., Magazzini, L., & Riccaboni, M. (2011). The productivity crisis in pharmaceutical R&D. *Nature Reviews Drug Discovery, 10*(6), 428–438.
Penders, B., & Vermeulen, N. (2015). *Collaboration across health research and medical care: Healthy collaboration*. New York: Routledge.
Rabeharisoa, V., & O'Donovan, Ó. (2014). From Europeanization to European construction: The role of European patients' organizations in the shaping of health-care policies. *European Societies, 16*(5), 717–741.
Rajan, K. S., & Leonelli, S. (2013). Introduction: Biomedical trans-actions, post-genomics, and knowledge/value. *Public Culture, 25*(3 71), 463–475.
Rheinberger, H.-J. (1997). *Toward a history of epistemic things: Synthesizing proteins in the test tube*. Stanford: Stanford University Press.
Rossi, E., & Spinelli, A. (2006). *Il manifesto di Ventotene*. Milano: Mondadori.
Said, E. (1978). *Orientalism*. New York: Pantheon.

Salter, B. (2013). Governing innovation paths in regenerative medicine: The European and global struggle for political advantage. In *The Global Dynamics of Regenerative Medicine* (pp. 194–216). London: Palgrave Macmillan.
Schot, J., & Geels, F. W. (2008). Strategic niche management and sustainable innovation journeys: Theory, findings, research agenda, and policy. *Technology Analysis & Strategic Management, 20*(5), 537–554.
Skelcher, C., & Smith, S. R. (2015). Theorizing hybridity: Institutional logics, complex organizations, and actor identities: The case of nonprofits. *Public Administration, 93*(2), 433–448.
Spiegel Online. (2015). *The people pay, corporations cash in.* Retrieved from http://www.spiegel.de/international/europe/imi-in-eu-project-citizens-count-corporations-cash-in-a-1025550.html.
Streeck, W. (2014a). *Buying time: The delayed crisis of democratic capitalism.* London: Verso Books.
Streeck, W. (2014b). Small-state Nostalgia? The currency union, Germany, and Europe: A reply to Jürgen Habermas. *Constellations, 21*(2), 213–221.
Wadmann, S. (2014). Physician–industry collaboration: Conflicts of interest and the imputation of motive. *Social Studies of Science, 44*, 531.
Webster. (2015). The health of collaborations. A reflection. In *Collaboration across health research and medical care: Healthy collaboration.* New York: Routledge.
Webster, A., & Eriksson, L. (2008). Governance-by-standards in the field of stem cells: Managing uncertainty in the world of "basic innovation". *New Genetics and Society, 27*(2), 99–111.

Authors' Biography

Luca Marelli is a post-doctoral fellow with the Research Unit on Science and Technology Studies at the European Institute of Oncology (IEO) in Milan. He completed (2016) a PhD in Foundations and Ethics of the Life Sciences at the European School of Molecular Medicine (SEMM) and University of Milan, and was a visiting research fellow in the Program on Science, Technology and Society at the Harvard Kennedy School (2013–2014).

Giuseppe Testa is a life scientist and STS scholar. A twice European Research Council (ERC) awardee, he is professor of Molecular Biology at the University of Milan and Principal Investigator at the European Institute of Oncology (IEO), where he heads the laboratory of Stem Cell Epigenetics and the Research Unit on Science and Technology Studies (STS), and where he co-founded the PhD program in Foundations and Ethics of the Life Sciences within the European School of Molecular Medicine (SEMM).

PART II

Making Value and Remaking Research in the Bioeconomies

CHAPTER 5

The Underworlds Project and the "Collective Microbiome": Mining Biovalue from Sewage

Luísa Reis-Castro

Introduction

It was a freezing morning on February 11, 2016, as I biked through the slushy streets of Cambridge, Massachusetts. I was on my way to meet a group of researchers at the corner of Portland and Washington Streets, just outside the borders of the MIT campus. We were meeting to take samples of the sewage flowing beneath the city streets. A worker from the traffic department also joined in. In fact, he had earlier closed the traffic on Washington Street and now opened up a manhole so that researchers could send sampling instruments down into the sewage beneath the streets. With my hands almost numb from the bitter cold, I helped these scientists tie their aprons and put on their silicon gloves; since I had not received any training in their designated tasks, this was the extent to which I could help. A cotton patch was laid just next to the manhole and a small electric pump placed on top. A light snow started falling.

L. Reis-Castro (✉)
HASTS, MIT, Cambridge, MA, USA

V. Pavone and J. Goven (eds.), *Bioeconomies*,
DOI 10.1007/978-3-319-55651-2_5

A long and thin hose attached to the pump was dropped inside the rusty manhole. Wastewater slowly started moving up the hose, filling a plastic sampling bottle…

This experiment was part of the *Underworlds* project, an interdisciplinary, cross-departmental research collaboration funded by a $4 million grant from the Kuwait-MIT Center for Natural Resources and Environment (CNRE). Its aim is to develop a system to collect and analyze sewage water in order to investigate urban health patterns (Underworlds 2015). The group at MIT now includes people in labs from Civil and Environmental Engineering to Computer Science and Artificial Intelligence, but the enterprise started early on as a conversation between two MIT professors, one from Urban Studies and Planning and another from Biological Engineering. The researchers I met on that cold February morning were almost all affiliated with the Biological Engineering lab. They are part of MIT's new Center for Microbiome Informatics and Therapeutics (CMIT) and focus on "computational and experimental models to engineer the human microbiome" (MIT News 2014; Engineering the Human Microbiome 2015). A former lab member, one of the project's conceivers, explains *Underworlds'* goal as aiming "to develop the science and technology needed to build a proof-of-concept system to track microbial indicators of health and disease in real time, through the city's "*collective gut*" (Turgeman 2015, 14) [emphasis added]. In the first semester of 2016, I accompanied lab members doing sampling procedures in the streets of Cambridge, watched as they transformed the wastewater into material for lab experiments, joined them on a visit to the Boston area Waste Water Treatment Plant facilities at Deer Island in Boston Harbor, attended talks where they were presenting their work, and conducted semi-structured interviews with project members.[1]

The *Underworlds* project aims to create what these scientists call a "smart sewage platform" to investigate human health and behavioral data by identifying the bacteria, viruses, and chemical compounds present in sewage waste. Such organisms and elements were once in the human gut and were shed through stool and urine; they aggregate in the sewers and compose what the researchers call a "collective microbiome" (Underworlds 2015). This microbiome is made up of the genes and genomes of the microorganisms present in the human body. Under the optic and analytic of the microbiome, the human body becomes a "microbial body"; the microbe-human relationship is posed not as one of adversarial relation, marked by potential disease, but rather as a largely symbiotic relation (Sangodeyi 2014). The microbiome defines the very constitution of the biological environment of the body. The *collective* microbiome, however, is a more distributed entity. It crosses many bodies and places. And it lurks in the sewers. Researchers from *Underworlds* argue that this subterranean collective microbiome can be mined for community health information: the underground sewage as a proxy for the world at street level and above.

The *Underworlds* website opens with these words: "A vast reservoir of information on human health and behavior lives in our sewage. And this resource is untapped." Mining the DNA of microbes living in the sewage as a way to inquire after human health can be understood as a mode of finding what medical sociologist Catherine Waldby (2000) has called *biovalue,* defined as being "generated wherever the generative and transformative productivity of living entities can be instrumentalized along lines which make them useful for human projects" (33). Feces and urine constitute a paradoxical category within the frame of biovalue, since they contain living matter (microbes) taken (or expelled) from the body, but are not themselves human tissue. Hence, the biovalue being "tapped" by *Underworlds* includes both the microbes (and chemicals) and the biological matter of those who have flushed away their bodily waste.

By investigating the *Underworlds* project, this chapter traces scientific practices of creating value from sewage, from metabolized chemicals and microbes. Although I argue that *Underworlds* produces a new form of value, I first recount how the project fits within a long history of trying to obtain information about a community in the sewers and an even longer history of trying to remake waste—as loss of value—into something once again valuable. I then outline the practices of the *Underworlds* research team members as they establish three proxies: (1) wastewater in the sewers as aggregated community information; (2) microbes and chemicals as data about human health and behavior; and (3) aggregated samples averaged out as biomarkers for bodily conditions and health of individuals in the community. This chapter describes how the *Underworlds* team must constitute the "collective microbiome" through these three proxies in order to mine biovalue from the wastewater in the sewers. In their study of tissue economies, Catherine Waldby and Robert Mitchell (2006: 114–115) have traced how the notion of waste is established through an active detachment and cast off from its sources, a separation between fragment and whole. The *Underworlds,* as it explores sewers in the form of "collective gut," simultaneously separates and rearranges: the wastewater as both an extension and a fragment of human bodies.

But the *Underworlds* team aspires to go beyond mining health information from the sewers of Cambridge. It also aims to create a smart sewage platform that enables implementation elsewhere and "scalability" (Turgeman 2015: 81). The project is also an effort to improve and develop methods and technologies to better understand the link between samples and bodies, between the structure and behavior of microbes and their relationship to humans, and between biomarkers and the diseases and conditions they indicate. Hence, while the project is grounded on the three proxies it aims at the same time to refine and tune the relations that sustain these proxies. The datasets and techniques developed and improved by the research team can be employed in other settings and capitalized in other endeavors. For example, some members of the *Underworlds* team started working on the establishment of the *Everybiome* startup. Through analysis of individual stools, *Everybiome* would offer "daily personal health analytics [that enable] better management of chronic diseases, diet and baseline health monitoring, early diagnosis, and real-time epidemiology" (Everybiome 2015). This chapter argues, then, that as it constitutes the "collective microbiome," *Underworlds* also contributes to the actualization and circulation of these bioeconomic proxies.

Shit and Sewage

The study of "shit" and sewage can reveal much about the humans who produce these substances, and it can also tell us much about the social relations into which ideas and flows of these substances enter. The psychoanalyst Dominique Laporte (1978) investigates the *History of Shit* as a path toward understanding the history of modern subjectivity. His book opens by recounting a French royal decree of 1539 regulating the disposal of human waste in Paris, marking new disciplines and boundaries for the individual and for the state. For Laporte, the negative reference to the odor of shit, particularly the odor of the shit of the "other," is a marker of the emergence of the individualized body and of the bourgeois state. Shit is transformed during this moment into something private, and public shit is redefined as sewage, though also marked as something to be shuttled out of sight and smell. This private/public distinction brings with it a new set of responsibilities for individuals and the state. For Laporte, waste becomes the ambivalent and inexorable dilemma of civilization: "The necessary outcome of socially profitable production, it is the inevitable by-product of cleanliness, order, and beauty" (14). The proper individual domestication of fecal matter and the state control of sewage become the maximum signifiers of civilization.

The ceaseless production of waste brings with it attempts to transform such matter into something valuable—a successful economy must make even its waste circulate (Cooper 2008: 45–49). Or, as Waldby and Mitchell (2006) put it, "Waste material which may appear as the very opposite of value in one context can become the starting point for the generation of significant degrees of value in another" (84). Waste, as loss of value, has also within it the possibility to become, once again, valuable. "But shit cannot be converted into cash through mere elimination" (26), Laporte reminds us.

In the 1850s, for example, with the population growth of London and the popularity of water closets amongst the upper and middle classes, the Thames River was transformed into an immense sewer with a foul stench during low tides. Concerns were raised about the appalling condition of the river, and the Royal Society of Arts started a debate about the disposal and utilization of sewage. Constructing a sewage system, however, was an elaborate and expensive proposition. The German chemist Justus von Liebig presented a proposal to not only solve the sewage issue, moving it away from Londoners' eyes and noses, but also to use it profitably. Liebig urged Europeans to implement a practice already long adopted in China of returning human and animal waste into the farmed soil.[2] Transforming London wastewater into manure would allow this practice to be done on a large scale. For Liebig, reusing the wastewater was not only a profitable solution to the sewage issue in London but also a necessity in the wake of population growth. Returning the waste to the soil would guarantee the fertility needed for the land to continue to produce (Brook 1997). Sewage as fertilizer transforms a noxious anti-commodity—money must be paid in order for it to be taken away—into profitable and essential manure. From value in its absence, the value of the waste would come from its potential to continue (food) production. As Laporte (1978: 34) says, "It can enrich the earth and sprout again as gold." Shit, like money, is made to circulate.

Move now, to thinking of the Boston area waste treatment facility on Deer Island, close to the city's harbor, which still, today, transforms wastewater into manure. *Underworlds* also hopes to retransform the waste, to add a new use and value to it—but the project seeks to use sewage to produce a quite different form of (bio)value than manure. This novel value-production was often foregrounded to me by the members of *Underworlds* when explaining the importance of the project. For example, on an April afternoon, I met with Xiaoying,[3] a postdoc who had just started her project with the *Underworlds* group. Sitting in a noisy café on the MIT campus, I asked if I could record the interview to save me from taking copious notes while we talked. As I was setting up the recording device, she took a book from her backpack. She said, "I brought this for you; it might help you understand what we are doing." The book was *Water 4.0*, by David Sedlak. She explained that, for Sedlak, Water 1.0 referred to the transport of water into cities from far away places—water distribution dating back to Roman aqueducts. Water 2.0 represented treating the water before drinking it—advancements in water treatment around the nineteenth century. Water 3.0 was the result of humans cleaning their discharge—the developments in sewage treatment during the twentieth century. Finally, Water 4.0 emerged with developments to transform sewage into useful material as fertilizer and energy. The *Underworlds* project, Xiaoying argued, would push Water 4.0 to yet another level (perhaps Water 5.0?), by transforming the wastewater into *information*. The information *Underworlds* hopes to gather concerns the patterns of human community health, diseases, and wellness. It would collect this data through the analysis of microbial and chemical components found in the biological matter that has converged to the sewage.

The group works to detect two types of information that correlate to the community's health: "metabolic signatures of human activity" and "pathogens of interest in the collective microbiome" (Turgeman 2015, 100). For the first type of information, they analyze the metabolites—small molecules identified as the intermediaries or product of metabolic activity. In this metabolomic analysis, or (as it has been described) the "systematic study of the unique chemical fingerprints that specific cellular processes leave behind" (Daviss 2005, 25), the research team hopes to be able to make inferences about the population's wellbeing, nutrition, or behavior (e.g., drug use). For the second type, pathogens of interest, the research team is working on creating a viral detection and a metagenomic platform. The lab has developed an algorithm called "Latent Strain Analysis," a method that "separates reads into biologically informed partitions and enables assembly of individual genomes" (Cleary et al. 2015, 1053). This method allows them to detect low-abundance strains that might be very diffused within the wastewater collected, and later collect genomic information on these strains (Preheim et al. 2013). The researchers describe the metabolic signatures and pathogen DNA, or biomarkers, as their "raw data," which can later be transformed into meaningful information (Turgeman 2015, 98). The idea of raw data, however, can be considered an oxymoron (see Gitelman 2013). After all, all data needs to be collected and in the act of collecting there are already put in motion assumptions about the correlations and representations such data can yield..

Collecting a "collective microbiome"—in the sense of bringing it together through systematic sampling and meticulous data analysis and modeling—was the goal of *Underworlds*. During my research, I was told several times how they planned to create a "smart sewage platform" (Underworlds 2015) that could enable the gathering, exploration, monitoring, manipulation, and communication of urban health information. They seek to establish

> a data platform that aims to guide public health policy, municipal strategy, urban planning and epidemiological science in a larger sense. The system offers the possibility to build rich databases that document, integrate and analyze fluctuations in urban health parameters over time. (Turgeman 2015, 83)

Hence, the platform aims not only to collect data but also to potentially transform how health practitioners and public health policy makers relate to data about communities. This process, in which data-driven analytics propose to not only refine practices but reconfigure institutions and the relations between and within these institutions is what Shreeharsh Kelkar (2014) has defined as "platformization." (See also Gillespie (2010) for the strategic and political uses of the term "platform" within media studies.) But how can flushed-stool-turned-sewage-turned-sample-turned-data be turned into health information about a community that can inform public health policies and practices?

Flushing Data

The idea of sewage not only as wastewater but also as wasted data, an untapped resource out there waiting to be mined, was almost ubiquitous during conversations with lab members. For example, Carmen, a grad student working in the project, remarked during a presentation of her work at MIT that, "every day we flush health data from the toilet." Xiaoying, during our interview explained, "All our poop goes into the sewage, every day we are generating this biological sample. We just don't know how to use it. We just don't know *yet* how to decompose this information." Stool becomes no longer waste, but a "noninvasive sample rich in health data"; and more, one that is already "submitted routinely" (Turgeman 2015, 77–78).

This informational rudder can steer understandings within epidemiology and public health management. Anthropologist Natasha Dow Schüll (2016) examines trends towards gathering and examining big data,[4] looking at how wearable technologies, with their continuous tracking capacities, transform what will count as personal and population data. Such monitoring devices move the management of disease and health into the hands of individuals as "quantified selves."[5] The constant flow of data produced by these gadgets transforms the clinical gaze (Foucault 1973) into "informational bits whose cumulative diagnostic value could never be detected by the eyes of a doctor in real-time" (Schüll 2016, 9).These continuous measurements can produce an overwhelming amount of data that needs to be rendered legible.

Underworlds aims to expand this quantification endeavor in order to gather data not about individual bodies, but about a collective. The constant stream of the sewage is transformed into a constant flow of epidemiological data, measured through "the study of different species of bacteria, viruses, and chemical compounds that live in the human gut and converge in the city's sewage—our collective microbiome" (Underworlds 2015). The data mined from the sewage needs to be rendered legible, something the group is still working on.

What, then, are the imaginaries and assumptions of those making this "big data" legible (Boellstorff 2013)? For example, some of the team members are developing a model of flow rates and mixing to inform how the biometrics collected should be interpreted based on "sewer network morphology," accounting for the distance the sewage travels as well as slopes and angles of pipelines (Turgeman 2015, 84). They are also working on a "geographic modeling of the sewage network," by surveying "geographic morphology, network topology, as well as demographic distribution, in conjunction with waste water loads over time, to propose and validate a model which informs waste water sampling and correlates to target population samples" (86). The research team also conducted a 24-h sampling experiment to gather data throughout the day, in order to identify differences in one sample location and infer relations about how certain samples might correlate with times of human behavior (e.g., going to the bathroom, cooking, or taking a shower) (Fitzgerald 2015). The research team has also developed an experiment to simultaneously sample ten different locations in the city of Cambridge (Turgeman 2015, 115).

The project is both based on, as well as hopes to contribute to, studies about the possibility of using biomarkers in the samples to estimate the number of people whose waste is included in sewage sampled (Daughton 2012; Chen et al. 2014; O'Brien et al. 2014). The *Underworlds* team will test human-produced biomarkers such as creatinine, coprostanol, cortisol, and androstenedione, as well as chemicals common in consumed products, such as caffeine, ibuprofen, and paracetamol. These biological estimates are then compared with census municipal estimates of the number of people served by each sewage catchment. Finally, the group also

> will be able to look for a stress biomarker in the sewage to gauge the *health status of a city*. In this respect, we will follow the workflow outlined by Christian Daughton on using isoprostanes as a *general biomarker of health*. Isoprostanes can be used to measure oxidative stress and are remarkable biomarkers since they are excreted in urine; are biostable; are not impacted by diet; are relatively stable on a day-to-day basis; and are detectable using low-cost immuno-assays. They can be used as a measure of public health since a diverse set of conditions give rise to elevated isoprostane levels: obesity, diabetes, cardiovascular disease, neurodegenerative disorders such as Alzheimers disease, smoking, alcohol consumption, among others. (Turgeman 2015, 90) [emphasis added]

What does it *mean* to talk about the health status of an entire neighborhood or city? And what does it mean to pinpoint health status through isoprostanes? The computed information is translated and displayed as thresholds of health, or the lack of it, promoting understandings of bodies and selves through numbers—what sociologist Deborah Lupton (2013) has called "quantifying the body." As a public health project, *Underworlds* measurements can create a *quantified collective self*, in which epidemiological health data can be defined through microbial numbers and thresholds. The microbes and biomarkers lurking in the sewage would become an approachable and measurable proxy for the city inhabited by humans. However, this epidemiological big data information can be valuable only if it is connected to a certain community being sampled. As STS scholar Michael Fortun (2008) observed in his study of corporate attempts in Iceland to gather genetic data from Icelanders and correlate these with existing medical records, there are complex and at times fraught questions of politics, property, and representation at stake in such efforts. The connection with a particular community, the "re-use" aspect of the sample, is what enables information/value to be mined from the sewage.

Anthropologist of science Nicole Labruto (2012), in her work with middle-class garbage rummagers in Cambridge, USA, and with working-class waste workers in Belo Horizonte, Brazil, introduces the notion of "re-use value" to highlight how value from recycled goods can be derived precisely from them being on the verge of becoming trash (see also Hawkins and Muecke 2003). Re-use value in the case of *Underworlds* does not come from the political and ethical possibility of expanding the usefulness of objects, as Labruto might argue, but rather through the recycling of biological material "shed" by a community. The *re-use biovalue* to be mined from the sewers comes from microbes and chemicals connected to a particular group of humans. The *collective microbiome* has to be connected to a specific *collective of humans*, who have shed biological data (i.e., fecal matter and urine) and from which health and behavioral data can be gathered. The "surplus value of [microbial] vitality" (Waldby 2000, 19) as flushed data, already being produced daily as individuals go to the toilet, can be transformed into epidemiological information. *Underworlds* can extend the pipelines of the bioeconomy, creating informational value from where there was none.[6] The "smart sewage platform" (Underworlds 2015) would convert communal microbes into health information, shit into epidemiological and epistemological gold.

Everybody Poops

Public health decisions and policies have long been defined through epidemiological (big) data. When I interviewed grad student Claire, she reminded me, however, how epidemiological data can be of poor quality, slow to be gathered and difficult to measure. Public health records can be collected only when people seek medical help and doctors report their diagnoses. However, not everyone goes or can afford to go to medical clinics; further, once this data is compiled, it may already be too late to create public health policies. But "everybody poops," Carmen commented once, and "our health data is naturally aggregated in the sewage." Using this logic, mining the sewage for information would account for everyone. "Passive" sampling, as lab-members described it to me, would also allow tracking "health data" even before a significant number of people are sick or go to the doctor, promising insights into epidemics yet to happen. Sewage could become not just a proxy for the aboveworld: it can become its oracle. And more current and comprehensive measurements could better inform public health guidelines.

This is not a completely new approach. In our first interview, Carmen outlined how successful sewage-mining has been adopted for particular public health endeavors. She noted how surveilling wastewater for poliovirus had been effectively used to control the disease, perhaps most famously in Israel (Manor et al. 2013). Since the virus can silently circulate in a population, tracing it through environmental surveillance of the sewage allows a response to be implemented even before cases are reported. She also mentioned how monitoring sewage has been used to track illicit drug use (see Frost and Griffiths 2008; Daughton 2011).

While hunting for poliovirus or illicit drugs involves targeting specific microbes and substances, *Underworlds* aims to create an all-encompassing "prototype smart sewage platform." This consists "of physical infrastructure, biochemical measurement technologies, and the down-stream computational tools and analytics necessary to interpret and act on our findings" (Underworlds 2016). The goal is to create the tools for a generalized monitoring, capable of detecting health conditions and behavioral patterns, which can later be deployed into different places. In a collaboration with Kuwaiti scientists, the six labs are involved in transferring expertise and plan to implement in 2017 a larger scale project in Kuwait City (MIT News 2015; Underworlds 2016).

To facilitate the gathering of such information, another lab within the project is designing robots that could process samples on their own and live-stream the data back to scientists. After all, as Sarah Franklin (2013, 56) has noted, while analyzing human embryonic cell lines, "[f]or the bioeconomy to become productive, it is necessary for new sources of capital stock to become more streamlined in order that they can be scaled up, banked, and used." In order to produce meaningful data, the *Underworlds* project needs to rely on a stream of wastewater samples. The first robot model developed was named "Mario" and a more recent and improved version is called "Luigi."[7] These bots—named after the famous duo of video game plumbers—would not only make the process of sampling easier, they would make it constant, offering up-to-date insights into the community being targeted. Or as their website explains, "We imagine a future in which sewage is mined for real-time information that can inform policy makers, health practitioners, designers, and researchers alike" (Underworlds 2016). The presence of Luigi, however, reminds us that although "everybody poops," the project needs to put in place an infrastructure capable of collecting the "spatio-temporal biological signatures from the city sewage" (Turgeman 2015, 91). The goal is to collect not only samples but also a "sewage sensor network" that includes information about the pH, temperature, water flow and level, and biological oxygen demand—a measure of the amount of oxygen used by microorganisms in the oxidation of organic matter.

Thus, in order to collect samples and data, *Underworlds* needs to put in place an infrastructure parallel to the sewage one: not only a sensor network and an automated sampler enabled by the robot Luigi (which they refer to as a "cyber-physical infrastructure"), but also a computational infrastructure.

> In addition to the cyber-physical infrastructure above, we are developing key computational components for a smart sewage system. While the data collected by Underworld's sensing apparatus will constitute a valuable asset in and of itself, multiple additional possibilities will be enabled by performing advanced data analysis over time and space, thus shedding new light onto the factors that give rise to the underlying patterns and variations. (Turgeman 2015, 93)

These infrastructures promise value in the samples they collect and in the data they produce, as well as in the operational tools they help develop (see Chun 2008, 19), offering a glimpse into the rationality underlying this technological project (Larkin 2013). These cyber-physical and computational infrastructures have embedded in them ideas about (possible) relations between human bodies and sewage water, between "collective microbiome" and community health, between "spatio-temporal signatures" and indications of a disease or bodily condition.

In a newspaper article discussing the *Underworlds* project, a journalist comments that when a person goes to the toilet, they flush away not just data but *free* data (Hines 2016). The content found in the sewers is the product of aggregated biological matter from different people, raising the question of who owns the sewage substance as well as the information mined from it. In an article entitled "Who owns your poop?", Alice Hawkins and Kieran O'Doherty (2011) outline questions around privacy, consent, and ownership on biobanked poop. Can something that was once discarded still belong to that person? Waldby and Mitchell (2006, 84) show how designations of waste allow body parts to be disentangled from individuals, and transformed into property to be manipulated for biomedical research, therapies, and products. They also comment that tissues shed or excreted are more likely to be classified as waste.

The material lurking in the sewers has yet another complicating aspect: as Laporte (1978) noted, sewage is considered public. Who owns the sewage, and more specifically the stool and urine (information) lurking on it? Does it still belong to those who went to the bathroom to get rid of it? Or is it now a property of the (public or private) organization that manages the sewer infrastructure? Or even, does it now belong to the researchers, who have collected, manipulated, and interpreted the samples? The answers to these questions might depend on *what* the material being collected is understood to be and, consequently, its connection to those who have flushed their toilets filled with biological data. Scholars like Paul Rabinow (1996) and Hannah Landecker (2007) have probed questions of continuity and fragmentation between cells and bodies when examining the development of the "Mo" and "HeLa" cell lines from tissues of the patients John Moore and Henrietta Lacks. But the material collected by *Underworlds* cannot be traced back to only one individual, like "Mo" and "HeLa." It refers to a multitude of people flushing a toilet in Cambridge: the sewers as a *collective* gut. And the metaphor of *gut* adds yet another complexity: is the gut part of the human body or is it a container for the non-human?

Microbes as (Not) Humans

Underworlds researchers often emphasize that connections between the data being gathered and individual human bodies cannot be established. For example, during a presentation of the project as part of the

Cambridge Science Festival events (Underworlds: "Using Sewers as a Diagnostic Tool for Public Health" 2016), the team was asked whether they were looking for human DNA. They answered no, the *Underworlds* project searches for microbial DNA in the sewage, not human DNA. As Carmen and I later talked about this matter, she reminded me that only 0.2% of all reads in their sequencing library align with human DNA. In a "microbial body," however, the difference between what is human microbial and non-human microbial might be a vexed issue.

While the project is trying to gather data to create a platform for *human* epidemiology, it does that through the analysis of DNA of microorganisms and traces of chemical substances that were once inside and metabolized by human bodies. Microbes and chemicals are made into representatives of human health and behavior. As anthropologist Stefan Helmreich (2009) concludes while examining the "genome of the Sargasso Sea," "microbes rise to the status of ecological representatives when their disembodied genes are treated as proxies" (172). In *Underworlds,* it is the human body and its ecology that are the object of inquiry understood through microbial data. Hence, to paraphrase Helmreich's questions: how do samples of the sewage come to be imagined as meaningfully summed up as the microbiome of a particular community? How does accumulated biological matter or aggregated samples come to represent a collection of people? How does the city sewer become a "collective gut"?

Microbes as proxies for human behavior and health are part of what Brigitte Nerlich and Iina Hellsten (2009) examine as "a new view of what it means to be human which extends over numerous genomes, from "our" genome to the genomes of microbes and beyond to interactions between microbes, humans and the wider environment" (23). The particularity of the *Underworlds* project, however, is their interest in the *collective.* A microbial understanding of the human body might prompt us to rethink the notion of a bounded biological individuality (Dupré and O'Malley 2007; Juengst 2009), perhaps through a "pluralized microbiomic self" (Fadok 2015). In the case of the *Underworlds,* however, this pluralized understanding of what is the human body can also allow the circumvention of criticisms of surveillance while still gathering data about the community. Through microbial genetic sequences acting as (potential) proxies for human health information, the data collected from the sewage simultaneusly becomes not human material *and* data about the human population.

Investigating the collective microbiome would not only offer epidemiological information about a community, it could feed back into studies of the microbiome. As Claire explained to me, "One of the current challenges in the microbiome sciences is that everyone is so different, so there is a lot of 'noise.' If there is a difference of signal in disease states, you either need a lot of samples of healthy vs. diseased, or you need very big differences in the disease state." One of the project's goals is to use the aggregated samples collected from the sewage to average out the noise.

Claire illustrated the process through the example of obesity, which is considered a major concern for their collaborators in Kuwait.[8] Obesity might have a really subtle signature, a biomarker, in a person's microbiome—whether that is a cause or an effect is yet another question. Since it is such a small signal, the difference between individuals will overshadow the possibility of comparing and tracing it. With aggregated samples, once the noise is averaged out, the signature of obesity might become more measurable. But aggregated samples can only be analyzed if researchers can make use of knowledge known a priori about the population—evoking, once again, the importance of connections between medical records and (microbial) genetic data (cf. Fortun 2008).

In this recursive mode of definition, the coupling of information about a community with the aggregated biological data of the "collective microbiome" helps determine more precise biomarkers, while at the same time these biomarkers and microorganisms associated with certain bodily conditions or diseases promise an insight into the health of communities. *Underworlds*' continuous flow of information carries with it the promissory condition of a more immediate recursive definition:

> with realtime feedback on the levels of microorganisms and biomarkers associated with obesity, that may allow for realtime evaluation of public health policy long before changes in population are observed. (Turgeman 2015, 105)[9]

Social scientists have pointed out that while biomarkers might be important new developments to assess the "propensity," prevent diseases and promote well being, they might at times offer a limited understanding of health when "elusive biomarkers" are made to stand as "the core, or essence, of the disease" (Lock 2007, 69; see also Clarke et al. 2009). In the case of the *Underworlds* project, the biomarker as an indicator (or possible indicator) of a biological state moves from the scale of the individual body to that of a community (cf. Montoya 2013). Biomarkers and microbes act as proxies for the "experienciable" human bodily condition and health, as they become the means to measure and assess public health policies (Dumit 2012).

Conclusion: The Collective Microbiome and Bioeconomic Proxies

In their endeavor to mine the informational biovalue from the sewage, the researchers at *Underworlds* establish three proxies: (1) sewage water as aggregated community information, (2) microbes and chemicals as data about human health and behavior, and (3) the average of aggregated samples as noiseless (or at least less noisy) biomarkers for bodily conditions and the health of individuals within a community.

The notion of the *collective microbiome*, I argue, as converged fecal matter, as microbes and chemicals metabolized by a specific community, and as averaged samples amalgamates and actualizes these three proxies. Social studies of microbes and the microbiome have already pointed to shifts in the understanding and representations of "the human," what Helmreich (2016) has called the *Homo microbis,* emerging amidst perhaps a "microbiomania." Scholars such as Amber Benezra et al. (2012) claim there has been an expansion of the "sense of personal identity," with humans "seeing [themselves] with increasing definition as a "supraorganism" composed of microbial and human cells, as well as human and microbial genes, with the number of microbial components vastly exceeding the number of human (Homo sapiens) components" (6378). Perhaps microbes are yet another caution against what science studies scholar and historian of biology Donna Haraway (2008) has called the "fantasy of human exceptionalism," that is, "the premise that humanity alone is not a spatial and temporal web of interspecies dependencies" (11). In an analysis of attempts to deploy parasitic worms as "gut buddies" to tackle autoimmune disease, Jamie Lorimer (2016) reminds us, however, that even within hyped talk about how the human body is not simply human but formed through human-microbial multispecies relations, the human is still central. "Securing this human as a multispecies microbial self involves defense, communication, and recruitment, differentiating the good, the bad, and the indifferent in all their microbial diversity" (72).

In its attempt to both identify and monitor harmful viruses and bacteria as well as use microbial data to inform and assess the public health conditions and policies, the *Underworlds* project understands human-microbial relations both in terms of potential enemies and allies lurking in the sewers (cf. Paxson 2008, 2013). Or, to use the terms from Heather Paxson and Stefan Helmreich (2013), the project combines the "idiom of peril" with the "idiom of promise." However, the *Underworlds* project not only distinguishes between 'good' and 'bad' microorganisms—as Paxson (2014, 118) describes on her account of cheese-making practices—but it also harnesses these microorganisms and substances in the sewage to become informants for human bodies and human health: microorganisms as "information-carrying" agents (Turgeman 2015, 25).

The collective microbiome, as it mobilizes the three different proxies together, represents reformulations of living processes that enable yielding vitality and produce expressions of biovalue (see Waldby 2002, 310). Since this collective microbiome is only accessible through aggregated bodily waste, the sewage water stands as an example of what Waldby and Mitchell (2006, 109) have defined as *constitutive waste*: waste resignified and redeployed, transformed into new forms of value. The project is still incipient in its implementation; hence, its impact on scientific understandings of the microbiome and on the practices of public health practitioners and policy-makers is still unclear. However, I argue, the mining of informational biovalue from the sewage is already grounded in and capitalizes on the (promissory) bieconomic proxies that constitute the collective microbiome.

Notes

1. I first want to thank the Underworlds research team for warmly welcoming the anthropology grad student from across campus, and patiently answering my sometimes-quirky questions. Many thanks as well to the editors, Joanna Goven and Vincenzo Pavone, for the invitation to contribute and for offering thoughtful comments and encouragements. I also want to thank David S. Jones and Stefan Helmreich for their support in the conception and development of my analytical framework. In particular, I am grateful to Stefan for caringly and carefully reading many iterations of this paper. I thank the HASTS community for supporting my work and Marilyn Levine for guiding me on the ways of the English language. And Philipp Rosenbaum, for always...

Two interviews were recorded and transcribed and a third was documented in notes. I also gathered field notes from all my interactions with the *Underworlds* team.

2. Laporte (1978) comments on such triumphant hygiene endeavors, reminding readers that the use of fecal matter to fertilize soil was a common practice among Romans and even in the contemporary countryside. "Rather, they found justification for the *nec plus ultra* of agricultural technology in the diaries of travelers who had journeyed to China. This pattern of repetition and revival helps us better understand the oscillations of civilization's anal imaginary: that which occupies the site of disgust at one moment in history is not necessarily disgusting at the preceding moment or the subsequent one" (32). For him, this example points to the historically particular ambivalences between the relationship between civilization and waste.
3. I asked my interlocutors what name they would like me to use to refer to them in this essay. The result is a mixture of pseudonyms and real names.
4. The OED defines big data as "data of a very large size, typically to the extent that its manipulation and management present significant logistical challenges; (also) the branch of computing involving such data." But as Schüll remarks, "'Big data' has come to mean many things. Typically the phrase characterizes the continuous collection of data streams and the convergence of multiple streams and types of data such that previously undetectable patterns can be discerned—with the right tools. Some definitions of big data include the novel analytic tools that are brought to bear on vast data sets, such as advanced mining techniques, predictive modeling, dynamic systems modeling and new machine learning algorithms" (Schüll 2016, 9).
5. The term "quantified self" is employed by users themselves and corresponds to a community of avid self-trackers. As Schüll explains, "Founded by two former editors of *Wired* magazine in 2007, Quantified Self currently claims 45,000 members in 40 countries. In online forums and in meetings around the world, quantified selfers share their attempts to experiment with diet and meditation, monitor drug side effects, correlate hormone levels with mood fluctuations and relationship dynamics, or even evaluate semantic content in daily email correspondence for clues to stress and unhappiness" (3).
6. Melinda Cooper (2008) has investigated the practices and promises of the new bioeconomies through mobilizing life to create value and overcome the limits of growth. She even comments, "the bioeconomy will take us beyond all limits, transforming even industrial waste into a source of surplus value" (47). As we have seen, however, this practice, and its promise, is far from new.

7. For a video of the *Underworlds* sample being tested in the lab, see here: https://vimeo.com/140844914.
8. Kuwait is among the countries with the highest obesity rates worldwide (see Karageorgi et al. 2013).
9. An example of this policy evaluation would be done by "monitoring obesity/inflammation biomarkers and how they change upon interventions in dietary intake, and monitoring phthalate levels and how they change upon banning specific personal care products or plastics to reduce exposure" (Turgeman 2015, 101).

Acknowledgments I first want to thank the Underworlds research team for warmly welcoming the anthropology grad student from across campus, and patiently answering my sometimes-quirky questions. Many thanks as well to the editors, Joanna Goven and Vincenzo Pavone, for the invitation to contribute and for offering thoughtful comments and encouragements. I also want to thank David S. Jones and Stefan Helmreich for guiding me through the conception and the development of my analytical framework. In particular, I am grateful to Stefan for caringly and carefully reading many iterations of this paper. I thank the HASTS community for supporting my work and Marilyn Levine for guiding me on the ways of the English language. And Philipp Rosenbaum, for always...

References

Benezra, A., DeStefano, J., & Gordon, J. I. (2012). Anthropology of microbes. *Proceedings of the National Academy of Sciences, 109*(17), 6378–6381.

Boellstorff, T. (2013). Making big data, in theory. *First Monday, 18*(10). http://firstmonday.org/article/view/4869/3750

Brook, W. H. (1997). Liebig and London: The chemistry of sewage. In *Justus von Liebig: The chemical gatekeeper* (pp. 250–272). Cambridge: Cambridge University Press.

Chen, C., Kostakis, C., Gerber, J. P., Tscharke, B. J., Irvine, R. J., & White, J. M. (2014). Towards finding a population biomarker for wastewater epidemiology studies. *Science of the Total Environment, 487*, 621–628.

Chun, W. H. K. (2008). *Control and freedom: Power and paranoia in the age of fiber optics.* Cambridge, MA: MIT Press.

Clarke, A. E., Shim, J. K., Shostak, S., & Nelson, A. (2009). Biomedicalising genetic health, diseases and identities. In P. Atkinson, P. Glasner, & M. Lock (Eds.), *Handbook of genetics and society* (pp. 21–40). London: Routledge.

Cleary, B., Brito, I. L., Huang, K., Gevers, D., Shea, T., Young, S., et al. (2015). Detection of low-abundance bacterial strains in metagenomic datasets by eigengenome partitioning. *Nature Biotechnology, 33*(10), 1053–1060.

Cooper, M. (2008). Life as surplus: Biotechnology and capitalism in the neoliberal age. In *Vivo: The cultural mediations of biomedical sciences.* Seatle: University of Washington Press.
Daughton, C. (2012). Real-time estimation of small-area populations with human biomarkers in sewage. *Science of the Total Environment, 414,* 6–21.
Daughton, C. G. (2011). Illicit drugs: Contaminants in the environment and utility in forensic epidemiology. *Reviews of Environmental Contamination and Toxicology, 210,* 59–110.
Daviss, B. (2005). Growing pains for metabolomics. *The Scientist, 19*(8), 25–28.
Dumit, J. (2012). Prescription maximization and the accumulation of surplus health in the pharmaceutical industry: The_BioMarx_Experiment. In K. S. Rajan (Ed.), *Lively capital: Biotechnologies, ethics, and governance in global markets* (pp. 45–92). Durham: Duke University Press.
Dupré, J., & O'Malley, M. A. (2007). Metagenomics and biological ontology. *Studies in History and Philosophy of Biological and Biomedical Sciences, 38*(4), 834–846.
Engineering the Human Microbiome. (2015). Retrieved from http://almlab.mit.edu.
Everybiome. (2015). *MassChallenge.* Retrieved from http://masschallenge.org/startups/2015/profile/everybiome.
Fadok, R. (2015). Human, all too nonhuman: Personal microbiomics and the dissolution of the sovereign self. Unpublished work.
Fitzgerald, M. (2015). What does Cambridge sewage say about residents? MIT plans to find out. *The Boston Globe.* Retrieved from http://www.bostonglobe.com/ideas/2015/01/09/what-does-cambridge-sewage-say-about-residents-mit-plans-find-out/qxBK9jbHYJFMcNGiTM2L2I/story.html?s_campaign=email_BG_TodaysHeadline.
Fortun, M. (2008). *Promising genomics: Iceland and deCODE genetics in a world of speculation.* Berkeley: University of California Press.
Foucault, M. (1973). *The birth of the clinic: An archaeology of medical perception.* New York, Pantheon Books.
Franklin, S. (2013). *Biological relatives: IVF, stem cells, and the future of kinship.* Durham: Duke University Press.
Frost, N., & Griffiths, P. (2008). Assessing illicit drugs in wastewater: Potential and limitations of a new monitoring approach.
Gillespie, T. (2010). The politics of 'platforms'. *New Media & Society, 12*(3), 347–364.
Gitelman, L. (Ed.). (2013). *"Raw data" is an oxymoron.* Cambridge: MIT Press.
Haraway, D. (2008). *When species meet.* Minneapolis: University of Minnesota Press.
Hawkins, A. K., & O'Doherty, K. C. (2011). 'Who owns your poop?': Insights regarding the intersection of human microbiome research and the ELSI aspects of biobanking and related studies. *BMC Medical Genomics, 4*(1), 72.

Hawkins, G., & Stephen M. (2003). Introduction: Cultural economies of waste. In *culture and waste: The creation and destruction of value*, (pp. ix–xvii). Lanham: Rowman & Littlefield.

Helmreich, S. (2009). *Alien ocean: Anthropological voyages in microbial seas.* Berkeley: University of California Press.

Helmreich, S. (2016). Homo microbis: Species, race, sex, and the human microbiome. In *Sounding the limits of life: Essays in the anthropology of biology and beyond* (pp. 62–72). Princeton: Princeton University Press.

Hines, N. (2016). Autonomous super mario-themed sewer robots live-stream. *Inverse.* Retrieved from https://www.inverse.com/article/15158-autonomous-super-mario-themed-sewer-robots-live-stream-neighborhood-microbiome-data.

Juengst, E. T. (2009). Metagenomic metaphors: New images of the human from 'translational' genomic research. In M. Drenthen, J. Proctor, & J. Keulartz (Eds.), *New visions of nature: Complexity and authenticity* (pp. 129–145). Berlin: Springer.

Karageorgi, S., Alsmadi, O., & Behbehani, K. (2013). A review of adult obesity prevalence, trends, risk factors, and epidemiologic methods in Kuwait. *Journal of Obesity, 2013,* 1–14.

Kelkar, S. (2014). The MOOC as laboratory: The implications of A/B testing in higher education. *American Anthropologist, 116*(4), 834–835.

Labruto, N. (2012). Re-use value and re-use economies. *Anthropology News, 53*(8), 14–15.

Landecker, H. (2007). *Culturing life: How cells became technologies.* Cambridge, MA: Harvard University Press.

Laporte, D. (1978). *History of shit* (N. Benabid & R. El-Khoury, Trans.). London: MIT Press. (2000).

Larkin, B. (2013). The politics and poetics of infrastructure. *Annual Review of Anthropology, 42,* 327–343.

Lock, M. (2007). The future is now: Locating biomarkers for dementia. In R. V. Burri & J. Dumit (Eds.), *Biomedicine as culture: Instrumental practices, technoscientific knowledge, and new modes of life* (pp. 61–87), New York: Routledge.

Lorimer, J. (2016). Gut buddies: Multispecies studies and the microbiome. *Environmental Humanities, 8*(1), 57–74.

Lupton, D. (2013). Quantifying the body: Monitoring and measuring health in the age of mHealth technologies. *Critical Public Health, 23*(4), 393–403.

Manor, Y., Shulman, L. M., Kaliner, E., Hindiyeh, M., Ram, D., Sofer, D., et al. (2013). Intensified environmental surveillance supporting the response to wild poliovirus type 1 silent circulation in Israel. *Euro Surveillance: Bulletin Européen Sur Les Maladies Transmissibles = European Communicable Disease Bulletin, 19*(7), 1–10.

MIT News. (2014). New interdisciplinary center at MIT to focus on the microbiome and human health. Retrieved from http://news.mit.edu/2014/new-mit-center-microbiome-and-human-health-1106.

MIT News. (2015). Real-time epidemiology from urban wastewater. Retrieved from http://news.mit.edu/2015/real-time-urban-epidemiology-from-wastewater-1102.

Montoya, M. J. (2013). Potential futures for a healthy city community, knowledge, and hope for the sciences of life. *Current Anthropology, 54,* 45–55.

Nerlich, B., & Hellsten, I. (2009). Beyond the human genome: Microbes, metaphors and what it means to be human in an interconnected post-genomic world. *New Genetics and Society, 28*(1), 19–36.

O'Brien, J. W., Thai, P. K., Eaglesham, G., Ort, C., Scheidegger, A., Carter, S., et al. (2014). A model to estimate the population contributing to the wastewater using samples collected on census day. *Environmental Science and Technology, 48*(1), 517–525.

Paxson, H. (2008). Post-pasteurian cultures: The microbiopolitics of raw-milk cheese in the United States. *Cultural Anthropology, 23*(1), 15–47.

Paxson, H. (2013). *The life of cheese: Crafting food and value in America.* Berkeley: University of California Press.

Paxson, H. (2014). Microbiopolitics. In E. Kirksey (Ed.), *The multispecies salon* (pp. 115–121). Durham, NC: Duke University Press.

Paxson, H., & Helmreich, S. (2013). The perils and promises of microbial abundance: Novel natures and model ecosystems, from artisanal cheese to alien seas. *Social Studies of Science, 44*(2), 165–193.

Preheim, S. P., Perrott, A. R., Martin-Platero, A. M., Gupta, A., & Alm, E. J. (2013). Distribution-based clustering: Using ecology to refine the operational taxonomic unit. *Applied and Environmental Microbiology, 79*(21), 6593–6603.

Rabinow, P. (1996). *Essays on the anthropology of reason.* Princeton: Princeton University Press.

Sangodeyi, F. I. (2014). The making of the microbial body, 1900s–2012 (Doctoral dissertation, Harvard University, 2014). *History of Science.*

Schüll, N. D. (2016). Data for life: Wearable technology and the design of self-care. *Biosocieties, 11*(3), 317–333.

Turgeman, Y. J. (2015). *Microbial mediations: Cyber-biological extensions of human sensitivity to natural and made ecologies* (Doctoral dissertation, Massachusetts Institute of Technology). http://hdl.handle.net/1721.1/99304

Underworlds. (2015). Statement on initiation status: Underworlds MIT researchers are developing a smart sewage platform to monitor urban health.

Underworlds. (2016). Retrieved January 27, 2016, from http://underworlds.mit.edu.

Underworlds: Using sewers as a diagnostic tool for public health. (2016). *Cambridge Science Festival*. Retrieved from https://www.cambridgesciencefestival.org/wp-content/uploads/2016/09/CSF16PB_Final_lr.pdf.

Waldby, C. (2000). *The visible human project: Informatic bodies and posthuman medicine*. New York: Routledge.

Waldby, C. (2002). Stem cells, tissue cultures and the production of biovalue. *Health: An Interdisciplinary Journal for the Social Study of Health, Illness and Medicine, 6*(3): 305–323.

Waldby, C., & Mitchell, R. (2006). *Tissue economies: Blood, organs, and cell lines in late capitalism*. Series: Science and cultural theory. Durham: Duke University Press.

Author Biography

Luísa Reis-Castro is a PhD student in the History, Anthropology, and Science, Technology, and Society (HASTS) program at MIT, USA. Her dissertation project investigates the development and implementation of modified mosquitoes for controlling mosquito-borne disease as a window to discuss science and public health policies in Brazil. Luísa's inquiries into the *Underworlds* project come from her underlying interest on valuation practices and notions of health.

CHAPTER 6

Bio-Identification, Value Creation and the Reproductive Bioeconomy: Insights from the Reprogenetics Sector in Spain

Vincenzo Pavone

Introduction

Supported by a strong interaction between pre-implantation genetic testing, stem cell research and in vitro fertilization, human assisted reproduction is emerging as a core technological platform (Throsby 2004; O'Riordan and Haran 2009) of rapidly growing tissue economies (Waldby 2002). In these tissue economies, the creation and appropriation of bio-value has recently been at the center of an intense scholarly debate, which has highlighted how these economies, ultimately, come to be constitutive elements of a larger, encompassing *reproductive bioeconomy* (Waldby 2014).

Addressing value creation in the reproductive bioeconomy, some observers argue that value is effectively generated through the reproductive work and the regenerative labor of the female body (Waldby and Cooper 2010). More specifically, they argue, value is generated through displacement of risk away from capital to progressively less organized and

V. Pavone (✉)
Institute of Public Goods and Policies (IPP), Consejo Superior Investigaciones Científicas (CSIC), Madrid, Spain

V. Pavone and J. Goven (eds.), *Bioeconomies*,
DOI 10.1007/978-3-319-55651-2_6

protected individuals, and through the incorporation into the market of reproductive practices, which had been hitherto confined to unpaid, kinship-related, domains (Waldby and Cooper 2014). As human reproduction practices are being incorporated into the regime of capital accumulation through a combination of technological devices, organizational practices and normative arrangements,

> the life science industry relies on an extensive yet unacknowledged labor force whose service consists in the visceral experience of experimental drug consumption, hormonal transformation, more or less invasive biomedical procedures, ejaculation, tissue extraction and gestation.

This is what Waldby and Cooper define as "clinical labor" (2014, 7). Focusing more on the regenerative medicine sectors of the bioeconomy, which are characterized by innovation clusters, spin-offs, patent regimes and venture capital, other scholars suggest that the immateriality of assets, such as intellectual property rights, stands at the core of the value creation process rather than the materiality of tissues, technologies and bodies (Birch and Tyfield 2013). To some of them, the new bioeconomy is just another example of the *financialisation* of the knowledge-based economy (Birch and Mykhnenko 2014; Petersen and Krisjansen 2015; Birch 2016).

This chapter aims at contributing to this debate by analyzing, conceptually and empirically, the process of creation and appropriation of value in the reprogenetics sector (Knowles and Kaebnick 2007)[1] of the Spanish reproductive bioeconomy. The reprogenetics sector is a key component of the Spanish national economy as a whole, and has substantially contributed to making Spain the leading international reproductive hub in Europe (Pavone and Arias 2012). Importantly for the conceptual analysis, the reprogenetics sector is inherently different from the sectors analyzed (respectively) by Waldby and Cooper (2014) and by Birch and Tyfield (2013): it does not directly involve the clinical labor central to gametes donation, clinical trials or surrogacy; and it is only related but not directly affected by the "financialized" value creation processes of the highly speculative sectors of the bioeconomy. The empirical analysis draws primarily on 28 semi-structured interviews, conducted in 2010–2012, with women undergoing PGD [pre-implantation genetic diagnosis] or PGS [pre-implantation genetic screening].

The chapter is organized as follows. The first section outlines and discusses the characteristics of the global reproductive bioeconomy. The second section addresses the variety of ways in which the processes of value creation, attribution and distribution have been conceptualized and analyzed in current debates about the bioeconomy. The third section describes, and identifies the key facilitating factors of, the reprogenetics sector in Spain. Drawing from 28 patient interviews and inspired by the concepts of bio-objectification and bio-identification (Holmberg et al. 2011) the fourth section analyses and discusses the creation, attribution and distribution of value in the reprogenetics area of the reproductive bioeconomy in Spain. The final section of the paper summarizes the main arguments and findings and sets them in context.

The Global Reproductive Bioeconomy

The reproductive bioeconomy is being constituted around both emerging and consolidated sociotechnical practices of human assisted reproduction, and it is growing rapidly (Kupka et al. 2014). It has emerged at the intersection of three biopolitical problems: an aging population, a scarcity of biological resources and a decline of both female and male fertility (Waldby 2002). First developed around a core set of Assisted Reproductive Techniques [ARTs], the reproductive bioeconomy has recently expanded, incorporating a variety of interconnected technologies and practices, mostly related to the emerging field of regenerative medicine (Franklin 2013).

The term *bioeconomy* also effectively serves to capture how, in this economy, a specific set of *biotechnologies* plays a constitutive role, enabling the transformation of otherwise inaccessible biological materials, such as oocytes and embryos, into *bio-objects*. The concept of bio-objects will be discussed at length in the next section; here it is simply necessary to note that the material agency of such "bio-objects" derives at least in part from the novel sociotechnical relations they generate (Webster 2012; Metzler and Webster 2011; Holmberg et al. 2011). These sociotechnical relations, which are constitutive of what Thomson defined the ontological choreography of reproductive technologies (2005), remind us that medical treatments, human tissues and reproductive processes are increasingly embedded in markets, but that their value cannot be restricted to the economic one. Public, philanthropic, and not-for-profit organizations also play a key role in their functioning and consolidation. Moreover, it would be impossible to explain the expansion of the reproductive bioeconomy without also taking into account heteronormativity, patriarchal kinship, religion and ethnicity (Roberts 2012; Franklin 2013).

The global reproductive bioeconomy is marked by other characteristics that make it different from other emerging bioeconomies. It features, for instance, a close interaction between clinical practice and cutting edge research laboratories in which the regenerative medicine research community occupies the position of end user and major beneficiary (Franklin 2006a, b; Waldby and Cooper 2010; Cooper 2008). In addition, it is a strongly gendered economy: due to a peculiar interaction between reproductive work and regenerative labor (Waldby and Cooper 2008), women endure the greatest share of social and economic burden, facing serious health, social and economic risks, but rarely constitute the main recipients of the economic benefits produced (Waldby and Mitchell 2006; Waldby and Cooper 2014).

However, it also shares with existing bioeconomies the constitutive role of hopes and expectations and a global nature. Whilst it is true that stem cell research using human oocytes is an integral part of the speculative investments of a biotech industry characterized by economic hype (Rajan 2006), in the reproductive bioeconomy hope is also strongly associated with the discovery of new treatments or a wider access to available techniques. In both cases, though, the reproductive bioeconomy clearly capitalizes on the construction and mobilization of expectations and hopes (Novas 2006; Brown 2007; Brown et al. 2011; Petersen and Krisjansen 2015).

Although national regulations do play a fundamental role, the reproductive bioeconomy has an increasingly global nature (Salter and Faulkner 2011; Salter 2013). In spite of national constraints—or, perhaps, thanks to them—in the reproductive bioeconomy, patients, tissues, capital, technologies, scientists and practitioners increasingly move across national borders. While different national cultures may encourage a more or less restrictive approach to given technologies and practices, it has been argued that there is a neoliberal tendency to construct recipients and donors as informed, empowered consumers who seek the best value for money when it comes to their reproductive choices (Thompson 2007; Waldby and Cooper 2008). Moreover, the European reproductive bioeconomy is characterized by a peculiar mix of public and private healthcare arrangements. As these two domains follow different ethical, epistemological and social criteria, which end up constructing different regimes of access and care (Martin et al. 2008; Brown 2007), their interaction plays a constitutive role in the reproductive bioeconomy.

Finally, the reproductive bioeconomy is not a monolithic phenomenon, and internal differences do exist. Traditional IVF practices, where no gamete donors or genetic testing are involved, address different types of users, deploy different marketing strategies and have different relations with research labs compared to the reprogenetics area or to donor-based assisted reproduction (Mamo 2007; Waldby 2002). In fact, the reproductive bioeconomy is constituted by different yet deeply intertwined tissue economies (Waldby and Mitchell 2006), some of which share the sociotechnical core of ARTs (Franklin 2013), while others, such as the ones that have emerged around cord blood preservation (Kent 2008; Brown et al. 2011; Martin et al. 2008) and fetal tissue research (Kent 2008), share only the technological and economic interfaces with regenerative medicine (Fig. 6.1). While each of these tissue economies shares important aspects with the others, each is organized around different tissues, techniques, normative rules and assumptions, as well as different actors.

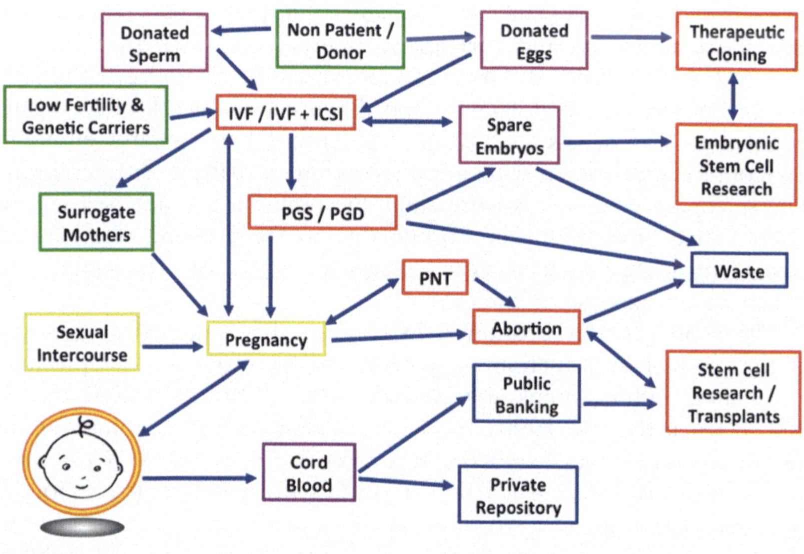

Fig. 6.1 The reproductive bioeconomy *Source* in *yellow*: social practices, in *red*: techniques, in *blue*: institutions, in *purple*: bio-objects, and in *green*: bio-subjects

The **traditional IVF economy** emerged directly along with the gradual implementation of assisted reproduction techniques, which have been increasingly framed as a technological response to the decline in fertility rates (Waldby and Cooper 2008). Whether provided by public or private health care, ARTs constitute the core technologies of a market that knows no decline and involves sociotechnical services, biological products and reproductive labor (van Niekerk and van Zyl 1995). The **embryo economy** and the **gametes economy** have emerged as a result of a surplus produced by ARTs. More specifically, they have been arranged around the extraction, procurement, storage and exchange of "spare" *embryos* and *gametes*, which are not used in IVF cycles and get frozen and stored (Almeling 2007, 2009; Shanley 2002). With the rise of embryonic stem cell research and somatic-cell nuclear transfer, existing procurement practices of embryos and gametes (insofar as human eggs are concerned) also face a growing demand from the research community; embryos and gametes are in demand from those developing embryonic stem cell lines (Waldby 2008; Waldby and Cooper 2008; Franklin 2006a) and from regenerative medicine researchers working with somatic nuclear transfer or parthenogenesis (Schultz and Braun 2013).[2]

More recently, and thanks to the gametes surplus, a new tissue economy has emerged around **surrogacy**, especially in India, the US, and Mexico (Waldby 2006; Waldby and Cooper 2008). Surrogacy combines IVF techniques, donated gametes and the gestational work of surrogate mothers to give access to biological parenthood to (wealthy-enough) gay couples and women with uterine conditions that prevent gestation. The surrogacy economy is increasingly becoming a relevant part of the reproductive bioeconomy in Europe due to the growing numbers of Europeans going abroad to access surrogacy, often with the support of European law agencies or clinics.[3]

The **reprogenetics economy**, in contrast, has emerged as a result of the gradual introduction of genetic testing into assisted reproduction (Silver 2000; Parens and Knowles 2003; Knowles and Kaebnick 2007). Since the late 1990s, advances in genetics and genomics allow the genetic screening of embryos in vitro before implantation. Currently, reprogenetics is mainly articulated around two very different techniques, pre-implantation genetic screening (PGS) and pre-implantation genetic diagnosis (PGD), which are both employed before the fertilized embryo is implanted in the uterus. PGS is usually employed to search for chromosomal abnormalities, and it has become the core technology of a new market targeting older women or women with a history of recurrent miscarriages or implantation failures (Pavone and Arias 2012). In contrast, PGD targets hereditary genetic diseases or predispositions, and is rather encouraging the extension of assisted reproduction techniques to fertile couples, who might have serious problems with the genetic profile of their offspring if they became parents through sexual reproduction.

Sharing what Sarah Franklin has recently defined as the technological platform of ARTs (Franklin 2013), these tissue economies (Waldby and Mitchell 2006) appear to be deeply intertwined. In the following figure, which illustrates the organization of the reproductive bioeconomy in Spain, it can be appreciated how these tissue economies have come to constitute a wider, and well-organized, reproductive bioeconomy. As the figure is immediately representative of the Spanish reproductive bioeconomy only, some key institutions, such as a variety of disease-specific charities that play a key role in other European countries, are not present in this graphic reconstruction. Conversely, an important and specifically Spanish actor represented here is the CNRHA, the assisted human reproduction committee, which authorizes PGD for diseases and conditions that do not explicitly meet the requirements of the 2006 Assisted Reproduction Act, as explained later in Sect. "The Reprogenetics Sector in Spain".

Value Creation in the Reprogenetics Bioeconomy

The reproductive bioeconomy located at the intersection of regenerative and reproductive medicine is changing fast, driven by several economic, technological, and social factors (Franklin 2013). New regenerative medicine and tissue engineering markets are emerging at the global level (Salter et al. 2014, 2015). Oocyte vitrification, single embryo transfer IVF and time-lapse blastocyst selection processes are also producing remarkable changes in the way reproductive medicine is evolving (Waldby 2014; Forman et al. 2012). A number of recent studies have addressed the economic and the political aspects of reproductive and regenerative medicine (Brown et al. 2011; Cooper 2008; Pande 2009; Braun and Schultz 2012, 2013). These studies have been very successful in shifting the focus away from the moral status of the individual body, the embryo and the oocyte, to the "*process and practices of 'doing bodies' in their specific context and circumstances*" (Schultz and Braun 2013, 121). They show how different social dynamics, institutional factors and economic driving forces, as well as neoliberal policy agendas and strategies, have all contributed to shaping value creation in a variety of different sub-sections of the reproductive bioeconomy (Waldby and Cooper 2014).

Recent scholarship on the reproductive bioeconomy, thus, has been focusing on the issue of value creation. Some authors suggest that value is being generated through the capitalist appropriation of the life surplus produced by the reproductive work and the regenerative labor of the female body (Waldby and Cooper 2010). This approach, which focuses on the materiality of the human bodies and tissues, has been criticized for not taking into due consideration the relevance of immaterial assets in the production of value and for suggesting that such value is inherent to the biological characteristics of the tissues and processes involved (Birch and Tyfield 2013). Responding to these critiques, Waldby and Cooper have clarified that surplus value is not intrinsic to the biological tissues and reproductive cycles, for they reject that "*the ultimate use value and source of all wealth could be located in the intrinsic generativity of living biology*". Rather, they link the creation, accumulation and appropriation of value to the role played by *clinical labor*, as a "*process of material abstraction by which the abstract, temporal imperatives are put to work at the level of the body*" (Waldby and Cooper 2014, 12). The emphasis, therefore, is not on the value of biological tissues and processes per se, but in the multiple ways in which these biological processes and tissues are put to work in the complex set of existing practices of capital accumulation.

Though crucial in the value creation process of some areas of the reproductive bioeconomy, such as IVF with gamete donation or surrogate gestation, clinical labor may not be as central to other areas of the reproductive bioeconomy, such as, for instance, reprogenetics. This paper, thus, tries precisely to explore and discuss the value creation process in areas, like reprogenetics, where clinical labor is not directly involved. What can we learn about value creation, appropriation and distribution in the reproductive bioeconomy when we focus our analysis on these other sectors of the reproductive bioeconomy? Based on empirical research discussed in the next section, my argument is that value is ultimately produced and appropriated through the integration of reproductive tissues and practices, whether proceeding from and involving clinical laborers or not, into a series of medical as much as socio-economic processes, organized at an industrial scale.

While this is partially due to the actual nature of Assisted Reproductive Technologies [ARTs], whose "*industrial logic breaks down reproductive biology into component processes*" (Waldby and Cooper 2014, 35), it is also due to the role played by some specific sociotechnical processes affecting reproductive materials as they face technological manipulation. Some recent studies on regenerative medicine have suggested, for instance, that human embryos and gametes undergoing ARTs manipulation become *bio-objects* (Metzler and Webster 2011). Bio-objects are "matter out of place," living entities that, as a result of technological manipulation, challenge our understanding of life and matter, and of the social, ethical and political networks of meanings and practices that are deeply intertwined with this understanding of life. In this process, known as *bio-objectification*, "*life-forms or living entities are first made into objects, become possible, through scientific labour and its associated technologies, and then come to be attributed with specific identities*" (Holmberg et al. 2011).

Defying mainstream categorization and definition, bio-objects tend to generate controversies, activating processes of standardization and stabilization usually aiming at resolving, silencing or overcoming such controversies. This process, known as *bio-identification*, re-defines, re-configures and re-categorizes bio-objects through legal standards, medical protocols, bio-ethical guidelines, organizational procedures and operating licenses, in order to allow for their circulation, trade, storage, exchange, donation and commercialization. Bio-identification, in other words, aims at a successful incorporation of these objects in the existing networks of socio-technical and economic practices, where they get stabilized into renewed regimes of regulation, standardization and commercialization, in which they acquire a new bio-identity (Holmberg et al. 2011).

For these reasons, bio-objectification and bio-identification are likely to play a crucial role in the process of value creation, attribution and redistribution in the global reproductive bioeconomy. Drawing from these concepts, this study examines the processes that have made living entities into bio-objects, and the assemblage of practices, rules and controversies that have intervened in the process of bio-identification thereafter, in the Spanish reprogenetics sector. The research suggests that, in the reprogenetics sector of the reproductive bioeconomy, bio-identification actually occurs through four simultaneous technical, normative, social and economic configuration steps, which help to standardize, stabilize and attribute value to the embryos as they move through the PGD/PGS cycle.

As outlined in the following section, these procedures of socio-technical configuration are arranged in a way that closely resembles a value-chain, as depicted in the business literature (Burns et al. 2002) The concept of value chain refers to the chain of all the technical and organizational steps and processes that make a product acquire value. At each of these steps, the product gains some value so that, by the time it reaches the market, the product has acquired a value that is *higher* than the value possessed in each of the previous stages. It is also applied at industry level, where it is also known as supply chain (Mentzer et al. 2001). This concept has been also applied to study how value is created in healthcare manufacturing and markets (Burns et al. 2002), and recent studies have argued that societal value and economic value are coproduced through what have been defined as "shared value chains" (Porter and Kramer 2011). According to Porter and Kramer (2011, 63), firms gain competitive advantage from how they configure the value chain, that is, the set of activities, processes and tools that are involved in creating, producing, selling, delivering and supporting their products and services. The value chain, thus, constitutes a valuable conceptual tool representing the various processes involved in producing goods and services, starting with raw materials and ending with a marketable product/service.

In the Spanish reprogenetics value chain, the gametes are first collected and then fertilized; the resulting embryos are screened, some are selected and implanted, and "spare" embryos are either discarded or donated to research, where they enter an entirely new domain, i.e. regenerative medicine (which is characterized by a totally different set of sociotechnical interventions). Each of these steps contributes to ensure the success of the treatment offered and provide the assisted reproduction cycle with an ultimate market value of ten to twelve thousand euros. As a result, a number of collateral business activities, such as those involved in the storage, use, exchange and procurement of human embryos for research, are indirectly sustained.

Operating similarly to an industrial value chain, these steps are sociotechnical interventions that reconfigure bio-objects into products and services, allowing them to be standardized objects of valuable economic exchanges, either into the open market or through public health-care services.

The Reprogenetics Sector in Spain

Although the reproductive bioeconomy could be observed and studied at any of its points of intersection, one of these points, i.e. the interface between IVF, reprogenetics and stem cell research (Franklin 2006b), has been more extensively studied (Fig. 6.2). Some scholars have explored this interface in relation to the higher propensity of women undergoing PGD or PGS to donate their "spare", genetically discarded embryos to research (Franklin et al. 2005, Svendsen and Koch 2008). More recent qualitative studies on women and couples undergoing PGD have highlighted the importance of reproductive choice, the desire to avoid abortion and the welfare of the future child as the main factors influencing the decision to use PGD (Ormondroyd et al. 2012). Other studies, however, have suggested that economic incentives and constraints should not be neglected (Drazba et al. 2014). It has also been outlined how the decision-making process, in which these factors are taken into account, involves a psychosocial journey through different phases, which operate as filters (Hershberger et al. 2012). Finally, it has been observed that restrictive legislation not only reduces the prescription of PGD, it also encourages many affected couples to go abroad in order to undergo IVF with PGD (Gianaroli et al. 2014). These studies, however, are essentially focused on PGD for hereditary genetic disorders, and little is known today about couples undergoing PGS for chromosomal abnormalities or related disorders. Yet, many of the fears of an extension of PGS to a wide variety of chromosomal and genetic characteristics (King 2013), have so far failed to materialize, at least in Spain (Pavone and Arias 2012).

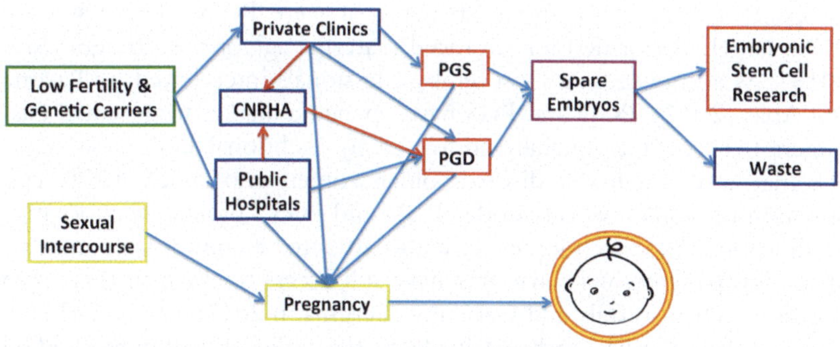

Fig. 6.2 The reprogenetics economy *Source* in *yellow*: social practices, in *red*: techniques, in *blue*: institutions, in *purple*: bio-objects, and in *green*: bio-subjects. The *red arrows* indicate the special procedure to be followed when hospitals and clinics want to perform a PGD for diseases and conditions that do not explicitly meet the requirements of the 2006 Assisted Reproduction Act

Among the European countries where pre-implantation genetic testing is allowed, Spain performs a little less than half of all the PGDs and PGSs in Europe (Kupka et al. 2014). In Spain, pre-implantation genetic diagnosis is performed by both private clinics and public hospitals. There exist regional variations, but the social security system usually subsidizes up to three treatments in public hospitals or private clinics collaborating with the Spanish NHS; however, it may take up to 3 years to go through the waiting list. In contrast, it is a relatively quick procedure in out-of-pocket private clinics, but it generally adds about 3000 euros to the average cost of an IVF cycle. The latter costs around 6000 euros in private clinics, but, when accessed through the social security system, is completely free or largely subsidized, depending on the regional healthcare system. In both private and public hospitals, PGDs that do not address "hereditary, early-onset, untreatable" diseases need to be authorized by the National Assisted Reproduction Committee (CNRHA). PGS, in contrast, is performed only by private clinics, and the Spanish NHS does not subsidize it. Private clinics usually offer it as an additional tool to 'improve' the chances of success in the cases of advanced maternal age, repeated implantation failure and miscarriages. PGS, however, is a controversial technique (Harper et al. 2010; Hardarson et al. 2008) and the medical literature is divided about its ability to improve success rates.

Out of the 2744 pre-implantation genetic testing cycles performed in Spain in 2012, only about 20 per cent were PGDs. The rest were PGSs, largely performed for advanced maternal age or miscarriages (SEF 2013). As acknowledged by several professionals interviewed by Pavone and Arias (2012), PGD and PGS have opened a market to potential customers that are not generally addressed by traditional IVF: fertile couples that have hereditable diseases, older women who prefer PGS to egg donation or women who consider PGD and PGS a preferable alternative to abortion. These techniques have also attracted a considerable number of foreign patients, who may not have full access to them in their own country, such as in Italy and Germany (Bergmann 2011; Salama 2014).

A recent study has analyzed in detail the political economy of PGD and PGS in Spain, showing how the large amount of pre-implantation genetic testing performed is the result of a peculiar combination of (1) a public that is in general very supportive of science and medicine in general, (2) liberal legal guidelines that tried to encourage both the consolidation of the IVF business and the emerging stem cell research community, and (3) a lack of effective, independent checks and controls on the functioning of the sector (Pavone and Arias 2012). These combined factors have encouraged a thriving private IVF industry. High technological levels and a vast, organized network of private IVF clinics make the supply side of the industry very competitive and technologically reliable, especially if compared with a public system negatively affected by lack of resources and recent austerity cuts (Adeces 2015). Thanks to a permissive and relatively non-transparent regulatory regime and the concentration of IVF clinics in centers of tourism, the reproductive bioeconomy in Spain seems to be characterized by a strong orientation toward the commercialization of reproductive tissues, services and practices. This orientation remains largely unchecked by either public authorities or civil society organizations. As a result of this specific political economy background, the Spanish reprogenetics sector constitutes an excellent case to study the organization, articulation and functioning of the value generation, appropriation and distribution in the reproductive bioeconomy.

Bio-Identification and Value Creation

The analysis of value generation, appropriation and distribution is based on qualitative data proceeding from 28 interviews with women from both private and public hospitals in 8 regional communities in Spain between late 2010 and late 2012. Originally gathered to shed light on the articulation of PGD and PGS in Spain from the perspective of the women involved, these interviews were collected following the principle of maximum variation sampling (Creswell 2013) in order to ensure access to a wide range of different experiences, especially with regard to the technique experienced (PGD and PGS), the region and the type of clinic or hospital involved (public, private or private with public subsidy). Approximately two thirds of the women interviewed had undergone either PGD or PGS, while the remaining women were considering this possibility at the time of the interview.

The interviews were organized around a semi-structured set of open-ended questions aimed at exploring different issues around PGD and PGS, such as: personal experiences, choice of IVF center, level of information, informed consent, access to the tests, psychological and genetic counseling, embryo donation, public versus private IVF settings and future prospects of the technique (Table 6.1). The interviews lasted between 45 min and one hour, were recorded and transcribed and, finally, analyzed through a combination of *thematic data analysis* and *discourse analysis.* Thematic data analysis was employed to identify the most recurrent topics and reconstruct the various organizational, medical and social steps of the PGD/PGS journey (Marshall and Rossman 2011). Discourse analysis was aimed at identifying, analyzing and interpreting the emotional, discursive and representational tools adopted by the interviewees to make sense of their personal experience and to frame and communicate it to themselves and to others, including the interviewer (Potter 1997). The interviews of the women who had not yet experienced PGD or PGS were used as a sort of control group, against which the characteristics of the other interviews were considered and assessed.

The analysis of the interviews casts light on the micro-level dynamics of the reprogenetics economy and explores the variety of sociotechnical interventions usually deployed in the trajectory of a PGD/PGS cycle. Operating like a value chain, these interventions submit the embryos to a variety of technical, normative, social and economic configuration steps and turn them into ultimately stabilized bio-objects (Holmberg et al. 2011), which acquire value as they get successfully incorporated into the market.

Technological Configuration: The Embryo as an Object for Quality Assessment

One of the most common experiences narrated by the women interviewed is their gradual exposure to a framing process in which they are encouraged to approach embryos as objects of technical manipulation and selection: "*They use hatching to help the embryo implant, they remove part of the skin, so to speak, like an onion, you know?*" (P18) "*We had no choice, it was either select the good one, the healthy one I mean, or take the risk and do it naturally*" (P17). As a result of manipulation and selection, the PGD and PGS embryos are generally subjected to a screening that is framed as a quality assessment process.

Many interviewees, though, are reluctant to consider their embryos as objects of quality assessment. This is especially true for PGS patients, because it was very hard for them to even reach the stage where they had some viable embryos. After such a long and uncertain process of fertilization, it was especially difficult to accept that the embryos should *also*

Table 6.1 Interview data

Interviewee	*IVF Centre*	*Age*	*PGS*	*PGD*	*Cycles*	*Diagnostic*	*Psychological support*	*Source of info*	*Spare embryos*	*County*
P1	Private	31	No	No	AI 3; IVF 1	Azoospermia	No	Internet, staff	2	Madrid
P2	Private	42	No	No	IVF + ICSI 4	Ectopic pregnancy	Yes (private)	Internet, staff	No	Madrid
P3	Private	29	No	No	AI 1; IVF 1	Tube failure	Yes (private)	Internet, staff	No	Madrid
P4	Private	31	No	No	IVF + ICSI 2	Tube failure	Yes (private)	Internet, staff	1	Madrid
P5	Private	37	No	No	IVF + ICSI 3	Early menopause	Yes (private)	Internet, staff	No	Madrid
P6	Private	41	Possible for the future	No	AI 1, IVF + ICSI 1	AMA + POS	No	Internet	No	Madrid
P7	Public–Private	39	No	Yes	IVF + ICSI 2; move to egg donation	AMA + Chromosome translocation	Yes (private)	Staff	No	Madrid
P8	Public–Private	39	Possible for the future	No	AI 2; IVF 2	AMA + Ectopic pregnancy	Yes (private)	Internet, friends	No	Basque Country; Valencia
P9	Private	Unknown	Possible for the future	No	AI 3; IVF + ICSI 2	POS + RIF	Yes (private)	Internet	6	Madrid; Valencia
P10	Public	Unknown	No	Yes	IVF + ICSI 1	Recurrent miscarriages + Inverted chromosome + Ectopic pregnancy	No	Internet, staff	No	Madrid
P11	Private	Unknown	No		IVF 1	Unknown	Yes (private)	Internet, staff	No	Madrid

(continued)

Table 6.1 (continued)

Interviewee	*IVF Centre*	*Age*	*PGS*	*PGD*	*Cycles*	*Diagnostic*	*Psychological support*	*Source of info*	*Spare embryos*	*County*
P12	Private	33	yes	No	IVF + ICSI 4	Recurrent miscarriages + Chromosome translocation	Yes (private)	Internet, staff	No	Catalonia
P13	Public–Private	Unknown	No	Yes	IVF + ICSI 4	Both brother and father with HCC	No	Internet, staff	No	Galicia
P14	Private	Less than 40	No		IVF 1	Recurrent miscarriages + Ectopic pregnancy	Yes (private)	Internet, staff	No	Madrid
P15	Private	35+	Yes	No	IVF + ICSI 5	Azoospermia + Chromosome anomaly	Yes (private)	Internet, staff, friends	No	Catalonia
P16	Private	35+	Yes	No	IVF 4	Chromosome Anomaly	No	Internet, staff	No	Aragon, Valencia
P17	Public	40	No	Yes	IVF + ICSI 3	Recurrent miscarriages + Chromosome traslocation	No	Staff, internet	No	Aragon
P18	Public–Private	39	Planned	No	IVF + ICSI 4	Azoospermia	No	Staff, internet	No	Catalonia
P19	Private (paid by SS)	35+	No	Yes	IVF + ICSI 1	Genetic condition	No	Internet	No	Galicia
P20	Public	Unknown	No	Yes	IVF 1	Spinal muscular atrophy	No	Staff, internet	No	Andalucía
P21	Private (paid by SS)	Unknown	No	Yes	IVF 2	Genetic condition	No	Staff, internet	No	Madrid
P22	Public–Private (paid by the SS)	Unknown	No	Yes	IVF 1	Hereditary retinoblastoma	No	Staff	No	Madrid

(continued)

Table 6.1 (continued)

Interviewee	*IVF Centre*	*Age*	*PGS*	*PGD*	*Cycles*	*Diagnostic*	*Psychological support*	*Source of info*	*Spare embryos*	*County*
P23	Private	Unknown	No	Yes	IVF 1	Genetic condition	Yes (private)	Staff	No	Baleares
P24	Private	30+	Planned	No	IVF 1	Recurrent miscarriages + Ectopic pregnancy	Yes (private)	Internet	No	Catalonia
P25	Private	30+	No	Yes	IVF + ICSI 1	Hereditary colon cancer + POS + RSM	No	Staff	No	Madrid
P26	Private insurance (but PGD paid out of pocket)	35+	No	Yes	IVF 3	Hereditary neurodegenerative disease	No	Staff, internet	No	Castilla Y Leon
P27	Public	30+	No	Yes	IVF 1	Genetic condition	No	Staff	No	Madrid
P28	Private insurance (but PGS paid out of pocket)	35	Yes	No	IVF 5; IVF + ICSI 2	No clear diagnostic	Yes (private)	Staff, internet	No	Valencia

Source
POS Polycystic Ovarian Syndrome
HCC Hereditary Hypertrophic cardiomyopathy
AMA Advanced Maternal Age
RIF Repeated Implantation Failure
RSM Reduced Sperm Motility

be subjected to a quality assessment: "*Moreover, the doctor said we had to take into consideration that the quality of the embryo was not good enough to take a risk with PGS*" (P18). PGD patients were more willing to accept this step, for they had decided to try the IVF process not so much for reason of infertility but precisely to *select* offspring not affected by hereditary diseases.

Eventually, though, the conceptualization of the embryo as a bio-object that can be submitted to a quality assessment procedure is generally accepted and constitutes a first step into the process of bio-identification. For both kinds of patients, it is then easier to frame embryos failing the quality assessment test into objects that can either be "donated" for medical research or simply considered waste: "*I always got good and bad embryos. With the bad ones, I thought, they* [the scientists] *can do whatever they want, if they do it for the good of other people. I do not know what they actually do, though*" (P16) "*The embryos were all bad, genetically bad, and I thought: at the end of the day it is just a microscopic cell, if it turns out well, okay, otherwise you can just get rid of it*" (P12).

Normative Configuration: Informed Consent as Paperwork

The bio-identification process operating in the reprogenetics IVF journey also involves a normative configuration step centered on the concept of informed consent. According to the contractual approach adopted by the regulatory guidelines, informed consent is the constitutive step bringing the contract to life. Yet, in the experience of several interviewees, informed consent works more as a budget quote or a risk waiver form: "*Yes, you have to sign about the treatment, the costs and so on*" (P15); "*Yes, I was informed that… you know, success is not guaranteed, and the risk of anesthesia…*" (P18).

This is directly related to the fact that PGD and PGS are also normatively configured as something outside or beyond the traditional moral boundaries, i.e. as something endowed with a logic of their own: "*Once you have decided to go for IVF, you need to put aside moral and religious concern*" (P17). PGS is also presented by the doctors as an end in itself: "[The doctor said] *You need to produce a minimum number of embryos, and so I was doing repeated cycles and getting one or two embryos each time and freezing them until I got 6 and they could perform the PGS*" (P16). To many interviewees, in contrast, PGS is a necessary step into what they hope is going to be a successful journey to parenthood. This legitimate aspiration extends its moral legitimacy also to the means employed to pursue it: *"I know that* [PGS] *analyses some chromosomes and then helps selecting those who are healthy, genetically healthy* [...] *given it was our last chance, we thought that selecting a healthy embryo would give us more chances* [to become parents]*"* (P8).

Normative configuration, thus, seems to operate in the clinical settings differently from how it is conceptualized by the legal framework. For instance, the Spanish 2006 Law on assisted reproduction introduces the concept of "pre-embryo", to allow spare embryos to be used for stem cell research: *"The pre-embryo is the embryo created* in vitro *constituted by a group of cells resulting from the subdivision of the oocytes between the fertilization and the 14th day"* (Art. 1, comma 2). However, no such distinction can be seen in the narrative of the women interviewed, who rather tended to distinguish between good and bad embryos, reproductively valid ones and disposable ones.

By the same token, the contractual approach underpinning the legal framework considers informed consent as the key act that brings the contract to life. In fact, the interviewees experienced informed consent as a necessary bureaucratic procedure to be fulfilled in order to have access to a treatment that had been chosen long before. Though differently from what is conceptualized by the legal guidelines, the normative configuration step plays a crucial role in the bio-identification process. Despite being considered a pro-forma bureaucratic hurdle, or precisely because of this, informed consent represents a constitutive step in the valuation process for it enables the legal inclusion of the to-be-tested and selected embryos and, consistently, of the set of sociotechnical arrangements that allow this selection to occur, into the reprogenetics value chain.

Social Configuration: Pre-implantation Genetic Testing as Reproductive Choice

STS and political economy literatures (Franklin 2006a; Waldby 2006) have successfully shown how, in embryonic and tissue economies, economic value and social value cannot be understood or analyzed separately, as they are simultaneously co-constructed. In many ways, the interviews not only confirm this mutually constitutive relationship, they also show how the bio-identification process not only operates through the progressive reconfiguration of the normative regime, it also involves the mobilization of social legitimacy, which encourages affected women to frame embryo selection in terms that positively resonate with their own ethical and social narrative. In the narratives of the women interviewed, pre-implantation diagnosis) and screening are, often, framed as an issue of reproductive choice.

The journey through the two different techniques, PGD and PGS, however, follows different framing routes and produces different outcomes. In the case of PGS, choice is often related to the opportunity to decide when, and in which socio-economic context, it was the right moment to have a baby: "*If I had known in advance, if they had told me when I was 20 I would have certainly vitrified my eggs, had it been possible…*" (P16). This narrative is widely shared by the women undergoing PGS, to the point that, sometimes, natural pregnancy dynamics are denied: "*I see in television that other women get naturally pregnant at 42, so if they had done it, why shouldn't I?*" (P14). In general, however, the actual statistics about the chances of late pregnancy and the related risks are often ignored, and infertility gets often framed as common disease: "*This is a health problem…people may believe this is a fancy thing, but this not fancy at all, this is a disease*" (P2). In this case, choice is reframed as the right to have infertility addressed by public health care treatment: "*It must be paid by the Social Security because we are talking about a disease*" (P10); "*Sure, it should be provided by the public health care system. I believe everyone has the right to have children, no?*" (P18).

The case is quite different when it comes to PGD. Here the women interviewed had no history of infertility, nor were necessarily close to the end of their fertile period. Rather, they often had their first baby diagnosed with some hereditary disease, and have opted for PGD to avoid having to go through the same experience again or to go through abortion in case the fetus were to show an affected genetic profile: "*My husband and I, we are carriers of a genetic disease, and our first baby developed the disease and died only five months after birth, so we decided to have our second baby with PGD…*" (P23); "*I did not want to wait for two months and then, if the diagnostic was bad, go through abortion*" (P21). In the case of PGD, thus, choice is much more related to the health of the prospective child. In fact, among PGD interviews the technique almost becomes morally imperative: "*You see, if you have a sick child because it was impossible to detect the illness, that is fine… but that you have this child while you knew he/she was sick all the time, and nowadays there are techniques that can do that... I don't know... I respect all moral choices but...*" (P20); "*Sometimes you are desperate and would do anything, but in my case it's different, in my case it's kind of compulsory, I mean, there is no other choice...*" (P19).

In both cases, however, the bio-identification process encourages the adoption of a narrative that positively resonates with the dominant narrative of reproductive choice and further contributes to the creation and attribution of both social and economic value to the practice of embryo testing and selection, and to the sociotechnical arrangements that enable such selection.

Economic Configuration: Patients as Pro-active Consumers

The analysis, so far, has highlighted how the bio-identification process, through the technological, the normative and the social configuration steps, stabilizes PGD and PGS embryos as bio-objects and contributes to their valuation process, as it endows to-be-selected embryos with a special biological, social and, therefore, economic value. The following quotations from the interviews, however, show how an economic reconfiguration process helps to transform, in both physical and semiotic terms, these embryos and their interwoven sociotechnical arrangements into valuable products and services, which successfully find their way into both public healthcare schemes and private reproductive markets.

The bio-identification process, first, encourages patients to approach ARTs centers in terms of professional reputation and price variations and to assess techniques and services offered in terms of a cost-benefit relationship: "*I went back home, because in this center the IVF cycle could cost me twice as much, for it was the most expensive center*" (P2) "*I do not care if they are pirates, the only thing I wanted was a baby. All I care about is results*" (P16). Similarly, some interviewees show how experienced social and economic differences need to be framed and accepted in terms of market differentiation: "*I mean, of course, some people cannot afford private clinics...but, on the other hand, some people have planes and others go by bike. That's life, right?*" (P16).

Consistently with what has been found elsewhere (Pavone and Funes 2014), not all patients are comfortable with the role of market consumers looking for an excellent price-quality relationship, they admit that often they had to behave *as if* they were consumers to improve their chances of successfully navigating through their IVF journey: "*I mean, it is really bad that we have to pay to have a child, but, then, when you are in this situation you have to be prepared to do anything to achieve what you desire most*" (P24). Indeed, several patients interviewed eventually got enrolled into new specialized markets, such as IVF psychotherapy and tailor-made IVF bank loans, especially because the public sector sets very strict admission criteria in terms of age (40) and diagnosis (proven medical infertility), and it only subsidizes three IVF cycles per person with waiting lists of up to 3 years. But what is common to all of them, whether in the public or in the private clinical setting, is their feeling of being 'numbers', of being, in a way, no more than the customers of a medical service: "*I changed because in the previous two centers I felt like a number, and I don't like feeling like a number*" (P10, similar arguments in P13).

Bio-Identification: A Key Aspect of Value Creation

In the reprogenetics field of the Spanish reproductive bioeconomy, value creation can be viewed as a process elsewhere described as bio-identification, which turns initially controversial *bio-objects* (Metzler and Webster 2011; Holmberg et al. 2011) into familiar, stabilized objects of established medical protocols and, then, into either highly sophisticated marketable products or highly restricted publicly subsidized medical treatments.

This process is based on four configuration steps, which are deeply intertwined and help the PGD/PGS trajectory to operate as a value chain. In the first step, the technological ability to extract, assess, manipulate and store embryos is a key passage, for it enables a variety of important sociotechnical changes in the way embryos are framed, addressed or constructed, such as, for instance the reframing of the embryos as objects for quality assessment, and the subsequent distinction between healthy, valuable ones and non-viable, disposable ones.

The normative and social configurations steps simultaneously provide legal and social legitimacy, allowing the newly created bio-objects to leave the experimental petri dish. We have seen, for instance how the regulatory regime frames and disciplines the variety of embryo manipulations enabled by the technology as part of a voluntary contract based on informed consent and how this is differently experienced in the clinical setting. This normative process is deeply performative and not only establishes applicable values and norms, it also defines the relevant actors and the social legitimacy of their related stakes (Faulkner 2012). The social configuration reinforces this process by encouraging women to frame the genetic selection of their embryos in terms that positively resonate with their own ethical and social narrative, which is often articulated around the concept of reproductive choice. Finally, the economic configuration step facilitates the deployment of market transactions by urging the interviewees to approach ARTs centers in terms of professional reputation and price variations, assess techniques and services offered in terms of a cost-benefit relationship, accept social and economic differences in terms of market differentiation and behave as if they were pro-active consumers. In the end, to-be-tested embryos become standardized, legally circulating and socially legitimate entities that can be treated as the commercially valuable outcomes of equally valuable, sociotechnically complex, medical procedures.

The interviews also show how the mutual articulation of the technical, the legal, the social and the economic configuration steps in the bio-identification process do operate as a value chain, in which each process of reconfiguration, apart from enabling biological materials and related products/processes to move a step further towards the final delivery to their potential customers, also attribute to them an important portion of their value, intended as both societal and economic value (Porter and Kramer 2011).

The interviews also show how, in the Spanish reproductive sector, the process of value creation is mostly embedded into private sector centers through a dense web of market-based transactions. In a context where patients feel encouraged to abandon their traditional moral benchmarks, the women interviewed often experienced their bodies and their tissues as part of a largely commercial set of value relations fundamentally oriented towards the market, which is not confined to private institutions: even in public health centers, the products and services that patients are offered are regulated by a logic of new public management, i.e. savings, austerity and resources optimization.

Though the process appears to be similar for both PGD and PGS, relevant differences between PGD and PGS exist. The technical reconfiguration associated with screening and quality assessment is initially resisted in the case of PGS, while it is actively sought in the case of PGD. Moreover, while PGD is initially associated with a sociotechnical process that is situated outside traditional moral boundaries, it becomes morally compulsory, because it is framed as a tool to avoid later abortion. Quite differently from PGS patients, the technique becomes morally legitimate because it boosts their chances of success in a moment in which they consider that the necessary socio-economic conditions to become parents have materialized.

These outcomes confirm that, whatever value may be latent in the regenerative ability of women's bodies (Waldby and Cooper 2010), it can only be harvested through the coordinated intervention of a complex set of sociotechnical steps of configuration, which appears to work, effectively, as a value chain. The deeply political nature of these sociotechnical interventions has been recently discussed by Waldby and Cooper (2014), who have emphasized the constitutive role played by clinical laborers such as egg donors or surrogate mothers. Focusing on a reproductive sector where clinical laborers are not involved, this chapter casts new light on how the organization of these sociotechnical interventions into a bio-identification value chain is a key aspect of the complex trajectory of incorporation of reproductive tissues and practices into the capitalist regime of accumulation and, thus, plays a key role in the process of value creation and attribution.

Conclusion

In recent years, a growing body of literature has turned its analytical gaze towards the economic aspects of reproductive and regenerative medicine. Instead of focusing on the ethical, legal and social implications of reproductive and regenerative sociotechnical practices, these works have approached the interactions between these two domains as constitutive elements of rising economies. Rather than separate economies organized around the commodification of different human tissues, these economies constitute a comprehensive reproductive bioeconomy. To address and study the latter, new and important concepts and tools have been elaborated, such as *biovalue*, *regenerative labor*, *life surplus* or *bioeconomy*, and the process of value creation, appropriation and distribution has been a key issue in this scholarly debate.

While some argue that, in these economies, value is generated through the reproductive work and the regenerative labor of the female body, others have emphasized that the immateriality of assets, such as intellectual property rights, is what stands at the core of the value creation process. More recently, rather than the intrinsic biological properties of the female body per se, it has been argued that the work of clinical laborers, such as egg donors and surrogates, stands at the core of value creation in the post-fordist reproductive bioeconomy. Yet, what happens to the process of value creation when clinical labor is not centrally involved? Here I suggest that the process of bio-identification at work in the reproductive bioeconomy is key to enabling the incorporation of human reproduction into the capitalist regime of accumulation. In other words, while the contribution of clinical labor cannot be underestimated, in general, the legal, technical, social and economic configurations involved in the bio-identification process operate like a value chain that actually generates, appropriates and distributes value.

Focusing on the reprogenetics economy in Spain, this chapter draws from 28 interviews with women undergoing PGD and PGS to outline and discuss how these configuration practices play a crucial role in the creation, attribution and appropriation of value. Technical, normative, social and economic configurations operate similarly to what has been defined, in the business literature, as a "value chain", where products acquire economic and social value by going through a series of valuation stages. At each of these stages, the product gains some value so that, by the time it is ready for commercialization, the product has acquired its final market value.

Thus, the socio-economic value associated with the latent (re) generative capabilities of the human body is extracted and appropriated by the capitalist modes of production and exchange through the combined deployment of four configuration steps that are part of a bio-identification process. The latter technically, legally, socially and economically reframes and reconfigures subjects, objects and their related sociotechnical practices of reproduction, care and medical consumption to the point that, first, it attributes them with a specific type of social and economic value and, second, it enables its appropriation by the existing regime of capitalist accumulation and commercialization.

Notes

1. The most common definition of reprogenetics proceeds from Knowles, L. P. & Kaebnick, G. E. 2007. Reprogenetics: Law, policy, and ethical issues, JHU Press. They define reprogenetics as "the field of research and application that involves the creation, use, manipulation or storage of gametes or embryos" (p. 255). While generally adopting this definition in this paper I have it narrowed down to the most important application in the field of reprogenetics in reproductive medicine, Prenatal Genetic Diagnosis and Screening (Knowles and Kaebnick 2007).
2. Both gametes and embryos economies, therefore, share a strong interaction with stem cell research laboratories. However, while the abundance of embryos combined with existing moral and legal rules prevents their procurement from being managed by market rules (Ehrich and Williams 2010; Svendsen et al. 2008; Zarzeczny and Caulfield 2009), egg procurement has been recently described as a growing industry with an expanding market (Braun and Schultz 2012; Spar 2007). Recent advances in egg vitrification, which now enables the correct preservation of eggs for personal (social freezing) or altruistic purposes and the development of egg commercial banks are very likely to dramatically change the overall organization of the egg economy. The commercial exchange of frozen eggs, in fact, allows for gametes to travel to the destination of the recipients, which no longer have to coincide in time and space with the donor. Before egg freezing, the economic value generated by an egg donation cycle was inextricably linked to the clinic providing the *in fresco* donor, whilst now it may actually follow the frozen eggs to wherever they are ultimately transferred to the recipients.
3. At the heart of a global market where class- and race-related stratification and economic exploitation issues are being fiercely debated (Parry 2015; Vora 2009; Waldby and Cooper 2014) surrogate mothers have come to represent the post-modern icon of reproductive laborers whose rights and risks are continuously ignored.

References

ADECES. (2015). *Técnicas de Reproducción Asistida La necesidad de invertir*. Madrid: Asociación pro Derechos Civiles, Económicos y Sociales.

Almeling, R. (2007). Selling genes, selling gender: Egg agencies, sperm banks, and the medical market in genetic material. *American Sociological Review, 72*(3), 319–340.

Almeling, R. (2009). Gender and the value of bodily goods: Commodification in egg and sperm donation. *Law & Contemporary Problems, 72*(3), 37–58.

Bergmann, S. (2011). Fertility tourism: Circumventive routes that enable access to reproductive technologies and substances. *Signs: Journal of Women in Culture and Society*, 36(2), 280–289.

Birch, K. (2016). Rethinking value in the bio-economy finance, assetization, and the management of value. *Science, Technology & Human Values, 42*(3), 460–490.

Birch, K., & Mykhnenko, V. (2014). Lisbonizing versus financializing Europe? The Lisbon Agenda and the (un) making of the European knowledge-based economy. *Environment and Planning C: Government and Policy, 32*(1), 108–128.

Birch, K., & Tyfield, D. (2013). Theorizing the Bioeconomy Biovalue, Biocapital, Bioeconomics or... What? *Science, Technology and Human Values, 38*(3), 299–327.

Braun, K., & Schultz, S. (2012). Oöcytes for research: Inspecting the commercialization continuum. *New Genetics and Society, 31*(2), 135–157.

Brown, N. (2007). Shifting tenses: Reconnecting regimes of truth and hope. *Configurations, 13*(3), 331–355.

Brown, N., Machin, L., & McLeod, D. (2011). Immunitary bioeconomy: The economisation of life in the international cord blood market. *Social Science & Medicine, 72*(7), 1115–1122.

Burns, L. R., Degraaff, R. A., Danzon, P. M., Kimberly, J. R., Kissick, W. L., & Pauly, M. V. (2002). The Wharton School study of the health care value chain. *The health care value chain: Producers, purchasers and providers* (pp. 3–26). San Francisco: Jossey-Bass.

Cooper, M. (2008). *Life as surplus: Biotechnology and capitalism in the neoliberal era.* Seattle: University of Washington Press.

Creswell, J. W. (2013). *Educational research: Planning, conducting, and evaluating quantitative and qualitative research.* Upper Saddle River, NJ: Pearson Education Limited.

Drazba, K. T., Kelley, M. A., & Hershberger, P. E. (2014). A qualitative inquiry of the financial concerns of couples opting to use preimplantation genetic diagnosis to prevent the transmission of known genetic disorders. *Journal of Genetic Counseling, 23*(2), 202–211.

Ehrich, K., & Williams, C. (2010). A 'healthy baby': the double imperative of preimplantation genetic diagnosis. *Health:, 14*(1), 41–56.

Faulkner, A. (2012). Law's performativities: Shaping the emergence of regenerative medicine through European Union legislation. *Social Studies of Science, 42*(5), 753–774.

Forman, E., Tao, X., Ferry, K., Taylor, D., Treff, N., & Scott, R. (2012). Single embryo transfer with comprehensive chromosome screening results in improved ongoing pregnancy rates and decreased miscarriage rates. *Human reproduction, 27*(4), 1217–1222.

Franklin, S. (2006a). Embryonic economies: The double reproductive value of stem cells. *BioSocieties, 1*(1), 71–90.
Franklin, S. (2006b). The IVF-stem cell interface. *International Journal of Surgery, 4*(2), 86–90.
Franklin, S. B. (2013). *Biological relatives: IVF, stem cells, and the future of kinship*. Durham: Duke University Press.
Franklin, S., Roberts, C., Throsby, K., Braude, P., Shaw, J., Lashwood, A., et al. (2005). Factors affecting PGD patients' consent to donate embryos to stem cell research. *Reproductive Biomedicine Online, 10,* 31.
Gianaroli, L., Crivello, A. M., Stanghellini, I., Ferraretti, A. P., Tabanelli, C., & Magli, M. C. (2014). Reiterative changes in the Italian regulation on IVF: The effect on PGD patients' reproductive decisions. *Reproductive Biomedicine Online, 28,* 125–132.
Hardarson, T., Hanson, C., Lundin, K., Hillensjö, T., Nilsson, L., Stevic, J., et al. (2008). Preimplantation genetic screening in women of advanced maternal age caused a decrease in clinical pregnancy rate: A randomized controlled trial. *Human Reproduction, 23,* 2806–2812.
Harper, J., Coonen, E., de Rycke, M., Harton, G., Moutou, C., Pehlivan, T., et al. (2010). ESHRE PGD consortium data collection X: Cycles from January to December 2007 with pregnancy follow-up to October 2008. *Human Reproduction, 25,* 2685–2707.
Hershberger, P. E., Gallo, A. M., Kavanaugh, K., Olshansky, E., Schwartz, A., & Tur-Kaspa, I. (2012). The decision-making process of genetically at-risk couples considering preimplantation genetic diagnosis: Initial findings from a grounded theory study. *Social Science & Medicine, 74*(10), 1536–1543.
Holmberg, T., Schwennesen, N., & Webster, A. (2011). Bio-objects and the bio-objectification process. *Croatian Medical Journal, 52*(6), 740–742.
Kent, J. (2008). The fetal tissue economy: From the abortion clinic to the stem cell laboratory. *Social Science & Medicine, 67*(11), 1747–1756.
King, J. (2013). Predicting probability: Regulating the future of preimplantation genetic screening. *Yale Journal of Health Policy, Law, and Ethics, 8*(2), 284–357.
Knowles, L. P., & Kaebnick, G. E. (2007). *Reprogenetics: Law, policy, and ethical issues*. Baltimore: JHU Press.
Kupka, M. S., Ferraretti, A. P., De Mouzon, J., Erb, K., D'Hooghe, T., Castilla, J. A., et al. (2014). Assisted reproductive technology in Europe, 2010: Results generated from European registers by ESHRE. *Human Reproduction, 29*(10), 2099–2113.
Mamo, L. (2007). Tissue economies: Blood, organs, and cell lines in late capitalism. *JAMA: The Journal of the American Medical Association, 297,* 414–415.
Marshall, C., & Rossman, G. B. (2011). *Designing qualitative research*. Thousand Oaks, CA: SAGE.

Martin, P., Brown, N., & Turner, A. (2008). Capitalizing hope: The commercial development of umbilical cord blood stem cell banking. *New Genetics and Society, 27*(2), 127–143.

Mentzer, J. T., DeWitt, W., Keebler, J. S., Min, S., Nix, N. W., Smith, C. D., & Zacharia, Z. G. (2001). Defining supply chain management. *Journal of Business logistics, 22*(2), 1–25.

Metzler, I., & Webster, A. (2011). Bio-objects and their boundaries: Governing matters at the intersection of society, politics, and science. *Croatian Medical Journal, 52*(5), 648–650.

Novas, C. (2006). The political economy of hope: Patients' organizations, science and biovalue. *BioSocieties, 1*(3), 289–305.

O'Riordan, K., & Haran, J. (2009). From reproduction to research: sourcing eggs, IVF and cloning in the UK. *Feminist Theory, 10*(2), 191–210.

Ormondroyd, E., Donnelly, L., Moynihan, C., Savona, C., Bancroft, E., Evans, D. G., et al. (2012). Attitudes to reproductive genetic testing in women who had a positive BRCA test before having children: A qualitative analysis. *European Journal of Human Genetics, 20*(1), 4–10.

Pande, A. (2009). Not an "Angel" not a "Whore": Surrogates as "Dirty Workers" in India. *Indian Journal of Gender Studies, 16*(2), 141–173.

Parens, E., & Knowles, L. P. (2003). Reprogenetics and public policy: Reflections and recommendations. *Hastings Center Report, 33*(1), 1–24.

Parry, B. (2015). Narratives of neoliberalism:'Clinical labour'in context. *Medical Humanities, 41*(1), 32–37.

Pavone, V., & Arias, F. (2012). Beyond the geneticization thesis: The political economy of PGD/PGS in Spain. *Science, Technology and Human Values, 37*(3), 235–261.

Pavone, V., & Funes, S. L. (2014). Pacientes, consumidoras o ninguna de las dos: narrativas y posicionamientos de mujeres en el caso de diagnóstico preimplantacional en el Estado español. *Revista de derecho y genoma humano, 1*, 289–300.

Petersen, A., & Krisjansen, I. (2015). Assembling 'the bioeconomy': Exploiting the power of the promissory life sciences. *Journal of Sociology, 51*(1), 28–46.

Porter, M. E., & Kramer, M. R. (2011). Creating shared value. *Harvard Business Review, 89*(1–2), 62–77.

Potter, J. (1997). Discourse analysis as a way of interpreting naturally-occurring talk. In D. Silverman (Ed.), *Qualitative research: Theory, method and practice*. London: SAGE.

Rajan, K. S. (2006). *Biocapital: The constitution of postgenomic life*. Durham: Duke University Press.

Roberts, E. F. (2012). *God's laboratory: Assisted reproduction in the Andes*. Berkeley: University of California Press.

Salama, M. (2014). Cross border reproductive care (CBRC): A global perspective. *Obstetrics & Gynecology International Journal, 1,* 00008.

Salter, B. (2013). Governing innovation paths in regenerative medicine: The european and global struggle for political advantage. In *The Global Dynamics of Regenerative Medicine: A Social Science Critique, edited by Andrew Webster* pp. (194–216). London: Palgrave..

Salter, B., & Faulkner, A. (2011). State strategies of governance in biomedical innovation: Aligning conceptual approaches for understanding 'Rising Powers' in the global context. *Globalization and Health, 7*(1), 1–14.

Salter, B., Zhou, Y., & Datta, S. (2014). Health consumers and stem cell therapy innovation: Markets, models and regulation. *Regenerative Medicine, 9*(3), 353–366.

Salter, B., Zhou, Y., & Datta, S. (2015). Hegemony in the marketplace of biomedical innovation: Consumer demand and stem cell science. *Social Science & Medicine, 131,* 156–163.

Schultz, S., & Braun, K. (2013). Procuring tissue: Regenerative medicine, oocyte mobilisation, and feminist politics. *The Global Dynamics of Regenerative Medicine: A Social Science Critique, edited by Andrew Webster* pp. (118–149). London: Palgrave.

SEF. (2013). *Registro de la Sociedad Española de Fertilidad: Técnicas de reproducción asistida (IA y FIV/ICSI).* Madrid: Sociedad Española de Fertilidad.

Shanley, M. L. (2002). Collaboration and commodification in assisted procreation: Reflections on an open market and anonymous donation in human sperm and eggs. *Law and Society Review, 36*(2), 257–284.

Silver, L. M. (2000). Reprogenetics: Third millennium speculation. *EMBO Reports, 1,* 375–378.

Spar, D. (2007). The egg trade—making sense of the market for human oocytes. *New England Journal of Medicine, 356*(13), 1289–1291.

Svendsen, M. N., & Koch, L. (2008). Unpacking the 'Spare Embryo'. Facilitating Stem Cell research in a moral landscape. *Social studies of Science, 38*(1), 93–110.

Thompson, C. (2007). Why we should, in fact, pay for egg donation.*Regenerative medicine, 2*(2), 203–209.

Throsby, K. (2004). *When IVF fails: Feminism, infertility and the negotiation of normality.* London: Palgrave Macmillan.

van Niekerk, A., & van Zyl, L. (1995). The ethics of surrogacy: Women's reproductive labour. *Journal of Medical Ethics, 21*(6), 345–349.

Vora, K. (2009). Indian transnational surrogacy and the commodification of vital energy. *Subjectivity, 28*(1), 266–278.

Waldby, C. (2002). Stem cells, tissue cultures and the production of biovalue. *Health, 6*(3), 305–323.

Waldby, C. (2006). Tissue economies. *The Wiley Blackwell Encyclopedia of Health, Illness, Behavior, and Society*, 2448–2450.

Waldby, C. (2008). Oocyte markets: Women's reproductive work in embryonic stem cell research. *New Genetics and Society, 27*(1), 19–31.

Waldby, C. (2014). 'Banking time': Egg freezing and the negotiation of future fertility. *Culture, Health & Sexuality: An International Journal for Research, Intervention and Care, 17*(4), 470–482.

Waldby, C., & Cooper, M. (2008). The biopolitics of reproduction: Post-fordist biotechnology and women's clinical labour. *Australian Feminist Studies, 23*(55), 57–73.

Waldby, C., & Cooper, M. (2010). From reproductive work to regenerative labour: The female body and the stem cell industries. *Feminist Theory, 11*(1), 3–22.

Waldby, C., & Cooper, M. (2014). *Clinical labor: Tissue donors and research subjects in the global bioeconomy.* Durham: Duke University Press.

Waldby, C., & Mitchell, R. (2006). *Tissue economies: Blood, organs, and cell lines in late capitalism.* Durham: Duke University Press Books.

Webster, A. (2012). Introduction. In N. Vermeulen, S. Tamminen, & A. Webster (Eds.), *Bio-objects: Life in the 21st century.* Farnham: Ashgate.

Zarzeczny, A., & Caulfield, T. (2009). Emerging ethical, legal and social issues associated with stem cell research & and the current role of the moral status of the embryo. *Stem Cell Reviews and Reports*, *5*(2), 96–101.

Author Biography

Vincenzo Pavone is tenured research fellow the Institute of Public Goods and Policies (IPP) of the Spanish National Research Council (CSIC). His research activities address the relationship between neoliberalism and emerging bioeconomies, with a special focus on reproductive bioeconomies.

CHAPTER 7

Making Value(s) Through Social Contracts for Biomedical Population Research

Erik Aarden

Introduction

The bioeconomy's promise of future health and wealth has generated an amalgam of policies, regulations, institutions, infrastructures, and technologies promoting advances in the life sciences. Prominent among them have been repositories of human biological material and data, commonly labeled "biobanks". In a 2009 article that describes them as one of the "ten ideas changing the world right now", *TIME Magazine* deftly employs the banking metaphor to contrast the potential of biobanks with the financial crisis of 2008 (Park 2009). The article argues that while monetary banks get by on government support, "the idea of creating yet another repository to safeguard your most valuable assets might seem downright ludicrous". Yet in language highly suggestive of the value attributed to biological materials and knowledge as a way out of the crisis, *TIME* argues that "it's not money they're after. It's your blood". In exchange for depositing tissue material, rewards consist of "medical

E. Aarden (✉)
Department of Science and Technology Studies, University of Vienna, Vienna, Austria

V. Pavone and J. Goven (eds.), *Bioeconomies*,
DOI 10.1007/978-3-319-55651-2_7

interest in the form of knowledge and therapies that grow out of that deposit". Attempts to build a national biobank in the United States, which follow similar initiatives in various other nations—so the article argues—may then help researchers "uncover the next generation of treatments for disease".

The promise of shared benefits, in the form of health and economic growth, is widely reflected in discourse on the value of biobanks and the ethical challenges they pose (Tutton 2010). Despite their diverse forms, biobanks may generally be defined as "structured collections of biological samples and associated data, stored for the purposes of present and future research" (Parodi 2015, 14; see also Kinkorova 2016), and are predominantly positioned as resources or infrastructures for research in genomics (Cambon-Thomsen et al. 2003; Chalmers 2011). Much of the ethical discussion around biobanks emphasizes traditional concerns about genetics, including consent and anonymity, at a larger scale, as well as concerns about ownership and privacy related to the economic potential attributed to biobanks. Yet the emphasis on an individual right to privacy and the economic and scientific promise that comes with ownership and use of samples (Sandor 2009) focuses the debate around biobanks on a relatively restricted set of issues. Wider social and political dimensions of biobanks' role in promised improvements in health, economic prosperity and national identity are thereby commonly obscured (Busby and Martin 2006; Kerr 2004; Mitchell and Waldby 2010).

In order to rectify this omission, here I will analyze governance *through*, rather than *of* biobanks (Gottweis and Petersen 2008). With governance through biobanks, Gottweis and Petersen refer to the ways tissue and data repositories reflect particular articulations and representations of human life that incorporate a set of norms for how to live. Analyzing governance through biobanks requires sensibility towards the assumptions of mutual commitments between biomedical researchers, donors of materials, and government or other funders. I develop this sensibility by employing the metaphor of a *social contract* to highlight the mutual expectations and commitments of the various parties engaged in producing "value(s)" in biomedical research. As we will see, value emerging from biomedical research collections in the form of benefits for a population can take various—financial and non-financial—forms, while simultaneously imposing values—in their normative sense—on populations (cf. Sunder Rajan (2012)).

Here I examine how contractual relations shape three population research projects that collect human life to produce and circulate particular forms of value(s): the Singapore Tissue Network, the Million Death Study in India, and the Framingham Heart Study in the United States. These three projects are located in three regions at the forefront of biomedical research—especially in connection with their economic potential—and link to prominent areas of concern in biomedical research. The analysis in this chapter is based on research carried out between 2012 and 2014, during which I interviewed 36 researchers and public administrators in connection with the three projects. In addition, I analyzed documents associated with these projects, such as protocols, manuals and publications, online resources and media reports. The chapter begins with an explanation of the social-contract metaphor and its application to science. This is followed by an analysis of each of the three research projects through the heuristic of the social contract. In the final section, I return to the question of value(s) to show how these three projects, while each pursuing different strategies and objectives, share a technocratic approach to public health problems, raising issues about justice in biomedical research that go beyond critiques of commercialization and the bioeconomy.

From Bioeconomies to Social Contracts

As the various contributions to this volume attest, the "bioeconomy" is a loosely defined concept employed to suggest the economic promise represented by the reproductive and generative capacities of life (Taussig et al. 2013). The economic potential of the life sciences is not a product of biological processes alone, and various scholars who have analyzed the defining role of capital in the contemporary life sciences have also pointed to the vital role of policies and regulatory interventions in facilitating the emergence of the life sciences as a sector of economic interest (Cooper 2008; Sunder Rajan 2006). The logics of capital are said to shape biomedical knowledge production as a result, although different authors' perspectives differ on whether economic value is located in biological materials or the rights to the derivative knowledge (Birch and Tyfield 2013; Waldby and Mitchell 2006). Yet the idea of the bioeconomy entails more than either the commercial exploitation of bioscientific materials and knowledge assets or a change in scientific practice to align with capitalist logics. It is simultaneously a performative construct that urges political action towards biomedical innovation and a form of public policy aimed at addressing a range of social problems through technical means (Goven and Pavone 2015; Petersen and Krisjansen 2015). Moreover, the idea that the production of value in the life sciences is predominantly framed in terms of contemporary logics of capital is biased towards a particular, market-based understanding of value (Nelson 2002) that obscures other ways in which biomedical research may be considered beneficial. Rather than to assume that the value of contemporary biomedical research lies primarily in its economic potential, I therefore investigate different ways in which knowledge produced in the life sciences may be considered to have value. I use the metaphor of a social contract for science to elucidate both what the value expected to be produced by the research projects consists of, and how this value is expected to be achieved.

The notion of a social contract is often used in science policy contexts to describe the mutual commitments and expectations research institutions and government funders have towards each other. In a classic example, Guston uses it to describe the relation between the United States Congress funding science on behalf of the American people and scientists committing to the production of useful outputs in return (Guston 2000). He succinctly describes this idea of the social contract as an arrangement where "the political community agrees to provide resources to the scientific community and to allow the scientific community to retain its decision-making mechanisms and in return expects forthcoming but unspecified technological benefits" (Guston 2000, 62). According to Guston, this particular contract allowed a strict but collaborative separation of roles between science and politics that lasted until a decrease in productivity and various scandals led to calls for more explicit accountability. While he thereby locates the notion of a social contract specifically in the decades after the Second World War in the US, other science policy scholars have adopted and expanded this notion. They use it to account for increasingly explicit demands for productivity and accountability as conditions for public funding of science (Elzinga 2012). However, across the science policy literature, the social contract is primarily used to examine how shifting contractual relations and demands affect perceptions of relevance (Hessels et al. 2009) and configurations of practice (Demeritt 2000) within *science itself*.

What such policy-centered perspectives do not account for is how the changing role attributed to science relates to changes in *society*. As Gibbons, Nowotny, and others have famously noted, the character of scientific research is subject to reconsideration as the social demands on science change. They argue this has led to the emergence of a "Mode 2 science [that] has developed in the context of a Mode 2 society" (Nowotny et al. 2001). While this thesis of scientific transformations has widely been criticized (see e.g. Hessels and Van Lente 2008), it suggests that social changes, and particularly the emergence of increasingly complex public problems in an increasingly uncertain environment, require a reconsideration of the purposes and practices of scientific research. Science and society thus need "a new social contract [involving] a dynamic process in which the authority of science will need to be legitimated again and again" (Gibbons 1999, C84). As traditional boundaries between institutions and structures of social authority dissolve, science needs to become more "socially robust"—that is, responsive to the demands and concerns of diverse social actors—to retain authority in the public arena. Yet, as Nowotny has indicated more recently, implying an alternative to the predominant discourse of science's economic productivity, responsiveness to social demands cannot be about rapid return on collective investment alone. Instead, science can be considered a public good for which "collective political imaginaries [need to] invoke citizens as the ultimate and factitious beneficiaries of all scientific and technological activities [...]" (Nowotny 2014, 17). She thereby suggests that one should think of social contracts for science explicitly in terms of the value that may be procured for (unspecified) publics at large.

This perspective on the value of science in terms of its responsiveness to the requirements of an ever-changing society raises several questions, which gravitate around the idea that knowledge ought to be "socially robust". The concept of social robustness does not attend to the power asymmetries that are of central interest to the social contracts that make up the bioeconomy (Sunder Rajan 2010). Rather than illuminating how scientific knowledge and its public valuation change, the idea of social robustness invites questions about the meaning of the social (i.e. for which collectives exactly is knowledge robust, who does this knowledge serve?) and in what ways knowledge is robust—in terms of credibility, (economic) value, or others? These questions are at the heart of feminist and postcolonial perspectives on the production and circulation of knowledge, which provide suggestive insights for analyzing biomedical research in terms of social contracts. Within these fields, various authors have pointed out how the authority of Western science has obscured and marginalized both the knowledge and the concerns of the disadvantaged

(Anderson and Adams 2007; McNeil 2005). Moreover, as Harding has argued, the production and circulation of scientific knowledge tends to perpetuate existing social hierarchies and asymmetries (Harding 2006). In marked contrast to the idea of socially robust knowledge, she thus sees social and political hierarchies, priorities, and exclusions reflected in scientific projects.

Approaching biomedical research projects in terms of social contracts from this angle thus raises the question how research affects justice (Benjamin 2013; Mamo and Fishman 2013). It forces us to consider for whom biomedical research produces value, and who is excluded. Below we will see how the answers to these questions are a product of the purposes, commitments, and expectations of the social contracts that frame particular research projects. The metaphor of the social contract suggests a three-part heuristic for analyzing core elements of the relations between researchers, funding agencies, and wider publics.[1] First, a contract has a *purpose* by virtue of being an agreement between two or more *parties* to exchange particular goods, services, or other items. Second, the agreement implies a *commitment* from the various parties to perform certain actions towards the contract's purpose. Third, contracts thereby document particular mutual *expectations* between the parties involved that may or may not be met and that can be (but are not necessarily) enforced. From this point of view, we thus may ask with what purpose parties engage in particular efforts to collect and understand human life; how they commit to actions oriented towards that purpose; and what mutual expectations the engagement in such an effort implies. In the next section, I will analyze these specific dimensions in three biomedical research projects.

Contracts for Making Value(s) in Three Population Research Projects

In this section I use the notion of social contract as a heuristic device to analyze the production and circulation of "value" in biomedical research. In each case, different actors committed to the establishment of research projects with the purpose of addressing particular social problems with the help of biomedical research. The particular purposes and actor commitments and expectations contribute to particular framings of these problems and their solutions, resulting in the exclusion of segments of the population as well as of alternative understandings of health and disease.

Building an Invaluable Resource for Biomedical Research in Singapore[2]

The commitments and expectations underlying the Singapore Tissue Network (STN), a centralized repository of human tissue, reflect the economic promise of the life sciences most closely among the biomedical research projects presented here. The social contract underlying the STN establishes a mutual commitment between biomedical researchers in Singapore and the country's government, through its basic research financier A*STAR (Agency for Science, Technology and Research) to contribute to the economic productivity of biomedical research. The STN was established in 2002 for "supporting the conduct of highest quality translational and population research in Singapore and improving healthcare for mankind".[3] It was part of the Biomedical Sciences Initiative, a government endeavor to "develop the Biomedical Sciences cluster as one of the pillars of Singapore's economy".[4] As Catherine Waldby has perceptively observed, this initiative "recapitulates [Singapore's] constitutive utopian vision in the mode of biological innovation" (Waldby 2009, 371) and thereby positions the life sciences as a vital sector to guarantee continued economic growth and the viability of Singapore as an independent nation (Lee 2007; Trocki 2009). On the other hand, the STN was described as a "boon for scientists" (Tan 2006), promising them unrestricted access to the highest quality research materials. The motivations for establishing the STN followed a globally established logic that biomedical research would produce economic value; this, over time, would override the idea that the tissue collection could improve population health.

For the STN to contribute to a vision of the life sciences as an engine of economic growth, A*STAR established the facility as a state-of-the-art collection of materials. The STN was located at the *Biopolis* research campus, a prestigious project bringing together various kinds of research institutes and biotechnology enterprises in close physical proximity. The STN offered fully automated tissue storage and retrieval, committed to international best practice standards and even developed a service for sample preparation and transportation to the repository. As one former STN administrator boasted in an interview, "all [researchers] had to do was to provide the nurse to take consent". Recipients of national research grants were expected to store their samples at the STN; in return, the STN provided this service to them without cost. Samples were thereby made available to other researchers, who could apply for access with the repository's steering board. In the word of one former high-ranking A*STAR official, the STN thus served as a "custodian" of samples and "facilitator" of biomedical research with economic potential.

However, getting researchers to commit to their end of the bargain proved a challenging proposition. When the Tissue Network was first established, clinicians and researchers collecting tissue proved reluctant to contribute to the STN collection. In various interviews, respondents attributed this to "Asian attitudes" among sample contributors (see also Huang and Tan 2010), by which they generally meant either the desire to receive a reward in exchange for contributing samples—as co-authors on resulting papers—or discontent with the major role of foreigners in biomedical research in general and the STN in particular. This included STN director Edison Liu, who had come over from the United States to lead various new research institutes and whose leadership was described by one respondent as "too American" for Singapore. Yet A*STAR found a way to increase the number of contributions to the repository, by obligating recipients of national grants to store their samples in the STN and by introducing the so-called 50/50-rule. This rule required that each sample that was contributed to the STN would be split in half. One half would be made available via the Network, while the other would be kept in what one former administrator metaphorically described as a "personal safety deposit box" for continued use in one's own research.

While this solution increased the size of the collection, it did not address the problem that samples stored in the STN were not extensively used. A*STAR had hoped to facilitate collaboration between researchers by making only basic data on age, gender, ethnicity, and minimal medical history available with the samples. This helped applicants identify samples they were interested in, but deliberately forced them to contact the person who brought the samples to the STN to get access to more detailed data. These people would often not make the data available without already having established a rapport with requesting scientists. As one researcher described his views on sharing samples, this process required familiarity, trust, and shared research interests as well as "sample bargaining" over the exact materials that would be shared. Since the STN did not take these needs into account for these procedures, researchers were unlikely to try and find collaborators through the network.

The STN was thereby largely unsuccessful in establishing collaborations that would contribute to economically valuable research outputs, but not because researchers objected to the economic motivations for funding research. Scientists rather saw the economic potential of biomedical research in a different timeframe than government officials did. From their point of view, the value of the tissue collection in both scientific and economic terms would develop over time. For researchers this trajectory would require the establishment of collaborations based on trust and familiarity first, followed by the production of knowledge about health and disease that, in some instances, would also have commercial potential. This vision conflicted with the timetable A*STAR was committed to, which expected economic benefits from biomedical research to begin flowing in after a decade. In that light, the agency critically evaluated the infrastructure it had developed and intervened where it thought it was necessary. In the sweeping reforms that followed in 2010 and 2011 (Edelson 2010; Fischer 2013), the STN was among the facilities that were shut down. While this decision was officially presented as a response to a lack of use of the available tissue materials (Chan 2012; Chang 2011), researchers rather described the repository as “a nice library” that was not very useful for research.

The fate of the Singapore Tissue Network can be seen as the result of the failure to align the different temporalities of expectation applied by researchers and administrators respectively. The social contract established with the STN thus broke down, which especially imperiled the repository’s potential value in terms of advancing scientific understanding of disease. While the repository was initially promoted in terms of its economic potential as well as its possible contributions to better health, it appears that its failure to deliver economic value in the short term was decisive in discontinuing centralized tissue storage. Since the simultaneous potential for advances in the economy and in health from biomedical research are often considered to depend on infrastructures that link together research and clinical care (see e.g. Coopmans et al. (2012) for an example specific to Singapore), the primarily economic motivation to close the STN raises questions about how economic and medical value expectations may be balanced in establishing research infrastructures.

Counting the Dead to Help the Living in India[5]

Most biobanks, like the one in Singapore, focus on the collection of snippets of human life, but the Million Death Study (MDS) in India is oriented towards gathering data about its end. The study started as a pilot with collaborators from the Office of the Registrar General of India (RGI) and the Center for Global Heath Research (CGHR) in Canada in 1998. After refining its methodology, it has been collecting information on causes of death in the Indian population since 2002. It uses a sample of less than 1% of the Indian population—which still totals eight million individuals—called the Sample Registration System (SRS). RGI uses this sample to keep track of population dynamics in between the full population censuses that take place once every decade. The SRS, described as the "largest demographic survey in the world" (Office of the Registrar General of India 2013, 3), allows RGI to register the number of births and deaths from a representative sample. For the MDS, RGI officials also collect data on causes of death. Most of the information that is available is not considered very reliable, since most people in India die without medical attention and statistics rely almost exclusively on hospital deaths. As one of the study leaders summarized the situation in India, "people who go to hospitals and die are very different from those who stay home and die [...]. If you base the national estimates on hospital statistics, you get really misleading numbers". Collaborators in the MDS therefore point to their study as an important public health project, since accurate cause-of-death information is considered vital in aiming policy interventions at the right targets (Jha 2012; Jha et al. 2006).

To collect cause of death information in India, researchers and RGI officials have developed a method called *verbal autopsy*, "which involve[s] a structured investigation of the circumstances and health symptoms leading to the death through an interview of a relative or associate of the deceased" (Jha 2012, 2). The method implies that census officials who visit households in their sample units every six months fill in a form to collect information in case a person has passed away. Central to this method, according to one MDS researcher, is that the official "should listen to the person [and ask]: just tell me how your grandfather died". Importantly, census officials are instructed in training videos not to determine cause of death themselves, but to elicit useful information from the person they interview.[6] They should gather that information in five steps that focus on the presence or absence, duration and order of twelve so-called cardinal symptoms (such as fever, weight loss, or jaundice). They are supposed to clearly indicate such information in a narrative they write on the form, which is collected and processed by the central RGI office in Delhi.

For analysis the forms are uploaded to a server for quality control, and the ones that meet necessary criteria are loaded onto the next software platform called Central Medical Evaluation. This platform distributes the forms over a panel of roughly 300 Indian physicians, depending on language (the verbal autopsy may be produced in any of 18 different Indian languages) and availability (as any one physician can only have ten cases on her or his desk at once). As one of the study initiators then describes the role of the physician, "if you are able to capture symptoms and events before the death, then the physician would be able to describe cause of death". Physicians are instructed to describe cause of death on the basis of the International Classification of Disease (ICD-10) and the principle "one death, one cause" (SRS Collaborators of the RGI-CGHR 2011, 9). As a means of quality assurance, two physicians code each death independently and usually arrive at the same conclusion. For the remaining 20% there is a procedure of, first, "reconciliation", where physicians see each other's diagnosis and are asked to reconsider. In the small percentage of cases where this does not resolve the disagreement a third, more senior physician is asked to "adjudicate". Physicians are further aided by the software platform, which uses what one MDS staff member describes as "intellectual know-how built into the system based on years and years of physician experience" to block or suggest particular diagnoses. As the same staff member explains, if "the person had a fever, the person was vomiting and you are saying they died of diabetes, that doesn't make sense. You should recheck your work". Per year, around fifty thousand deaths are autopsied, of which around forty thousand can be coded. The study involves various expert groups that prepare scientific publications on the basis of aggregate data, but these groups only get access to the data after the RGI has processed it. This shows how the study's policy purposes are given priority over its scientific outputs.

That the main purpose of the study is to contribute to policy change is a point that nearly all interview respondents emphasized. They would reference various examples of study findings that instigated policy change, ranging from tobacco control to the availability of anti-snake venom. As one researcher summarizes this perspective, "each common disease is rare somewhere in India, [which means] it is something we can avoid." This focus on policy change also has implications for how data collection and analysis are treated. According to another former study researcher, the central question is not whether the numbers are "right". Instead, "for a [health] program implementer, it is more important [to] do the same study after a few years and see what happens." Thus, comparing the effects of interventions over time is more important than applying diagnostic "gold standards"— various researchers I interviewed agreed that these could not be applied within the limits of the MDS. Physicians coding verbal autopsy narratives were consequently also instructed to "think from a public health perspective" and to recognize that "common causes are common" (SRS Collaborators of the RGI-CGHR 2011, 30). These instructions reflect the expectation that more reliable data would bring about policy change by "changing the mindset of the administrators", as one researcher argued—or in the words of study director Prabhat Jha in an interview with the *WHO Bulletin*, "the best advocacy is serious objectivity" (World Health Organization 2010).

Collaborators in the Million Death study thus see the value of their efforts in the impact cause-of-death information may have on health policy in India. While the information the study produces is undeniably of great importance for knowing the scale of health problems in India, there is a problematic assumption of equivalence in the idea that statistics informing policy makers about what Indians die from can directly be implemented in the form of revised policies. This problem becomes especially evident when one considers the so-called dual burden of disease in India. Even though the study has found that non-communicable diseases have become the "predominant cause in the overall mortality pattern in the country" (Office of the Registrar General of India 2009, vi), the MDS also finds that 38% of deaths are still due to infectious diseases—with the percentages even higher for women, in rural areas, and in economically disadvantaged states—and that one in five children does not survive childhood. These are all signs of health problems that persist—as a report on child mortality due to diarrhea argues—"more due to inaction or inadequate effort" (PHFI and Unicef 2012), whereas non-communicable diseases require investments in medical services. While cause of death statistics thus paint a particular picture of how deaths in India are distributed, the question of which courses of action should be pursued requires consideration of the mechanisms *behind* the numbers. In other words, cause of death statistics may be a necessary precondition for improving public health in India, but may not be sufficient for confronting what causes these causes.

Identifying the Coronary-Prone Individual in the United States[7]

While one of the core findings of the Million Death Study is that India is edging towards the "epidemiological transition" from infectious diseases to chronic, non-communicable diseases as the major cause of mortality, that same transition, which occurred in the United States sixty years ago, is often mentioned as one of the motivations for establishing the Framingham Heart Study (FHS). The history of "probably the most celebrated study in the history of medicine and public health" (Roberts 2000, 106) is usually told in terms of a sharp rise in cardiovascular disease mortality around the end of the Second World War (see for example Gehlbach (2005), Levy and Brink (2005) and Oppenheimer (2005, 2010)) on the study's history). In response, the US federal government committed to a twenty-year study of the causes of cardiovascular disease. On the basis of organizational considerations and estimates of the required size of the study sample, it was established in the town of Framingham, close to Boston. Study organizers invested considerable effort in securing commitment from the local community in order to ensure continued participation of research subjects. This close relation to the community remains a distinctive feature of the study today, as does the particular, predominantly white and middle aged, constitution of the FHS sample.

The study started in 1948 with 5127 participants, who were examined on a bi-annual basis. The aim of the study was to identify the "stresses and insults of various types" (Gordon and Kannel 1968, 1b-3) that researchers believed to result in cardiovascular disease. To identify the relative contributions of various characteristics, the study developed innovative approaches to investigating a set of hypotheses that referred to several potential causes (including smoking, alcohol consumption, diabetes, etc.) and defined precise criteria for diagnosing various manifestations of cardiovascular disease such as myocardial infarcts, cerebral embolism, or sudden death (Dawber 1980). By the early 1960s, study researchers proposed the notion of risk factors, described as "the outstanding achievement of CHD [coronary heart disease] epidemiology" (Epstein 1992, 25), to describe the (behavioral) characteristics increasing the likelihood of cardiovascular disease (Kannel et al. 1961).

The FHS had thereby delivered on its initial promise well within the original timeframe of twenty years, resulting in possible discontinuation once the National Institutes of Health (NIH) started evaluating what further benefits it expected from its existing studies (Patel 2012). Yet the study initiators considered the study "not as a unique effort, but rather as a continuing resource for the study of cardiovascular disease" (Gordon and Kannel 1968, 1b-1) and made two significant changes to the social contract of the FHS to secure the study's future. The first was a commitment to new directions of research by recruiting a second cohort of 5124 individuals, children of the original participants and their spouses. The research aim with this second cohort was to study transformations in risk factor profiles, with investigation of family patterns serving as "a post-facto justification more than a primary aim", according to one study researcher. This was different for a subsequent third generation of 4095 individuals, which was explicitly recruited for genetic research in the 1990s. The second major change was that the study officially transitioned to an institution external to NIH, Boston University (BU) School of Medicine, which would subsequently run the study on the basis of contracts with the NIH. These contracts are established every seven years and define core responsibilities and deliverables for the study, including continuous registration of cardiovascular disease cases, investigation of causes, and maintenance of the study infrastructure and (data and blood) repositories. NIH and BU can further agree on supplementary agreements, "which provide extra money to the contractor to do extra things", as one NIH official explains. These extra things may include new technological possibilities such as whole genome sequencing, which was introduced halfway through a contract cycle.

Both the original social contract between biomedical researchers and the Framingham community and its revision in the late 1960s continue to significantly shape how the study is conducted. This is evident in the study's efforts to understand the genomic dimensions of cardiovascular disease. Since the late 1970s, the FHS has progressed from the study of familial correlations of risk, to the study of linkages between genes and phenotypes of disease and risk, and on to genome-wide studies of associations between genomic variants and phenotypic expressions (so-called GWAS) and further sequencing techniques (Govindaraju et al. 2008; Myers and O'Donnell 1999). The study uses blood samples from its three cohorts that one researcher dates back to the 1990s, since "the study was transitioning around that time and became seriously active in the domain of genomics". These samples are processed for genetic analysis in the study's own DNA lab at BU and sent out for further processing into cell lines or for full sequencing at biotech companies like Illumina. But even while involving external companies, genomics research in Framingham remains distinctive, as its collection "provides a unique setting in which many clinical traits have been collected at approximately the same age in family members across three generations" (Splansky et al. 2007, 1335).

While for genomics in particular the value of the Framingham cohort, which is relatively small in number, lies in the family relations between the three generations of participants, one researcher argues that the value of the longitudinal information of phenotypes is of more general epidemiological significance. He describes the FHS as "a very different animal from many other epidemiological studies" due to the presence of sequential samples (that is, samples from the same individuals collected at different points in time). The study thus continues to investigate the causes and progression of cardiovascular disease both within and beyond the field of genomics. The study thereby treats genetics as a "co-factor", in the words of one researcher, which "will identify new pathways in risk promotion we did not know about". Just as this perspective on genetics is consistent with the study's broader understanding of risk factors as interactive, so the study's conduct remains consistent with its historically close connections with the Framingham community. Researchers describe this as both a blessing and a curse. On the one hand, the strong voice of the local community is said to be expressed in high levels of trust and consent to contributing to new research directions among participants. On the other hand, the strong engagement with the local community imposes limits on particular domains of research (e.g. related to reproduction or intelligence) and on commercial development of study resources.

Due to its historical legacy and the significance thereof in how the study engaged in genomics research, the FHS has thereby come to be somewhat of a hybrid between a traditional epidemiological study and one that is at the forefront of new developments in genomics (Jaquish 2007). While the study does not pursue commercial development of its results, it uses the services of commercial biotechnology firms and supplies its data to consortia and databases such as the NIH Database on Genotypes and Phenotypes (dbGaP), where—after an embargo period—they become part of a large data collection that at least in part serves the commercial endeavors of US-based biomedical research. Moreover, one may argue that the FHS has a significant role in the emergence of contemporary biopolitics focused on individual responsibility for health, where the patient is increasingly positioned as a consumer (Rabinow and Rose 2006; Sunder Rajan 2011). Framingham researchers themselves have occasionally advanced the argument that heart disease is ultimately a consequence of "a faulty life-style" (Kannel 1987, 218). Its reliance on "discrete, mechanism-driven risk factors" (Aronowitz 1998, 108–109)—meaning, factors tied to particular individual behaviors—thereby obscures how, for example, socioeconomic circumstances contribute to cardiovascular disease risks (Shim 2014). Since genomics in the context of the Framingham Study further confirms rather than challenges this individualized perspective on disease risks, we may consider its legacy in terms of a politics of individualized responsibility for health.

Contracting Knowledge and Value(s) in Biomedical Populations

In this chapter I investigated the mechanisms of value production in three population-based biomedical research projects in terms of the social contracts established through these projects. My analysis aligned with current perspectives on the economic objectives and dynamics of capital in the life sciences that emphasize how biomedical knowledge and value(s) (in both the financial and ethical sense) are coproduced (Sunder Rajan 2012). But in employing the notion of social contract, I moved beyond capital as the primary factor shaping biomedical knowledge making today to explore how other kinds of value(s) may be produced in research. To a degree my analysis thereby echoes Foucault's perspectives on the roots of modern government, which he locates in the emergence of the notion of population as an object both of governmental power and of scientific knowledge about the effects of government (Foucault 2007). In Foucault's terms, the exercise of mechanisms of power and knowledge making over a population was intended "to improve the condition of the population, to increase its wealth, its longevity, and its health" (Foucault 2007, 105). In a way, the three studies I analyzed in this chapter focus on one of these domains of improvement each and thereby show how the means *by* which and population *for* which conditions improve are shaped by the particulars of the purposes, commitments, and expectations reflected in the various social contracts. In conclusion, we may draw three observations about the dynamics of biomedical research and its value for populations from these cases.

First, each of the projects presented here attributes to different groups of actors or parties to the contract a particular set of commitments and expectations regarding the production and distribution of value. In each of the projects, the contract was a tripartite agreement between government funders, research practitioners and a population sample. Yet the population sample mostly figured in the contract as an object of knowledge-making. The populations were objectified in different ways that served the purposes of knowledge-making in each project. Different material and informational dimensions of the populations thereby came to matter in the various projects, ranging from the quality of tissue in the Singapore Tissue Network, to symptomatic stories told after death in the Million Death Study, and linkages between blood samples, information from examinations, medical histories and family relations in the Framingham Heart Study. Social contracts in biomedical research are thus not only about abstract agreements, but about the materiality of producing knowledge, and hence of making value. In contrast to perspectives that argue that the economic value of biomedical research or research collections lie primarily in (rights to the exchange of) extracted knowledge (Birch and Tyfield 2013; Turner et al. 2013), a focus on social contracts thus shows how the biological matters—exactly because it focuses on other forms of value than those amenable to exchange in a market.

Second, the different ways in which materials are collected, questions are asked, and knowledge gets produced in these three projects not only reflect different areas of biomedical science, but also the context in which these projects are situated. These projects cannot reasonably be understood as representative of national "cultures" of science-society relations (Felt et al. 2010; Jasanoff 2005), due to specific features that are idiosyncratic to the particular purpose, design, and operation of each study. Nevertheless, the juxtaposition of these three does reflect particular social concerns related to biomedical knowledge-making at particular moments in time and in particular sociopolitical contexts. These particularities may in part be found in the specific domains to which these projects seek to contribute—which, to adopt the categories named in my quotation of Foucault above, would be wealth in Singapore, longevity in India, and health in the United States. Yet these purposes take particular forms in the way value is produced in these studies. While the Singapore

Tissue Network was supposed to result in high-quality collaborations with economic potential, the Million Death Study was expected to produce reliable cause-of-death information for improving policy interventions, and the Framingham Heart Study provided insight into individual risk factors for purposes of health education and cardiovascular disease prevention. Furthermore, the particular shape of these different kinds of value also reflects specific values, in a normative sense, in making research efforts beneficial. In Singapore, biomedical research was supported insofar as researchers would direct their efforts towards efficient and industrious production of economic wealth; in India, research efforts were tailored towards correcting incorrect assumptions in health policy-making; and in the US, the project was oriented toward identifying particular areas of individual responsibility for health.

Third, the different forms of value(s) these research projects produce for specific contexts provide insight into the expectations associated with biomedical research and how these may reproduce existing inequalities. In each of these cases—as well as in the broader bioeconomy discourse—life becomes an adjudicator of value that seemingly works independently of the complexity of social life. All three of the projects presented here present knowledge about the inner workings of life, health, and disease as the solutions to the problems for which their social contracts were designed, reducing social problems to biological terms (cf. Goven and Pavone 2015). Even the social contracts addressed here, which were only partially designed in bioeconomic terms, thus share a logic of biomedicalization, where health problems become the domain of individualized, medical-technical, and entrepreneurial perspectives (Clarke et al. 2003). Such perspectives share a relatively narrow medical understanding of health problems and thereby obscure the importance of structural dimensions in health, and of coordinated collective action as a way of improving population health (Deaton 2013; Marmot 2015). Understanding population research projects in terms of social contracts may illuminate some of these asymmetries and highlight the need to act to avoid a situation where the collection of bits and pieces of human life for the purpose of making knowledge happens at the expensive of collective improvements in health.

Notes

1. I draw these elements of a contract from definitions retrieved from the Merriam-Webster online dictionary, http://www.merriam-webster.com/dictionary/contract (Accessed 3 October 2016) and the Legal Information Institute of Cornell Law School https://www.law.cornell.edu/wex/contract (Accessed 3 October 2016).
2. This characterization of the Singapore Tissue Network is taken from a newspaper report on the establishment of the repository (Soh 2002).
3. Cited from http://www.asiabiotech.com/08/0824/1393_1394.pdf (Accessed 3 October 2016).
4. Cited from https://a-star.edu.sg/About-A-STAR/Biomedical-Research-Council/BMS-Initiative.aspx (Accessed 8 August 2016).
5. This characterization of the Million Death Study is taken from a newspaper report on the study (Nolen 2012), although variants of the phrase are used widely to describe the study. See also Vogel (2012), World Health Organization (2010).
6. See http://vatraining.vm-host.net (Accessed 4 October 2016).
7. This characterization of the Framingham Heart Study is taken from one of the earliest publications of its results (Kannel et al. 1961), to which the introduction of the notion of "risk factor" is attributed.

Acknowledgment Research for this chapter was funded by a European Commission Marie Curie Fellowship, grant number 272996. The author would like to thank interview respondents across the three studies for their time and insight; Wiebe Bijker, Sheila Jasanoff and colleagues in Cambridge, Maastricht and Vienna for fruitful discussions; and Ashawari Chaudhuri, Catelijne Coopmans, Mike Fischer, Poonam Pandey and Ambuj Sagar for their fieldwork support. None of these people bear any responsibility for my shortcomings in writing this chapter.

References

Anderson, W., & Adams, V. (2007). Pramoedya's chicken: Postcolonial studies of technoscience. In E. Hackett, O. Amsterdamska, M. Lynch, & J. Wajcman (Eds.), *The handbook of science and technology studies* (3rd ed., pp. 181–204). Cambridge: The MIT Press.

Aronowitz, R. (1998). *Making sense of illness: Science, society and disease*. Cambridge: Cambridge University Press.

Benjamin, R. (2013). *People's science: Bodies and rights on the stem cell frontier*. Stanford: Stanford University Press.

Birch, K., & Tyfield, D. (2013). Theorizing the bioeconomy: Biovalue, biocapital, bioeconomics or…what? *Science, Technology & Human Values, 38*(3), 299–327.

Busby, H., & Martin, P. (2006). Biobanks, national identity and imagined communities: The case of UK biobank. *Science as Culture, 15*(3), 237–251.

Cambon-Thomsen, A., Ducournau, P., Gourraud, P.-A., & Pontille, D. (2003). Biobanks for genomics and genomics for biobanks. *Comparative and Functional Genomics, 4*(6), 628–634. doi:10.1002/cfg.333.

Chalmers, D. (2011). Genetic research and biobanks. In J. Dillner (Ed.), *Methods in biobanking* (Vol. 675, pp. 1–37). Dordrecht: Springer.

Chan, T. W. (2012). The closure of the national biobank in Singapore. *Asia-Pacific Biotech News Journal, 16*(1), 40–43.

Chang, A.-L. (2011, June 4). National biobank to close. *The Straits Times.*

Clarke, A., Shim, J., Mamo, L., Fosket, J., & Fishman, J. (2003). Biomedicalization: Technoscientific transformations of health, illness, and U.S. Biomedicine. *American Sociological Review, 68*(8), 161–194.

Cooper, M. (2008). *Life as surplus: Biotechnology and capitalism in the neoliberal era*. Seattle: University of Washington Press.

Coopmans, C., Graham, C., & Hamzah, H. (2012). The lab, the clinic, and the Image: Working on translational research in Singapore's eye care realm. *Science Technology & Society, 17*(1), 57–77.

Dawber, T. R. (1980). *The Framingham study.* Cambridge, MA: Harvard University Press.

Deaton, A. (2013). *The great escape: Health, wealth and the origins of inequality.* Princeton: Princeton University Press.

Demeritt, D. (2000). The new social contract for science: Accountability, relevance, and value in US and UK science and research policy. *Antipode, 32*(3), 308–329.

Edelson, S. (2010). Singapore's New Tune. *SciBX: Science-Business eXchange, 3* (42), 1.

Elzinga, A. (2012). Features of the current science policy regime: Viewed in historical perspective. *Science and Public Policy, 39*(4), 416–428.

Epstein, F. (1992). Contribution of epidemiology to understanding coronary heart disease. In M. Marmot & P. Elliott (Eds.), *Coronary heart disease epidemiology: From aetiology to public health.* (pp. 20–32). Oxford: Oxford University Press.

Felt, U., Fochler, M., & Winkler, P. (2010). Coming to terms with biomedical technologies in different technopolitical cultures: A comparative analysis of focus groups on organ transplantation and genetic testing in Austria, France, and the Netherlands. *Science, Technology and Human Values, 35*(4), 525–553.

Fischer, M. (2013). Biopolis: Asian science in the global circuitry. *Science Technology & Society, 18*(3), 379–404.

Foucault, M. (2007). *Security, territory, population: Lectures at the College de France 1977–1978*. New York: Picador.

Gehlbach, S. (2005). Searching America's heart: The Framingham study. In *American plagues: Lessons from our battles with disease*. New York: McGraw-Hill.

Gibbons, M. (1999). Science's new social contract with society. *Nature, 402*(Supp2), C81–C84.

Gordon, T., & Kannel, W. B. (1968). *Section 1. Introduction and General Background*. Bethesda: National Institutes of Health.

Gottweis, H., & Petersen, A. (Eds.). (2008). *Biobanks: Governance in comparative perspective*. London: Routledge.

Goven, J., & Pavone, V. (2015). The bioeconomy as political project: A polanyian analysis. *Science, Technology and Human Values, 40*(3), 302–337.

Govindaraju, D. R., Adrienne Cupples, L., Kannel, W. B., O'Donnell, C. J., Atwood, L. D., D'Agostino, R. B. et al. (2008). Genetics of the Framingham Heart Study Population. *Advances in Genetics, 62*, 33–65.

Guston, D. (2000). *Between politics and science: Assuring the integrity and productivity of Research*. Cambridge: Cambridge University Press.

Harding, S. (2006). *Science and social inequality: Feminist and postcolonial issues*. Urbana: University of Illinois Press.

Hessels, L., & Van Lente, H. (2008). Re-thinking new knowledge production: A literature review and a research agenda. *Research Policy, 37*, 740–760.

Hessels, L., Van Lente, H., & Smits, R. (2009). In search of relevance: The changing contract between science and society. *Science and Public Policy, 36*(5), 387–401.

Huang, A., & Tan, C. (2010, September 17). Achieving scientific eminence within Asia. *Science, 329*, 1471–1472.

Jaquish, C. E. (2007). The framingham heart study, on its way to becoming the gold standard for cardiovascular genetic epidemiology? *BMC Medical Genetics, 8*(1), 63.

Jasanoff, S. (2005). *Designs on nature: Science and democracy in Europe and the United States*. Princeton: Princeton University Press.

Jha, P. (2012). Counting the dead is one of the world's best investments to reduce premature mortality. *Hypothesis, 10*(1).

Jha, P., Gajalakshmi, V., Gupta, P., Kumar, R., Mony, P., & Dhingra, N. (2006). Prosepctive study of 1 million deaths in India: Rationale, design, and validation results. *Plos Medicine, 3*(2), e18.

Kannel, W. B. (1987). New perspectives on cardiovascular risk factors. *American Heart Journal, 114*(1), 213–219.

Kannel, W. B., Dawber, T. R., Kagan, A., Revotskie, N., & Stokes, J. (1961). Factors of risk in the development of coronary heart disease—Six year follow-up experience. *Annals of Internal Medicine, 55*(1), 33–50.

Kerr, A. (2004). *Genetics and society: A sociology of disease.* London: Routledge.
Kinkorova, J. (2016). Biobanks in the era of personalized medicine: Objectives, challenges, and innovation. *EPMA Journal, 7*(4), 1–12.
Lee, S. A. (2007). *Singapore from place to nation.* Singapore: Prentice Hall.
Levy, D., & Brink, S. (2005). *A change of heart: How the framingham heart study helped unravel the mysteries of cardiovascular disease.* New York: Alfred A. Knopf.
Mamo, L., & Fishman, J. (2013). Why justice? Introduction to the special issue on entanglements of science, ethics, and justice. *Science, Technology & Human Values, 38*(2), 159–175.
Marmot, M. (2015). *The health gap: The challenge of an unequal world.* London: Bloomsbury.
McNeil, M. (2005). Introduction: Postcolonial technoscience. *Science as Culture, 14*(2), 105–112.
Mitchell, R., & Waldby, C. (2010). National biobanks: Clinical labor, risk production, and the creation of biovalue. *Science, Technology and Human Values, 35*(3), 330–355.
Myers, R., & O'Donnell, C. J. (1999). Genetics. In D. Levy (Ed.), *50 years of discovery. medical milestones from the National Heart, Lung, and Blood Institute's Framingham Heart Study.* (pp. 549–550). Bethesda: Center for Biomedical Communication.
Nelson, R. (2002). The problem of market bias in modern capitalist economies. *Industrial and Corporate Change, 11*(2), 207–244.
Nolen, S. (2012, December 13). How counting the dead helps the living in india. *The Globe and Mail.*
Nowotny, H. (2014). Engaging with the political imaginaries of science: Near misses and future targets. *Public Understanding of Science, 23*(1), 16–20.
Nowotny, H., Scott, P., & Gibbons, M. (2001). *Re-thinking science: Knowledge and the public in an age of uncertainty.* Cambridge: Polity Press.
Office of the Registrar General of India. (2009). *Report on causes of death in India 2001–2003.* New Delhi: Office of the Registrar General of India.
Office of the Registrar General of India. (2013). *Sample registration system statistical report 2011. Report no. 1 of 2013.* New Delhi: Office of the Registrar General of India.
Oppenheimer, G. M. (2005). Becoming the Framingham study 1947–1950. *American Journal of Public Health, 95*(4), 602–610.
Oppenheimer, G. M. (2010). Framingham Heart Study: The first 20 years. *Progress in Cardiovascular Diseases, 53*(1), 55–61.
Park, A. (2009, March 12). 10 ideas changing the world right Now. Biobanks. *TIME.*
Parodi, B. (2015). Biobanks: A definition. In D. Mascalzoni (Ed.), *Ethics, law and governance of biobanking* (pp. 15–19). Dordrecht: Springer.

Patel, S. S. (2012). Methods and management: NIH administrators, federal oversight, and the Framingham Heart Study. *Bulletin of the History of Medicine, 86*(1), 94–121.

Petersen, A., & Krisjansen, I. (2015). Assembling "the bioeconomy": Exploiting the power of the promissory life sciences. *Jounal of Sociology, 51*(1), 28–46.

PHFI, & Unicef. (2012). *Neonatal & child health in India: The what works series 2011–2012*. New Delhi: Public Health Foundation of India.

Rabinow, P., & Rose, N. (2006). Biopower today. *Biosocieties, 1(2),* 195–217.

Roberts, W. C. (2000). *Facts and ideas from anywhere*. Armonk, NY: Futura Publishing Company.

Sandor, J. (2009). From private to public? Legal concepts of the right to privacy and ownership in the regulation of biobanks. In K. Dierickx & P. Borry (Eds.), *New challenges for biobanks: Ethics, law and governance* (pp. 123–135). Antwerp: Intersentia.

Shim, J. (2014). *Heart-sick: The politics of risk, inequality, and heart disease*. New York: New York University Press.

Soh, N. (2002, December 3). National tissue bank to be set up. *The Straits Times*.

Splansky, G. L., Corey, D., Yang, Q., Atwood, L. D., Cupples, L. A., & Benjamin, E. J., et al. (2007). The Third Generation Cohort of the National Heart, Lung, and Blood Institute's Framingham Heart Study: Design, recruitment, and initial examination. *American Journal of Epidemiology, 165*(11), 1328–1335.

SRS Collaborators of the RGI-CGHR. (2011). *Prospective study of million deaths in India: Technical document no VIII: Health care professional's manual for assigning causes of deaths based on RHIME reports*. Toronto: Center for Global Health Research.

Sunder Rajan, K. (2006). *Biocapital: The constitution of postgenomic life*. Durham: Duke University Press.

Sunder Rajan, K. (2010). The experimental machinery of global clinical trials: Case studies from India. In A. Ong, & N. Chen (Eds.), *Asian biotech: Ethics and communities of fate*. (pp. 55–80). Durham: Duke University Press.

Sunder Rajan, K. (2011). Two tales of genomics. In S. Jasanoff (Ed.), *Reframing rights: bioconstitutionalism in the genetic age* (pp. 193–216). Cambridge: The MIT Press.

Sunder Rajan, K. (Ed.). (2012). *Lively capital: Biotechnologies, ethics and governance in global markets*. Durham: Duke Univeristy Press.

Tan, J. (2006, January 13). Singapore tissue bank a boon to scientists. *The Straits Times*.

Taussig, K. S., Hoeyer, K., & Helmreich, S. (2013). The anthropology of potentiality in biomedicine. *Current Anthropology, 54*(S7), S3–S14.

Trocki, C. (2009). *Singapore: Wealth, power, and the culture of control.* London: Routledge.
Turner, A., Dallaire-Fortier, C., & Murtagh, M. (2013). Biobank economics and the "Commercialization Problem". *Spontaneous Generations: A Journal for the History and Philosophy of Science, 7*(1), 69–80.
Tutton, R. (2010). Biobanking: Social, Political and Ethical Aspects. *Encyclopedia of Life Sciences.* Chichester: John Wiley & Son.
Vogel, G. (2012, June 15). How do you count the dead? *Science, 336,* 1372–1374.
Waldby, C. (2009). Singapore biopolis: Bare life in the city-state. *East Asian Science, Technology, and Society: An International Journal, 3(2–3),* 367–383.
Waldby, C., & Mitchell, R. (2006). *Tissue economies: Blood, organs, and cell lines in late capitalism.* Durham: Duke University Press.
World Health Organization. (2010). Save lives by counting the dead. Interview with Prabhat Jha. *Bulletin of the World Health Organization, 88*(3), 171–172.

Author Biography

Erik Aarden is a postdoc at the Department of Science and Technology Studies of the University of Vienna, Austria. His research focuses on relations between scientific and technological developments, social institutions and the distribution of resources in biomedicine, primarily from a comparative perspective. Previously he has been a postdoc at RWTH Aachen University in Germany and a Marie Curie fellow with the Program on Science, Technology and Society at Harvard University, US and Maastricht University in the Netherlands.

PART III

Bioeconomies as Sites of Power and Struggle

CHAPTER 8

Data-Sharing Politics and the Logics of Competition in Biobanking

Aaro Tupasela

Introduction: The Promissory Discourse of Biobanks

The political and policy pre-occupation with the possibilities that biotechnology offers society is undeniable and apparent. The knowledge-based bio-economy (KBBE) has become an important political project through which the bio-economy has been made to matter economically, socially, and culturally. Concomitantly, since the turn of the millennium a number of new theoretical insights have been developed to explain the rapid growth and expansion in the use of human tissue and data in biomedical research. Concepts such as *tissue economies* and *biovalue* (Waldby 2002), as well as *biocapital* (Rajan 2006) have been deployed and elaborated upon in an attempt to describe the operation of what Cooper (2008) has described as a post-Fordist mode of production. According to Cooper, emerging biotechnologies reflect a new form of post-Fordism where life (microbes, bacteria, genes, etc.) become central elements in the processes of capitalist value creation and productivity and are supposedly meant

A. Tupasela (✉)
Department of Public Health, Centre for Medical Science and Technology Studies, University of Copenhagen, Copenhagen, Denmark

V. Pavone and J. Goven (eds.), *Bioeconomies*,
DOI 10.1007/978-3-319-55651-2_8

to offer new forms of sustainable development derived from scientific research.

Biobanks reflect a new promissory discourse related to bioeconomies by making populations productive in new ways (Tupasela 2017). In a similar vein, Waldby and Mitchell (2006) have argued that within the context of national biobanks, clinical labor—the work performed by people participating in and donating to large national biobanking projects—forms an important component of new forms of bioeconomies which rely on the participation of large parts of the population. Some Nordic commentators have gone as far as to suggest that what we are witnessing in some contexts is a process through which whole countries become national cohorts for research (Frank 2000). More recently, Hoeyer (2016: 73) has suggested that we are witnessing the emergence of *intensified data sourcing* where there is an increase at "attempts at getting more data, of better quality, on more people." In relation to biobanking, populations are emerging as new sites of expectations and practices of generating value. On the one hand, populations emerge as an important source of samples and data which can in turn be used in medical research (Tupasela et al. 2015). At the same time, researchers and policy-makers are seeking to "brand" the populations from whom they have collected samples and data in order to promote themselves and this resource on international research markets (Tupasela 2017). Biobanking operates at an important nexus in these developments. According to the OECD (2001):

> Biological resource centres (BRCs) are an essential part of the infrastructure underpinning life sciences and biotechnology. They consist of service providers and repositories of the living cells, genomes of organism, and information relating to heredity and the functions of biological systems. BRCs contain collections of culturable organisms (e.g. micro-organisms, plant, animal and human cells), replicable parts of these (e.g. genomes, plasmids, viruses, cDNAs), viable but not yet culturable organisms, cells and tissues, as well as databases containing molecular, physiological and structural information relevant to these collections and related bioinformatics.

Biobanks are important forms of infrastructure which seek to re-make research practices through the development of new ways of collecting, sharing and studying information on the human body. Biobanks serve as an important tool for making new types of research possible, and as a result, they are also expected to make scientific knowledge productive in new ways.

Fochler (2016) has termed this development *epistemic capitalism*, where he refers, not only to capitalism as a system of circulation, but also a "cultural way of producing, attributing, and accumulating specific forms of worth".

In this chapter I will examine the challenges associated with the competition logics of large prospective cohort (LPC) biobanks within the European context by identifying the challenges associated with the sharing of samples and data. LPC biobanking involves large collections of human tissue samples from specific populations (usually national). The collections are, usually, from healthy donors who are then tracked over decades to see what diseases they develop over the course of their lifespan. LPC biobanks, therefore, also collect large amounts of health and lifestyle information. LPC biobanks reflect only one type of biobank and it is important to recognize that there are many different types of biobanks ranging from clinical collections stored in hospitals to small collections collected by individual research groups. My interest in LPC biobanks stems from the fact that they are usually publically funded and maintained, and more importantly, within the European context, they have served as a platform to develop sharing practices among researchers. As such, LPC biobanks present a good case study of the challenges associated with competition logics within the biobanking community in general. In Europe, and elsewhere, LPC biobanks have struggled to live up to the political and commercial expectations laid out by policy discourse and to the role given them by social-scientific theorizations of biotechnological capitalist accumulation. Scientifically, however, LPC biobanks are extremely useful and valuable in studying disease in populations and represent a significant resource.

This chapter seeks to evaluate the emerging tension between policies and practices in LPC biobanking with reference to the sharing of samples, *and* to re-consider the applicability of social-scientific theories that have sought to explain the emerging bioeconomy. Under the auspices of the Biobanking and Biomolecular resources Research Infrastructure (BBMRI)—which aims to set up a European research infrastructure—there has been a pilot project focusing on developing sample- and data-sharing practices among LPC biobanks (BBMRI-LPC). I argue that existing policies relating to population biobanks are problematic in that they encourage biobanks to compete, on the one hand, while at the same time expect them to provide open access to competitors. I argue that this tension proves problematic in relation to long-term sustainability of biobanks because it expects biobanks to navigate competing logics of operation. Furthermore, I argue that current theories of the bioeconomy are not critical enough in relation to the actual practices related to competition logics. Concepts such as *biocapital* (Rajan 2006), *surplus value* (Cooper 2008), and *biovalue* (Walby and Mitchell 2006) have provided important insights into the ways in which biotechnology and economic thinking have become intertwined in policy and practice. In this chapter, however, I want to argue that social scientific theories of biovalue need to look increasingly at market inefficiencies and dilemmas within bioeconomies. There is a need to develop a more nuanced theory of biobanking politics, where the interests of scientists who control samples and data are better understood and recognized in relation to the more normative political expectations associated with sample- and data-sharing set out in policies. These developments suggest a need to revise both policy discourse and social-scientific theories of the bioeconomy.

Taking as the empirical starting point developments in the Nordic countries, I will contrast these developments to supranational biobank policies in Europe. Drawing on 14 semi-structured interviews with policy makers, researchers, and biobankers conducted in numerous European countries during the past years, dozens of informal discussions over the past decade, as well as policy paper analysis I will focus on the ways in which biobanks have been framed as generators of different forms of value: scientific, economic, and symbolic. I will compare the perspectives of these actors to more recent discussions in specialist journals such as *Biopreservation and Biobanking*, in which numerous articles have appeared on the topic of biobank sustainability. These articles discuss the challenges that biobanks face in relation to identifying ways in which value can be generated and acquired through existing frameworks. According to Vaught et al. (2011) in one such article, biobankonomics relates to calculating the "Total Life Cycle Cost of Ownership (TLCO)" [...] "to estimate all costs arising from owning, operating, and maintaining a large centralized biobank." Here, however, I would like to develop the notion of biobankonomics to explore the politics associated with controlling and sharing resources within scientific research markets. These articles provide a sobering contrast to the political visions and rhetoric related to knowledge-based bioeconomies. This difference in policy and practice will provide the basis for a theoretical discussion on the usefulness of post-Fordist theories of value creation in relation to biobanking. A critical and contextualized analysis of the emergence of biobankonomics is relevant to the bioeconomy discussion in that it identifies the practical challenges that biobanks face in the translation of information to different forms of value.

Biobanks and Capitalist Accumulation

The hype and expectations of biobanks associated with their potential to produce new forms of value chains related to human health emerged at the turn of the millennium (Tutton 2007; Douglas 2005). These expectations were particularly aligned with new forms of capital accumulation and ownership related to patenting and licensing of human genes, but, more importantly, with providing new ways of overcoming the limits of economic growth. This model, however, has become increasingly problematic due to the ensuing discrediting of the one gene-one disease model initially assumed by part of the genetics community, as well as the legal challenges associated with patenting of naturally occurring genes in the human body, such as with the case of BRCA genes and Myriad genetics' patent claims over them (Kendall and Bravin 2013). A further tension associated with this model has been the challenge of introducing new scientific practices in genetic research based on sharing of samples and data. Blumenthal et al. (2006) have noted that data withholding is common within the biomedical sciences and takes numerous forms.

It is important to notice that the development and growth of hype around biobanks has effectively occurred during the emergence and consolidation of a bioeconomy policy framework, which represents simultaneously a political project as well as a scientific agenda, based on neoliberal ideals of production. Within this model, life itself emerges as a target of neoliberalism, where individuals are expected to take increasing responsibility for their health, by using genetic tests provided by companies to explore their health risks, become consumers of personalized pharmaceuticals, as well as make reproductive decisions supported by high-tech medical decisions. Under a bioeconomy framework, state authorities across Europe have sought to find new ways in which market forces can be introduced into the scientific research system, as well as the healthcare sectors (cf. Birch and Siemiatycki 2015; Goven and Pavone 2015).

One of the major proponents of the idea of the bioeconomy has been the OECD, which has had a long-term interest in developing the idea that knowledge production serves as the basis of economic growth and commercial output (OECD 2001). These policies have then entered the national policies within a number of European countries, including the Nordic countries, such as Finland (see Academy of Finland 2002). According to the OECD, the bioeconomy reflects "that part of the economic activities which captures the latent value in biological processes and renewable bio-resources to produce improved health and sustainable growth and development" (OECD 2009: 9). The bioeconomy concept became a pervasive organizing principle within the European context as well (see European Commission 2002), where the European Commission (2005) further elaborated the notion of knowledge-based bioeconomy (KBBE) as a policy term to describe the economic, social and environmental expectations associated with the biotechnology sector. These concepts continue to find new forms within the *Horizon 2020* strategies, which specifically focus on sustainable growth strategies (European Commission 2010). Both independently and in association with this KBBE approach, Nordic countries have engaged in a long-term attempt to mobilize and utilize the populations, the samples collected from them, and their related health information for medical research, through a multitude of health registries and databases (Bauer 2014; Nordforsk 2014; Frank 2000).

Public and private funders have provided start-up funding for many biobanking ventures in the hope that they would eventually be able to generate enough revenue from their own activities to fund their own operation. This orthodox view, however, has proven very problematic in almost all biobanking cases, and evidence suggests that some biobanks have actually gone bankrupt or had difficulty maintaining their operations (Tupasela and Stephens 2013; Rose 2006). One of the main challenges facing biobanks today is the lack of adequate consideration of the actual cost-recovery models that they ought to have in place (cf. Henderson et al. 2015). This problem appears to undermine the policy rhetoric and vision outlined in supranational policy documents generated by organizations, such as the OECD (2006), which have suggested that knowledge-based bioeconomies will provide a more sustainable mode of production than existing resource-based industries. With the growth in the number of biobanks and their increased need to source samples and data from patients and the general public, the financial sustainability of biobanks is increasingly predicated upon social sustainability. As a result, biobanks need to engage in a multitude of practices to secure the public funding, political support, and social acceptance that are necessary to ensure long-term sustainability (Tupasela et al. 2015).

In the case of population-based biobanks, however, the bioeconomy model is further challenged by the policy strategy which places emphasis on data-sharing as a fundamental principle of biobanking activities. The notion of data-sharing as a major principle derives from the idea that publicly funded science should be open (OECD 2015). Furthermore, data-sharing has been seen as a new form of economic activity which is supposed to replace more traditional forms of economic thinking. Most recently, for example, the EU Competitive Council highlighted the importance of open science by noting that:

> [O]pen science has the potential to increase the quality, impact and benefits of science and to accelerate advancement of knowledge by making it more reliable, more efficient and accurate, better understandable by society and responsive to societal challenges, and has the potential to enable growth and innovation through reuse of scientific results by all stakeholders at all levels of society, and ultimately contribute to growth and competitiveness of Europe. (Council of the European Union 2016: 3)

The problem, however, is that there is no clear guidance as to how biobanks are supposed to operate financially under the data-sharing model. This has become even more salient with recent austerity measures in a number of countries. As Vaught et al. (2011) very aptly point out with regard to the huge public investments into biobanks, "despite these sizable investments, the economics of biobanking initiatives is not well understood." Data-sharing, alongside knowledge-based bioeconomic thinking, emerged as another major political pre-occupation in which the role of biobanks was seen, not just as generating new forms of capital, but also as serving as the "gold mine" from which others could derive valuable information for their own innovations. The problem, however, has been that the production of capital value for biobanks has become somewhat challenging alongside their expectations to share samples and data rather than capitalize on them. In this sense, data-sharing policies placed biobanks in a somewhat tenuous position, where they were expected to compete by collecting a rare and limited source of information, but at the same time share it with competitors and other actors within the biomedical research field (European Union 2016; OECD 2005, 2009). In this sense, the idea of the bioeconomy also contained an important principle, namely data-sharing (Piwowar et al. 2008), with which current biobanks have to contend (see also Colledge et al. 2013; Pereira 2013).

In the case of large prospective cohorts and other population-based biobanks in Europe, it is not clear what the political expectations regarding biobanks are: open resources meant to be shared or resources for competition? This reflects a general tension in political aspirations for science, as it is being re-aligned to partly reflect ideals of open science, while at the same time being expected to compete for resources.

Rajan (2006: 7) has argued that "[b]iocapital is one vantage point from which to view the complexities of capitalism(s), and like all situated perspectives, it contains within it both its specificities as well as its diagnoses of more general structural features of capitalism." If we examine the specificities of LPC biobanks and see them as one situated perspective among several forms of capitalisms that have developed and emerged around various forms of bioeconomies, it can be said that LPC biobanks have been thrust into a particular type of productive limbo where they are expected to compete and share at the same time. In this sense, the notion of the bank as a model for what we call biobanks is misleading since the politics of sample and data-sharing lend themselves more to the idea of a library or a public resource. At the same time, however, public funding for biobanks has been limited, and they are expected to generate enough of their own income through the sale of their services to support their own activities.

I now turn to address biobank policies the notion of sustainability.

Biobanks and Sustainability

I begin my analysis by identifying what I argue to be an important turning point within the large prospective cohort biobanking community. In September 2012 the HandsOn Biobanks conference was held in Uppsala, Sweden. The conference was one of the first in what has become a series of important conferences organized by the BBMRI-ERIC network, which is a biobanking research infrastructure for European biobanks. Within the conference there was a session which focused specifically on sharing the experiences of European biobanks in relation to the development of costing models for their activities. The main concern of the session organizers, as well as the participants, related to the ways in which biobanks could capture the value and cover the costs of their activities in the long run. On the one hand, biobankers are expected to keep costs as low as possible so that university researchers can afford to use the resources—the majority of researcher grants do not offer the possibility to cover market-based costs of sample handling. On the other hand, biobankers are expected to achieve financial independence and sustainability by selling their samples and services. It was clear from the session that the practices and experiences of the biobanking community were at odds with the political ideas and visions of capitalist accumulation through biobanking.

These discussions brought issues relating to the sustainability of biobanks to the fore. Vaught et al. (2011) have developed the term biobankonomics to describe the financial calculations involved in running and maintaining a biobank financially. This may include developing cost recovery models for sample handling, distribution and storage, personnel costs, costs associated with collecting more samples, as well as costs associated with public engagement. The concern with biobankonomics is not a singular concern of a few biobankers, but has become a broader discussion point among the biobanking community. This has developed to such an extent that the journal *Biopreservation and Biobanking*, for example, regularly publishes articles on the topic of biobank sustainability and even had a special issue devoted to the costs of running biobanks (see Henderson et al. 2015; Seiler et al. 2015). These articles, I argue, are a reflection of the challenges associated with cost recovery models in biobanks and a general concern with the sustainability of biobanks. Although biobanks are very different, and may define sustainability in different ways, they all face a similar challenge: biobanks need to balance cost recovery with ensuring optimal use of the samples and data. Underuse will not generate enough scientific output and would be seen as a waste of a collection, whereas using samples for research that may not be scientifically of the highest caliber would also be seen as a problem. Although biobankonomics is generally understood in economic terms, I would like to expand this to encompass the social and normative issues related to sharing and monetization.

Although it has been recognized that population-based collections should be seen as a common resource from which all research endeavors should be able to draw (ESHG 2003), the political will and public financial support for these activities suggest that population biobanks are also expected to become self-sustaining. According to Cooper (2008: 19), "When capital mobilizes the biological, how do we theorize the relationship between the creation of money (surplus from debt; futures from promise) and the technological re-creation of life?" With LPC biobanks, however, the issue is not so much the re-creation of life using biotechnological means, but rather the study of disease in humans using tissue samples and related health data so as to prevent disease in the future. The impact on life itself is further down-stream where it is difficult for LPC biobanks to extend their monetization activities and thus gain access to financial stability. The main fundamental questions of how do we theorize the creation of money or biocapital from these practices remains a challenge, however.

A further challenge that has emerged in relation to biobank sustainability and competition logics relates to the post-2007 global financial crisis. In response, numerous European governments have been undertaking austerity measures to cut public funding in an attempt to curb public spending. Given that the majority of LPC biobanks have been publicly funded and are reliant on public funds for their operation, many have felt the pressure that has been exerted by these cuts. The pharmaceutical industry has been proposed as an obvious candidate for new partnerships providing new markets and sources of funding (Editorial 2009). An emergent challenge in this, however, relates to pricing of services. Many biobanks would like to charge the pharmaceutical industry higher fees than researchers working in the public sector in order to improve their financial stability. Indeed, it could be argued that companies also have the resources to pay higher prices. Under many national competition regulations, however, charging one set of actors higher prices as opposed to others constitutes discriminatory pricing practices and would be met with anti-competition laws. As a result biobanks are finding themselves in a situation where they are unable to extract higher market prices from private companies due to competition legislation. This reflects another way in which the biobanks must navigate a political-economic landscape shaped by the precepts and contradictions of neoliberalism.

In the following I will look at some of the responses and discussions that emerged in my interviews with biobankers as they relate to developing and maintaining long-term sustainability of their activities. The interviews I conducted reflect a common perspective among European biobankers that although their activities are seen as an important over-all resource for biomedical research, the resources allocated to their development and maintenance reflect an expectation that those activities are sustained through the provision and distribution of services and products, namely samples and information. Furthermore, tensions emerge between different imperatives that are set forth for biobanks, namely a competitive imperative to attract income to ensure sustainability, but also the imperative to share with potential competitors. The discussions surrounding this changing terrain also reflects different preferences among biobankers as to which model of sharing they wish to undertake.

Managing LPC Biobanks in Practice

The networks in which biobanks operate are an important infrastructure that supports their activities through providing a forum in which both sharing and sustainability are discussed and debated. On one hand, the network forms a support mechanism, while on the other hand it is a source of competition among its members. For European biobanks the major supporting and facilitating network is the BBMRI-ERIC network. All countries who become members agree to develop national biobanking resources and further work towards common standards for sharing, as well as other standards which facilitate the exchange and movement of samples within Europe. The BBMRI-ERIC infrastructure seeks to develop and identify ways of making access easier. Within the broader BBMRI-ERIC network a smaller and more focused project is under way, namely the BBMRI-LPC, which seeks to develop and standardize activities among selected European Large Prospective Cohorts, or population based biobanks. The LPC project is to provide insights into the challenges associated with sharing from LPC biobanks, which the broader network can then utilize in developing more common biobank platforms for sample and data-sharing.

The move within biobanking to foster and develop practices of data-sharing reflects in some ways a new normative demand among biobankers. Although traditional norms within the scientific community have drawn on a discourse derived from the Mertonian norms (Merton 1973), such as communalism (scientific goods as a communal property), the actual practices within scientific communities have in general reflected what are called counter-norms. Although difficult to pin down, data- and sample-sharing can be seen to emerge within the European context in relation to the GenomEUTwin project (2002–2006), which sought to compare samples and data from twin registries in a number of countries (Peltonen 2003). An important feature of the project was to identify and develop a technological solution to facilitate sharing among partners so as to avoid free riders. They achieved this by using a method known as database federation, where access to data only became possible by contributing (Tupasela 2010; Muilu et al. 2007: 2).

In the interviews that I conducted this issue came up several times. In relation to the development of a new culture among biobankers, it was often noted that biobankers were not accustomed to sharing samples in the way that policy documents have envisioned. One biobank manager noted:

> In the medical research community, we have essentially relatively small groups. And they have to compete with other groups to get funding. Now we will have to go to the leading groups that successfully compete against each other and get them to work together. And that is a complete new culture in a medical research community. [...] If someone does not feel properly recognized for his or her contribution, they are not happy. Very simple. And this might result in active or passive resistance in supporting projects and providing access. (Interview with biobank manager 2015)

The notion of active or passive resistance among biobankers to the dominant narrative in policy discourses, which pursue open access and sharing policies, was a common theme among many of the interviews. Many people felt that they either did not gain any recognition or there was no real benefit from sharing their samples with others, especially competitors. Although the idea of sharing samples is seen to provide a general benefit for the scientific community, it fails to address the need among scientists who control samples to acquire merit from sharing activities. Much of this concern relates to the lack of individual benefits that can be attributed to sharing practices. Although sharing might meet certain Mertonian criteria for traditional scientific norms, such as communalism, there is very little benefit to biobankers and the scientists running those biobanks from sharing their samples. Sample collections are in many cases not institutionalized, and thus are always tied to the interests of individual scientists or research groups. Historically, the collection of large prospective cohorts, as well as other types of sample collections, has resulted from the interest of individual scientists or research groups in collecting samples and data. It is rather a recent phenomenon that institutions, such as hospitals and universities, have begun to show an interest in the collections, and thus the politics of sharing samples and data (Hoeyer 2004).

This concern was often related to how sharing was evaluated within the biomedical research community. As one researcher noted in an interview regarding performance evaluation:

> All our performance evaluation is essentially toxic to collaboration. We have no incentives if someone enables the success of someone else. We only collaborate if it helps our own success. Then we collaborate. That's reality. (Interview with biobank manager 2015)

Despite policy calls for sharing and openness, the academic merit system is still very much based on competitive criteria, which biobankers find difficult to reconcile with sharing practices. Although there have emerged examples of trying to develop ways of recognizing and awarding merit for sample sharing through the use of citation systems such as the Bioresource Research Impact Factor (BRIF) (Cambon-Thomsen et al. 2011), biobankers and researchers controlling collections still seek to serve as gatekeepers of their collections. As a result, notions of productivity and the development of tissue economies within current biobank networks are guided and generated on the basis of competitive, not sharing, principles. Although biobankers recognized the importance of sharing, the reality of career building and publication pressure forced them to act strategically. This tension was highlighted in one interview where a researcher noted:

> If everybody contributes, everybody could benefit from the general resource that is established. I think we are still too much influenced by the classical deal-making type of arrangement – I give you something, I get something back – which helps, but I think we should have a much more general approach: we are all contributing to a common resource. There is not an immediate return. (Interview with biobank manager 2015)

Part of the reason for this cultural clash within the biobanking community is the lack of common European-wide policies related to compensation and credit for biobankers who share their samples. As a result, each biobank develops and maintains its own criteria for sharing, which best meet the performance evaluation system to which it is accountable.

The development of common access criteria across biobanks in Europe is also receiving attention, but these efforts, too, are coming into conflict with individual interests related to individual biobanks. As one biobank manager noted regarding the process of trying to develop common access protocols for all LPC biobanks:

> As a personal opinion I think that developing Access SOPs [standard operating procedures] is challenging because so much of the access issues depend on internal issues within individual cohorts, which we cannot necessarily influence. (Interview with biobank manager 2015)

This problem was generally referred to as "data hugging" or the process of working as a type of gatekeeper to control access to samples and related information. Data hugging emerged as a major challenge in implementing common sharing policies and gaining access to European sample collections. Data hugging was attributed to many different causes, such as lack of scientific quality of the application for samples or the small number of samples left in the collection. In some instances, it was noted that some researchers or biobankers may have a vested interest in what types of research are done with specific sample collections, and thus have an interest in protecting their scientific reputations by not allowing, for example, researchers who they considered to be doing poor research access to their collections. For others it might be about preventing competition. A challenge for some biobankers therefore was that in order to facilitate long-term sustainability, it was not just a matter of sharing samples, but rather sharing samples with the "right kind" of collaborators. This meant developing strategic alliances using their sample collection, which would further their career, as well as possibly ensure future funding by collaborating with partners who had access to more funding or resources. Biobankonomics, therefore, was not just a financial issue, but a scientific world-view or cultural issue where competitors were blocked out and like-minded researchers or research agendas were supported through access.

In addition to data hugging as a major challenge to facilitating access to samples, there also appeared to be structural challenges in running a biobank with decreasing institutional funding. In one interview a researcher with a long career in the biobanking sector noted the historical changes that had been affecting the biobank research community:

> It used to be that we would give out samples to researchers for free when we had a lot of state funding, we did not charge researchers. [...] But now we have to, and surprisingly researchers have been willing to pay thus far, even though we have had to continue to increase the prices as our budget funding is decreased. [...] If we do a state-funded cohort collection then it is a one-time cost of about one million or two. But what happens after the collection, that's what we need external money for. (Interview with biobank manager 2015)

Many biobank managers noted that problems arose not with the collection of samples, for which funding was often available, but with long-term maintenance of sample collections. Many countries do not provide funding for day-to-day operations of the biobanks themselves, but banks are not necessarily able to cover their costs from the sharing of samples. In many interviews people noted that there does not exist a financing model that can support not only the collection of samples and health information, but also long-term storage and facility maintenance. As one interviewee noted:

> If you employ people you have to pay them their salary continuously. So we are really lacking the business model – how to finance, how to maintain

> the operation to continue when this primary investment ends. (Interview with biobank manager 2015)

Such comments were quite common in the interviews where biobank managers would reflect on the challenge of long-term sustainability. In part this problem has been created or exacerbated through the implementation of neoliberal state policies where basic institutional funding is continuously decreased and institutions are forced to find new sources of revenue. At the same time, they are limited in the business models that they can employ since they are quite often publicly funded in the first place.

Finally, it should be noted that there are considerable differences between countries and biobanks as to how samples and data are shared. This is, perhaps, one of the greatest challenges in fostering a common culture of sharing in Europe. From the interviews it became apparent that some biobanks were more inclined to see sharing as a competitive advantage within biomedical research, whereas others saw gatekeeping as a more useful competitive strategy. One cannot say which perspective is more salient without taking a normative stand as to what approach is more useful, since European policies seem to provide grounding and basis for both. Publicly funded biobanks are expected to operate according to the notion of open science where samples are shared while at the same time giving rise to new forms of monetization of the human body and health. Although, these two approaches may not necessarily be mutually exclusive, they seem to generate tensions and challenges among European biobankers.

Examining this from a situated perspective related to biocapital (cf. Rajan 2006), we can say that LPC biobanks and other population biobanks are placed in a strenuous operating environment, where on the one hand they are expected to take on business-like practices in gaining outside funding, while at the same time living up to the traditional Mertonian norms of scientific openness and data sharing. There is no obvious way to reconcile these expectations.

Conclusion: The Catch-22 of Biobanking

The emergence of a strong policy discourse on the knowledge-based bioeconomy, as well as policies for data sharing and open access, has garnered a great deal of social-scientific attention. In particular, theories related to tissue economies (Waldby and Mitchell 2006) and biocapital (Rajan 2006) have reflected the interest that has emerged in describing and theorizing emerging economies. This chapter has sought to identify two challenges in these trends.

First, I have sought to identify how within the LPC and population biobank community there have emerged concerns regarding long-term financial stability. It is clear from scientific articles and interviews that I have conducted that biobankers harbor a variety of concerns in relation to being able to maintain long-term sustainability without continued public funding. Biobanks are seen as a resource for other research, but due to cutbacks in public funding they are finding themselves in a difficult position to maintain their operations. Within this context, there has emerged a competition logic among biobanks which is not conducive to sharing in all instances. The competition logic operates, both at the national level, where countries seek to compete with each other over infrastructure development and scientific reputation, as well as the personal or institutional level, where specific biobanks and research groups compete with others for research funding and collaborators. This further complicates the role of biobanks within society as general resources for research and brings into question the sharing logic held within the open science movement ideology.

Second, I have sought to highlight the inability of social-scientific theories to critique this development in relation to sharing and competition practices which tend to be at odds in relation to LPC biobanking. Looking at these developments from a situated perspective, LPC biobanking shows the difficulty that some research areas within biomedical research have in monetizing their activities. It also highlights problems with policies which seem to apply a one-size-fits-all approach to biotechnological research and development. Theories, such as tissue economies, need to account in a more robust way for market failures and inefficiencies generated by political programs, which seek to undertake science-like actions to bolster economic development. This would allow for political interests to become more visible in the operations of tissue economies, as well as the underlying assumptions which operate in relation to notions of biocapital, for example. This also challenges some of the normative assumptions surrounding sharing economies (OECD 2015), whereby, sharing is seen as a neutral practice, devoid of political interests to control and manage collections. In this sense, tissue collections need to be better understood as sites of power and governance, which may seek to resist the enforcement of sharing logics from above or through networks, such as BBMRI.

Politically the bioeconomy is presented as a neutral, interest-free field of operation. In reality it is fraught with social, cultural and economic interests which clash with traditional conceptions of scientific activity, such as data sharing policies. EU policies regarding the bioeconomy seem to have a number of tensions built into them. On the one hand, they encourage hard-core economic values of competition, while at the same time expecting actors to value sharing practices, which could undermine their competitive advantage. The combination of these two approaches has proven to be difficult to apply in relation to large prospective cohorts. Researchers and biobankers operating within the LPC community employ a range of approaches to sharing their samples, ranging from perspectives which espouse traditional Mertonian norms of sharing, to strategies of selective access which have been termed "data hugging" among the biobanking community. At the same time, however, those who share samples are confronted with the market logic which does not provide adequate rewards or recognition for sharing with the scientific community, and in many cases competitors. In addition, they have had to deal with dwindling public funding to support the long-term operation of their facilities, placing further burdens in trying to make their collections more available to the research community. As such they appear to be trapped in a type of policy-practice prison where multiple conflicting expectations constrain everyday activities in developing sharing practices within European biobanking.

Acknowledgments This project has received funding from the European Research Council under the European Union's Horizon 2020 research and innovation programme (grant agreement number 682110), as well as the *Global Genes, Local Concerns* project funded by the University of Copenhagen's 2016 Excellence Programme for Interdisciplinary Research. I am also grateful to Vincenzo Pavone, Joanna Goven and Klaus Hoeyer for insightful comments on earlier versions of the text.

References

Academy of Finland. (2002). *Biotechnology in Finland. Impact of Public Research Funding and Strategies for the Future*. Evaluation Report 11. Helsinki: Publications of the Academy of Finland.

Bauer, S. (2014). From administrative infrastructure to biomedical resource: Danish population registries, the "Scandinavian laboratory," and the "epidemiologist's dream". *Science in Context, 27*(2), 187–213.

Birch, K., & Siemiatycki, M. (2015). Neoliberalism and the geographies of marketization: The entangling of state and markets. *Progress in Human Geography*. doi:10.1177/0309132515570512.

Blumenthal, D., Cambell, E. G., Gokhale, M., Yucel, R., Clarridge, B., Hilgartner, S., et al. (2006). Data withholding in genetics and other life sciences: Prevalences and predictors. *Academic Medicine, 81*(2), 137–145.

Cambon-Thomsen, A., et al. (2011). The role of a bioresource research impact factor as an incentive to share human bioresources. *Nature Genetics, 43*(6), 503–504.

Colledge, F., Elger, B., & Howard, H. (2013). A review of the barriers to sharing in biobanking. *Biopreservation and Biobanking, 11*(6), 339–346.

Cooper, M. (2008). *Life as surplus—Biotechnology and capitalism in the neoliberal era*. Seattle: University of Washington Press.

Council of the European Union. (2016). *The transition towards an open science system—Council conclusions (adopted on 27/05/2016)*. Brussels: Council of the European Union.

Douglas, C. (2005). Managing HuGE expectations: Rhetorical strategies in human genome epidemiology. *Science Studies, 18*(2), 26–45.

Editorial. (2009). Biobanks need pharma. *Nature, 462*, 448.

ESHG (European Society of Human genetics). (2003). Data storage and DNA banking for biomedical research: Technical, social and ethical issues. *European Journal of Human Genetics, 11*(Suppl 2), 8–10.

European Commission. (2002). *Life Science and Biotechnology—a Strategy for Europe*. Luxembourg: Office for Official Publications of the European Union.

European Commission. (2005). *New perspectives on the knowledge-based bio-economy—Transforming life sciences knowledge into new, sustainable eco-efficient and competitive products*. Conference Report. Retrieved from https://ec.europa.eu/research/conferences/2005/kbb/pdf/kbbe_conferencereport.pdf.

European Commission. (2010). *EUROPE 2020—A strategy for smart, sustainable and inclusive growth*. Brussels: European Commission.

European Union. (2016). *Open innovation, open science, open to the world—A vision for Europe*. Brussels: European Union.

Fochler, M. (2016). Variants of epistemic capitalism: Knowledge production and the accumulation of worth in commercial biotechnology and the life sciences. *Science, Technology and Human Values*. doi:10.1177/0162243916652224.

Frank, L. (2000). When an entire country is a cohort. *Science, 287*(5462), 2398–2399.

Goven, J., & Pavone, V. (2015). The bioeconomy as political project—A polanyian analysis. *Science, Technology and Human Values, 40*(3), 302–337.

Henderson, M., Simeon-Dubach, D., & Monique, A. (2015). Finding the path to biobank sustainability through sound business planning. *Biopreservation and Biobanking, 13*(6), 385–386.

Hoeyer, K. (2004). The emergence of an entitlement framework for stored tissue: Elements and implications of an escalating conflict in Sweden. *Science Studies, 17*(2), 63–82.

Hoeyer, K. (2016). Denmark at a crossroad? Intensified data sourcing in a research radical country. In B. D. Mittlestadt & L. Floridi (Eds.), *The ethics of biomedical big data.* Switzerland: Springer.

Kendall, B., & Bravin, J. (2013). Justices Strike Down Gene Patents. *Wall Street Journal.* Retrieved February 25, 2016, from https://www.wsj.com/articles/SB10001424127887324049504578543250466974398.

Merton, R. K. (1973). *The sociology of science—Theoretical and empirical investigations.* Chicago and London: The University of Chicago Press.

Muilu, J., Peltonen, L. & Litton, J-E. (2007). The federated database—A basis for biobank-based post-genome studies, integrating phenome and genome data from 600,000 twin pairs in Europe. *European Journal of Human Genetics,15,* 718–723. doi: 10.1038/sj.ejhg.5201850.

Nordforsk. (2014). *Joint Nordic registers and biobanks—A goldmine for health and welfare research.* Nordforsk policy paper 5. Oslo: Norden.

OECD. (2001). *Biological resource centres: Underpinning the future of life sciences and biotechnology.* Paris: OECD.

OECD. (2005). *The Bioeconomy to 2030: Designing a Policy Agenda. OECD International Futures Programme.* Available at http://www.oecd.org/dataoecd/47/19/35532457.pdf

OECD. (2006). *Creation and governance of human genetic research databases.* Paris: OECD.

OECD. (2009). *The Bioeconomy to 2030: Designing a Policy Agenda.* OECD International Futures Programme. Retrieved from http://dx.doi.org/10.1787/9789264056886-en.

OECD. (2015). *Making open science a reality.* OECD Science, Technology and Industry policy papers no. 25, Paris: OECD.

Peltonen, L. (2003). GenomEUtwin: A strategy to identify genetic influences on health and disease. *Twin Research, 6*(5), 354–360.

Pereira, S. (2013). Motivations and barriers to sharing biological samples: A case study. *Journal of Personalized Medicine, 3,* 102–110.

Piwowar, H. A., Becich, M. J., Bilofsky, H., & Crowley, R. S. (2008). Towards a data sharing culture: Recommendations for leadership from academic health centers. *PLoS Medicine, 5*(9), e183.

Rose, H. (2006). From hype to mothballs in four years: Troubles in the development of large-scale DNA biobanks in Europe. *Community Genetics, 9,* 184–189.

Seiler, C. Y., Eschbacher, J., Bowser, R., & LaBaer, J. (2015). Sustainability in a hospital-based biobank and university-based DNA biorepository: Strategic roadmaps. *Biopreservation and Biobanking, 13*(6), 401–409.

Sunder Rajan, K. (2006). *Biocapital—The constitution of postgenomic life*. Durham: Duke University Press.

Tupasela, A. (2010). Semantic interoperability and biobanking—The politics of setting technical standards in tissue economies. In A. Bammé, G. Getzinger & B. Wieser (Eds.), *Yearbook 2009 of the institute for advanced studies on science, technology & society*. Wien: Profil.

Tupasela, A., & Stephens, N. (2013). The boom and bust cycle of biobanking—Thinking through the life cycle of biobanks. *Croatian Medical Journal, 54*, 501.

Tupasela, A., Snell, K., & Cañada, J. A. (2015). Constructing populations in biobanking. *Life Sciences, Society and Policy, 11*, 5.

Tupasela, A. (2017). Populations as brands in medical research—Placing genes on the global genetic atlas. *BioSocieties*. doi:10.1057/s41292-016-0029-9.

Tutton, R. (2007). Banking expectations: Reflections on the promises of biobanks. *Personalised Medicine, 4*(4), 463–469.

Vaught, J., Rogers, J., Carolin, T., & Compton, C. (2011). Biobankonomics: Developing a sustainable business model approach for the formation of a human tissue biobank. *Journal of National Cancer Institute Monograph, 42*, 24–31.

Waldby, C. (2002). Stem cells, tissue cultures and the production of biovalue. *Health: An Interdisciplinary Journal for the Study of Health, Illness and Medicine, 6*(3), 305–323.

Waldby, C., & Mitchell, R. (2006). *Tissue economies: Gifts, commodities, and bio-value in late stage capitalism*. Durham: Duke University Press.

Author Biography

Aaro Tupasela works as an Associate professor of ethical, legal and social aspects of biobanking at the Department of Public Health, University of Copenhagen. As a sociologist with an interest in STS, his research has focused on the social aspects of biomedical use of human tissue sample collections and the information that can be gained from them. He also has an interest in the work of population geneticists and the ways in which human populations are characterized and described.

CHAPTER 9

Reproducing the Border: Kinship Legalities in the Bioeconomy

Sonja van Wichelen

Introduction

In 2012 a boy named Dev, who was born out of a cross-border surrogacy arrangement, stayed behind in India while his twin sister left with her intended parents to live in Sydney, Australia. While international surrogacy is prohibited in New South Wales, the state where the Australian couple lived, The Department of Foreign Affairs had issued the girl with Australian citizenship based on the genetic link between her and the intended father. Australian diplomats handling the case in New Delhi believed the boy was abandoned because of gender selection and made allegations that money exchanged hands over the domestic adoption of the boy. Despite public criticisms from prominent Australian judges aimed at the Australian government for not intervening in this matter, the boy stayed in India. According to officials in New Delhi the boy was adopted by a wealthy Indian couple under the appropriate Indian adoption laws (*ABC Foreign Correspondent*, 23 June 2015).

S. van Wichelen (✉)
Department of Sociology and Social Policy, The University of Sydney, Sydney, Australia

V. Pavone and J. Goven (eds.), *Bioeconomies*,
DOI 10.1007/978-3-319-55651-2_9

Six years earlier a 5-year old girl called Lovette who came from war-torn West Africa was denied an application for a humanitarian visa which would allow her to be reunited with her mother in Australia. The rejection was based on the fact that they did not have proof that the child belonged to her mother. While the girl's birth certificate was destroyed in a fire caused by rebels in 2003, which had also killed hundreds of people in her village in the Ivory Coast, her mother was not able to pay the $7000 costs attached to DNA testing. After a push from the newspaper that made this story public, the Department of Immigration and Citizenship in Australia agreed to accept a second application and pay for the DNA test. In 2010, after genetic parentage was established through the test, the now 8-year old girl was reunited with her mother (*Herald*, 27 February 2010).

What do bioeconomies have to do with the Australian stories just presented here? Their stories are significant from a number of perspectives, but in this chapter, I will focus my attention on how economies emerging from biotechnological advancements are shaping (and are shaped by) the regulation of families in globalization. The chapter addresses two sets of questions. The first set pertains to the relationship between bioeconomies and the kinds of troubled kinship relations depicted above: To what extent do bioeconomies produce stateless and parentless children?, How do bioeconomies shape new formations of legal kinship and legitimate families? And how do these kinship legalities in turn shape new economies as well? The second set of questions addresses the relationship between cross-border reproductive practices and migration regimes: What does Dev's story have in common with Lovette's story? What do these different configurations of reproductive politics tell us about current shifts in kinship legalities brought about by biologization and globalization? As I will demonstrate, migration regimes are a constitutive part of the bioeconomy, as well as of the neoliberal order. The two stories illuminate the range as well as the differential operations of kinship legalities couched in different but entangled bioeconomies.

Kinship relations are changing in globalization. Biomedical technologies, such as genetic testing and assisted reproduction, challenge the constitution of families. They influence how kin members relate to each other but also how the nation-state defines which families are legitimate and illegitimate, worthy and unworthy. The bioeconomy can be seen as encompassing the development and promotion of these technologies and has significant implications for the kin relationships formed or disturbed by them. My chapter focuses on these implications and looks at the question of kinship legality in the bioeconomy. I take the site where family life is most at stake: the borders of citizenship. Hence, it analyzes how the bioeconomy affects families at the borders of the nation state and engages with what I call the "reproduction-immigration nexus". To do this I compare different practices pertaining to family life—namely cross-border surrogacy and family reunion in immigration—that are not typically seen in conjunction but that taken together can illuminate how bioeconomies are implicated in the global regulation of families and reproduction.

The chapter places these two different phenomena, reproduction and immigration, side by side. Although reproduction and immigration are distinct areas of concern and occupy unique positions in legal and social science scholarship, I demonstrate the importance of comparing practices that have different internal configurations. The comparison acts as a necessary methodological strategy. Bringing them together as a "framing device" will allow for an examination of the biomedicalisation of parenthood as well as the role of the bioeconomy in border politics. Functions and jurisdictions regarded in formal law as separate are increasingly intertwined and networked in the globalized world. By looking at how the bioeconomy is caught up in regulatory and legal practice, my aim is to disclose how biotechnology and globalization are implicated in defining what kinship is and how kinship legalities are produced.

Cross-Border Surrogacy in the Reproductive Bioeconomy

The phenomenon of global surrogacy is the latest addition to what has been termed cross-border fertility tourism. The transnational exchange of gametes, embryos, and babies through artificial reproductive technologies, adoption, and surrogacy, fulfill an increasing demand for reproductive services in situations where domestic possibilities are either legally, technologically or financially unattainable.[1] As is well-documented, global reproductive markets are highly stratified, based on global inequality, and formed in uneven formations of demand and supply mechanisms (Storrow 2005; Vora 2009). In the case of surrogacy, the practice emerges in countries with high-end medical facilities but also with a population living in conditions of economic hardship who can provide cheap biological or clinical labor (Cooper and Waldby 2014).

Moreover, reproductive markets can only emerge and exist within certain formations of legality. The resulting child needs to be legally accommodated by the nation-state of the commissioning parents, whether it involves the provision of citizenship or legal parentage.

Currently, several states in the US (including New York, California, Arkansas, and New Hampshire), India, Russia, Cambodia, Georgia, and the Ukraine allow certain forms of commercial surrogacy. As a result of the availability of high-end medical facilities and having a reputation for providing excellent services in reproductive technologies (together with the UK, India was the first to deliver IVF children), India is emerging as a leader in fertility services (Banerjee 2014; Pande 2016). Until recently, it was one of the most popular destinations for infertile couples and individuals from the Global North because of the relatively low cost (Australian commissioning parents pay a fraction of what they would pay in the US).[2] Moreover, before the ban on global surrogacy, Indian clinics remained competitive in relation to other countries that provided the service (such as Thailand), not just in the pricing, but in the hiring and retention of Indian females as surrogates. When countries shut down agencies that cater to surrogacy services—often following widely publicized scandals such as the story of Dev presented in the beginning of this chapter—new countries spring up to fill in the gap. Thus, after Thailand and India either banned the practice, or restricted it to domestic consumers, Georgia, Ukraine, and Cambodia emerged as new frontiers.

As scholars have argued, gestational surrogacy involves forms of capital, labor, and the arbitration of reproductive labor value (Cooper and Waldby 2014: 83; Vora 2009). High costs at home are pushing fertility seekers across borders to search for more affordable options (Whittaker and Speier 2010). Fertility tourism, then, demonstrates the "relations between travel, purchasing power and reproductive service labor that constitute the core of cross-border surrogacy" (Cooper and Waldby 2014: 63). As with previous debates about globalized domestic work or sex tourism, the debate on global surrogacy oscillates between structure and agency, between the exploitation and victimization of third world women and the reproductive and intimate labor of these women who actively choose this option out of the (albeit grim) options available to them. Nevertheless, as Amrita Pande's work on Indian surrogates illustrates vividly, the global practice of surrogacy feeds into neo-eugenic dynamics of contemporary reproductive biopolitics, in which women of the Global South (or working class women more generally) are expected to restrict reproduction while women of the Global North (or middle class women more generally) are encouraged to increase their fertility, echoing older histories of race and empire (Pande 2016).

An essential component of the (cross-border) surrogacy industry—that intimately ties markets to the legal playing field—is the place of genetics. Surrogacy markets are based primarily on *gestational* surrogacy, meaning that it does not allow the birth mother or surrogate to be the genetic mother at the same time. A donor egg needs to be part of the arrangement and can be an anonymous donor egg or the egg of the intending mother. It is often, then, a requirement that at least one of the intending parents is the genetic parent of the child. In this way, genetic parenthood can be established on the basis of a DNA test, and the child can be granted citizenship based on this genetic relationship.[3] While Australia, for instance, automatically grants citizenship based on the right of blood (*jus sanguinis*), France and Germany have not automatically done this in the past. In the latter situation, if the country of the surrogate does not grant citizenship on the basis of the right to soil (*jus soli*)—that is, by virtue of being born in the country—children born there though surrogacy contracts can be rendered stateless. This was the case with Germany and India regarding the so-called *Balaz twins*. Toward the end of 2009 two children, commissioned by a German man (the genetic father) and his German partner, were denied German citizenship because the practice was a criminal offense in the country. Nor were they granted Indian citizenship because India does not grant automatic citizenship if the child is not genetically related to an Indian national. In the absence of clarity, the Supreme Court of India had to intervene on humanitarian grounds.[4]

Legal disjuncture also occurs at the sovereign border of the country into which the child is brought. A parent can legally bring to Australia her or his genetic child from a commissioned contract pregnancy in India or the US as long as this genetic relationship has been legally established in the country of the child's birth. However, the child's entrance into Australia does not necessarily imply a natural conversion to legal parenthood. Usually, parents wanting to establish such a relationship have to go through a domestic court process. In the event that commercial surrogacy, and in particular overseas commercial surrogacy, is banned (such as is the case in New South Wales), courts and their judges need to justify going against domestic law in order to confirm parentage or parenting responsibilities.

Judges in the Global North feel they are in a conundrum. In the story we opened with, the girl was given Australian citizenship in India based on the genetic link with her father. Commercial surrogacy is prohibited in three Australian states: New South Wales, the ACT, and Queensland. In these states entering into a commercial surrogacy arrangement, regardless of where it takes place, constitutes a criminal offense that includes a jail term of a few years. Yet, as we have seen, the legal prohibitions do not stop Australian individuals and couples from commissioning surrogacy arrangements overseas. On the contrary, the practice appears to be growing (Keyes 2013). So far, there have been no criminal prosecutions following illegal overseas surrogacy arrangements. Instead, when presented in court, the legal doctrines of the "right to family life" and "the best interest of the child" are used to confer legal parenthood or parental responsibilities onto the genetic parent or intending parents (Van Wichelen 2016; Harland and Limon 2016). Pleas have been made by judges and scholars to allow commercial surrogacy domestically in order to prevent overseas arrangements that cannot be properly regulated and overseen.[5] Human rights are often cited in these pleas.

A similar human rights stance is evident in Europe where the European Court of Human Rights (ECHR) recently ruled against national decisions not to grant citizenship to children resulting from overseas commercial surrogacy arrangements.[6] The cases involved French citizens who commissioned surrogacy in the United States; the European Court ruled that surrogacy—while respecting France's ethical position to ban the practice—was compatible with human rights. Moreover, the court emphasized that children's rights as defined in the Convention of the Rights of the Child were not sufficiently protected when parents were not able to obtain legal parenthood. Based on Article 8 of the European Declaration of Human Rights, which is the right to respect for private and family life, the ECHR ruled that intended parents who can give evidence of biological relatedness should be recognized as legal parents. Such interventions reveal a pragmatic stance toward the practice of overseas commercial surrogacy.

Such rulings do several things at the same time. First, they tolerate the moral stance of the sovereign nation-state. Rather than seeing this as an inconsistent move, the legal scholar Britta Van Beers argues that in fact it has an essential symbolic function: domestic assisted reproductive laws "have an important communicative, expressive, and anthropological meaning and function which surpass these laws' practical effectiveness"

(Van Beers 2014: 5). As such, while the European Court can rule pragmatically, tolerating surrogacy not out of ideal reasons but out of practical realities, states can keep to their normative principles.

Second, although morally states can abide by their stance, the European court rulings restrict the legal consequences of such a stance. In the two surrogacy cases that were at issue, the Court ruled that France had to acknowledge paternal relationship. A liberalization of banned practices is taking place which challenges central principles of international law. Australia seems to represent a microcosm of what happens in Europe on a regional scale. Similar to the situation in Europe, a liberalization of domestic law and state policy is occurring in allowing commissioned children to be with their parents in states where commercial surrogacy is banned.

Third, such rulings allow global reproductive markets to evolve. The bioeconomies informing reproductive markets, then, do not only progress depending on the legalities and politics in the treatment countries, but also on the legalities and politics in the destination countries, as well as on international human rights regimes. Although facilitated by the interests and pressures of an elite minority, who use geopolitical differences to break open domestic politics on surrogacy, it is ultimately the implementation of several "rights" instruments in the international realm that makes it possible for global surrogacy markets to advance.

Family Reunion in a Global Migration Industry

Theorists of globalization have pointed to the paradoxical way in which globalizing economies have at once softened as well as hardened borders (Andreas and Biersteker 2003; Mezzadra and Neilson 2013). While globalization "expedites the flow of commodities and finance, [it is also] conditioning the circulation of peoples within that space" (Feldman 2011: 78). This trend applies similarly to the bioeconomy. While on the one hand, we witness how the globalization of the life sciences allows for collaboration, research, and trade across and beyond national borders, the implications of the very same life sciences are also affecting how states are protecting, surveilling, and managing their sovereign borders. DNA kinship testing in immigration applications is a recent illustration of this trend.

Stringent measures have been introduced to lower the number of people entering nation-states of the Global North. Restrictions to family reunion are key in regulating this influx, and increasingly, DNA kinship testing is required to give evidence of identity and biological relatedness. Family reunification policies were brought into life to accommodate migrants and resettled refugees to legally bring their families to their new country of residence. The right to family reunification has been a long-standing principle in immigration policies of many countries throughout the world, but it has recently witnessed a backlash in North America, Europe, and Australia (De Leeuw and Van Wichelen 2012; Castles et al. 2013). While the influx of people in developed nations is increasingly premised upon the financial or human capital they bring, "non-productive" people are excluded from entry through biomedical and biotechnological means.

More than 17 countries have implemented DNA kinship testing in their immigration policies, and its use in on the increase (Weiss 2011). The increase should be placed in the context of a globalizing migration-apparatus (Feldman 2011; Gammeltoft-Hansen and Sørensen 2013), and it cannot be understood without addressing how the bioeconomy is becoming more and more integral to this apparatus. By placing at the center the genetic model of family life, DNA kinship testing facilitates the state's construction of the il/legitimate family in a way that makes possible the application of double standards to the determination of who and what is "legal kin".

Comparisons across countries are illuminating. Although formally working under the same European Council directive that aligns family reunion policies across member states, European countries operate under their own distinct immigration regimes (Heinemann et al. 2015). The geneticization of the family in Germany, for instance, reveals a practice of DNA analysis in cases of family reunion that strictly adheres to a biological model of family life (Heinemann and Lemke 2015) that is counter to more social models in other countries. Here, DNA evidence of a biological relation between assumed kin trumps all other (documentary and interview) evidence in establishing family relation. In Germany, where blended families (through divorce), LGBT families, or adoptive families have become more and more commonplace, such an explicit biological model, Heinemann and Lemke argue, creates a double standard for the recognition of family, to the detriment of immigrants (ibid.). It also relies on the idea that DNA evidence promises "true" knowledge of the body and relatedness, an idea much debated by social theorists of science and technology (ibid.: 30). In contrast to the more controlled situations in Finland and Austria, where testing agencies are state regulated and centralized into one or a limited number of institutions (Heinemann et al. 2015), agencies in Germany are privatized and are therefore in competition with each other, producing a situation where immigrants and their lawyers need to "shop around" to find a cost-effective lab.

While in Germany DNA kinship testing is used to exclude non-genetic relatives from the nation-state, the situation in Finland, as Tapaninen and Helén (2016) show, is somewhat more complicated. The DNA test is seen as helping officials with the overall decision-making around the applications for family reunion. Rather than an absolute truth for authorities to form a decision, the test is valued as accommodating a human right for the applicant (i.e., ensuring that kin are not unintentionally excluded). Contrary to the privatized situation in Germany, this humanitarian motivation provides the justification for a highly state-regulated practice segregated from the logic of profit, as witnessed by a strict adherence to a government-assigned testing lab at the University of Helsinki (ibid.: 98).

Australia seems to occupy an intermediate position. DNA kinship testing is not compulsory. According to the Department of Immigration and Border Protection, DNA testing is suggested "as a means to prove a claimed biological relationship when the department is not satisfied with available evidence of a claimed relationship".[7] The actual practice of DNA kinship testing in immigration was initially relatively rare.[8] However, according to the Immigration Advice and Rights Centre (IARC) the practice of offering genetic testing is increasingly used as a "blanket response" to suspicions of fraud in certain countries.[9] The costs associated with the DNA test are borne by the applicant (except in some refugee applications) and can range from AUD $1000 to $1600 per single parent-child test and increases with each additional party. Costs also run higher when tests need to be performed outside of Australia and when travel costs are involved.[10] While counseling is advised preceding and following testing, this too is the responsibility of the applicant.

Most of the emerging scholarship in this field rightfully points to the biopolitics of the state in defining the right kind of family worthy of entry into the respective country (Murdock 2008; Heinemann et al. 2013; Heinemann and Lemke 2013, 2014; Helen 2014). This family is premised on the heteronormative, nuclear, and biological model. Thus, they argue, liberal democracies are enacting double standards: while domestic families are increasingly defined in more inclusive terms (comprising adoptive families, blended families, and LGBT families), families of immigrants are more and more narrowly defined as including only the biological, and more specifically, the genetic family unit. Moreover, the explicit targeting of certain "suspect populations" for DNA testing indicates continuing racialization and discrimination in immigration practice.

In light of this biopolitics of the management of family reunion using DNA testing, Heinemann and Lemke (2014) propose that the concept of biological citizenship should be revisited. Existing scholarship on biological citizenship has usually defined the concept as enabling an extension of rights, stressing the "biological body as the basis of claims about social inclusion, recognition, and democratic deliberation" (ibid.: 502). However, in light of the recent literature on DNA testing, this idea of an "extension of rights" is actually questionable and questioned. People are not calling upon knowledge of their bodies to claim particular access or rights within the polity, instead, the uses of kinship testing in immigration indicate the continuing role of the state in determining which biological criteria are sufficient to convey citizenship rights within the boundaries of the sovereign nation-state. In view of these practices, the claim that biotechnology produces a new form of "molecular biopolitics" that displaces traditional forms of classification and exclusion (Rose 2009) should be reexamined. One could argue that the biological citizenship of Rose's work in fact involves the capacity and organizing power of a certain elite who, with their wealth and influence, are able to effectively lobby for common biological interests (disease, disability, genetic dispositions). However, this is often at the expense of other less organized or powerful collectives. The various studies on DNA testing show that genetic relatedness, rather than helping to extend rights, is actually restricting rights of citizenship.

It is essential to place the various new articulations of biological citizenship, and the involvement of DNA testing, medical screenings, and biometrics, against the backdrop of an ever-increasing biologization and geneticization of a global migration management regime that is increasingly privatized (Btihaj 2013; Feldman 2011; Menz 2009; Pugliese 2010). Increasingly, the management of (offshore) detention centers, passenger screenings at airports, and the organization of deportations are becoming the responsibility of profit-oriented organizations. The state outsources these tasks to global companies such as Groups 4 Securicor (G4S), Geo, Kalyx and Serco, which then subcontract a large number of (often unknown) organizations (for instance Broadspectrum, formerly known as Transfield in Australia). According to George Menz—who studied the privatization and outsourcing of migration management in the US, Europe, and Australia—migration management across the Western world is increasingly affected by macro-economic dynamics. This includes the rise of New Public Management in public policy design, but also, he argues, financial and political blame-avoidance strategies (2013). In the outsourcing of state duties and accountabilities, politics and economies are central in shaping a neoliberalized management of migration.

Biotechnologies are a recent addition to security systems, which in the past typically relied upon bureaucratic systems and procedures. In his work on the biopolitics of biometrics, Joseph Pugliese (2005, 2010, 2014) describes how biometric systems are increasingly being used across institutions and organizations, in particular following the terrorist attacks on the twin towers in New York City in 2011. Pugliese questions the assertion that "biometrics offer a natural convenience and technical efficiency in ensuring the production of identity authentication and assurance" (2005: 3). Addressing the spread of facial-, iris- and finger-scan technologies, he contends that the "natural convenience" and "technical efficiency" of these biometric technologies are far from assured for non-white subjects. Instead, the coming together of surveillance techniques and biometrics produces a biopolitics of pursuing racially profiled subjects. Similarly, DNA testing for immigration purposes, presumes the idea that it promises a more effective measure of determining kin. Apart from the fact that such an idea is not compatible with the more expanded idea of legal kin practiced domestically in many western states, empirical studies of DNA testing regimes reveal it to be considerably less accurate than assumed (Heinemann and Lemke 2013). In the meantime, profits are being made from the increasing use of these technologies (particularly in countries where DNA testing facilities are privatized), and companies are actively approaching governments as well as migration agents and migrants themselves to "sell" their technologies and services.

In sum, bioeconomies are increasingly informing migration industries, and it is by engaging the bioeconomy *as a political project* (Goven and Pavone 2015) that it becomes apparent how the promise of DNA is shaping kinship legalities.

Kinship Legalities

Regardless of their unique locations in the legal domain, comparing practices that have different internal configurations like reproduction and immigration can be revealing. Most prominently, the comparison makes evident the different standards applied to the constitution of family. Fertility tourism concerns the formation of family, and many cases in Australia and beyond illustrate that in the legal domain it is pushing the notion of family beyond the natural, biological, and heteronormative (Margalit et al. 2014; Van Wichelen 2016). Family reunion in immigration, on the other hand, occurs in a legal space which narrows the notion of family. It is a space that requires of applicants to give evidence of biological relatedness and demands of them a stricter adherence to (legal) definitions of family life that are either no longer used or at the cusp of being reconfigured in many western states. Hence, refugees and migrants are expected to adhere more forcefully to the biological model of the nuclear family than citizens are who seek to *make* families overseas.

Beyond the tool of comparison to illustrate differences, or, more normatively, to demonstrate double standards, my aim in invoking the analogy is also to analyze what biotechnologies and their accompanying economies are *doing* to their respective legal contexts: what is DNA *doing* to family law? What is it *doing* to immigration law?[11] Each environment evokes a particular configuration of the legal, the moral, and the biological, and represents a particular figuration (idea/discourse) about what a "family" actually is or ought to be. As such, it is not so much the ideologies of genetics that are under investigation here, but rather, the way in which genetic knowledge is currently being reworked in different legal contexts.

As argued by Sheila Jasanoff, scientific truth is not free of morality, nor is law merely applying scientific technologies to come to a reasoned conclusion (2011). As demonstrated above, the decisions as to whether to use DNA testing and what meaning to confer on it are political ones. These decisions include whether the determination of biological parentage should decide what constitutes a legitimate or illegitimate family. For example, in the surrogacy and family reunion cases with which I opened this chapter, DNA testing was determinative in the decision as to whether or not a commissioned child or a family member of a resettled refugee could lawfully live with those whom they regard as, or who regard them as, kin. The fact that the request for DNA testing can be offered or made compulsory makes applicants and their relations always already implicated by the institution of genetic parenthood. Here, law, whether family law or immigration law, is an active agent in producing social mores and constructing new families. In their consequences, biotechnologies and the geneticization of family life can therefore be seen as what Jasanoff calls "bio-constitutional":

> Radical shifts in the biological representation of life thus necessarily entail far reaching re-orderings in our imagination of the state's life-preserving and life enhancing functions – in effect, a repositioning of human bodies and selves in relation to the state's legal, political, and moral apparatus. (2011: 4)

Biotechnology is re-articulating family life through law, and one can argue that such new configurations produce new biolegalities (the biolegality of surrogacy/family reunion) and biolegal bodies (the embodied subjectivities formed from such biolegalities).[12]

Let me illustrate this idea of biolegality and biolegal bodies by going back to the two stories of Dev (surrogacy) and Lovette (family reunion). Dev's twin sister, who was brought to Australia, represents the promise of reproductive technology. Her figuration overcomes the limitations of biological fertility as well as what is called "social infertility" (when referring to reproductive desires of male gay couples or single men). Although her legal kin existence is premised on the genetic relationship with the intended parent that is also the genetic parent (this can be the intended mother who donated her eggs, the father who donated sperm, or both), genetic knowledge is not overruling legal knowledge. In other words, it is not genetic knowledge that is translated into automatic legal parenthood. Instead, legal knowledge is being reconfigured to include a new legal notion, namely that of "intentionality" and "intending parenthood". This notion of "intending parenthood" is gaining more and more traction in legal discussions of reproductive technologies. The biolegal body of Dev's sister, then, represents the legalization and "bioconstitution" of a progressivist idea of the "recombinant family" (Strathern 2005: 22).

Similarly, the figure of Lovette in the story of family reunion can, on the one hand, represent the determinant role of genetics in the legal realm. Without DNA technology she would not have been reunited with her mother. It echoes the celebratory reception of DNA in forensics when cases were solved through DNA evidence that enabled prisoners who were charged in error to be freed.[13] However, her reunion was precarious. If the Department of Immigration and Border Protection—responding to publicity around the case—had not paid for the DNA test, she would not have been able to be reunited with her mother. The exorbitant costs attached to the test represent a significant additional hurdle built into the promise of the DNA test, which, in this space, represents not only the identification of a genetic relationship as such, but more importantly, the lawful entry into the Global North.

Moreover, if the test had come out negative for Lovette, she would not have been able to be reunited with her mother. The Australian security regime (similar to European regimes in France and Germany) hesitates to accommodate families formed in ways not practiced in Australia, but common in countries of the Middle East, South America, Asia, and Africa: customary adoption, Islamic forms of fostering (known as *kefala*), and polygamy. This does not mean that the intention of the regime is to "withhold family life"; rather, DNA technologies (as with biometric technologies) are another *means* through which the border is protected from an "influx of immigrants." Such devices, built in and through bio-economies, with their claim of scientific authority and precision, "neutrally" police global structural inequality by blocking in-migration of the poor. They are in effect giving legal expression to what Žižek (2002: 149) calls "unabashed economic egotism". In quoting Žižek, Gregory Feldman argues that, "these new developments in the security of borders [take] racism to an even more brutal form of reasoning that is based neither on naturalism (the West as the best) nor culturalism (the West must also preserve its cultural identity) but rather on 'unabashed economic egotism' in which the fundamental divide is between those in the sphere of prosperity and those excluded from it" (2011: 79).

While Lovette might easily have been prevented from reuniting with her mother, the boy left behind in India by his contractual parents can lawfully be provided with Australian citizenship.[14] Biolegalities and bio-legal bodies are unevenly formed and produced in unequal worlds (Van Wichelen 2015). While biolegalities reinstate and reconfigure natural and biological formations of kin, they also function as devices in apparatuses of security and containment. The structural inequalities they reinforce are obscured through the practices of individualization that they facilitate. So while the globalized, gendered, and racialized practice of commercial surrogacy feeds off of this longing, and demand, for "one's own" children, the structures of inequality that make it possible to fulfill this desire go unacknowledged in the new biolegalities of surrogacy. As the genetic attributes of particular individuals are brought to the fore, the structural inequalities that produced them recede into invisibility.

Conclusion

It was the purpose of this chapter to examine the complex interplay between biotechnology and legal governance by looking at the economies informing assisted reproduction and migration. Comparing cases pertaining to global reproduction on the one hand and family reunification on the other revealed the normative dimension of this interplay but it also emphasized that this normativity is not necessarily linked to a preference of one truth system (law) over the other (science).[15] As Alain Pottage argues, the links between law, science, and kinship have rarely been scrutinized but it is in this relationship that one can study the complex effects that biotechnologies have on the making of legal institutions (2007: 340). However, the complexity of globalizing processes has often been overlooked in this relationship. By looking at the implications of the bioeconomy for the legal regulation of reproduction and immigration, I proposed in this chapter to consider the relationship between science, capitalism, and law as sustaining modes of governance.

Through the reproduction-immigration nexus, particularly looking at the ways in which the promissory power of DNA technologies in the bioeconomy are intervening, we can thus see how new legalities and legitimacies of family life are formed. These new legalities point to a rescaling of legal institutions. While the bioeconomy of immigration is imposing divisions and devaluing the worth of migrants, the bioeconomy of reproduction is redrawing borders and accommodating commissioned lives. Such rescaling suggests the movability of borders beyond the nation-state and calls out for a legal sensitivity toward the proliferation of new borders. It also suggests a multi-scalar practice of legal governance that accommodates modern legitimacies as well as bio-legitimacies in their interactions with scientific cultures, market cultures, and human rights cultures.

Notes

1. For a discussion on how adoption can be approached as a reproductive technology, see Van Wichelen (2014).
2. Following a number of scandals, including the story of the boy left behind in the beginning of this chapter, there has been a call for stricter regulation of surrogacy in India. Proposed guidelines set out in the long-awaited "Assisted Reproductive Technology (Regulation) Bill" have recently been passed (November 2015) and at the time of this writing are under inter-ministerial consultation. The Department of Health Research has also recently issued instructions conveying, among others, that the import of human embryos is banned except for research purposes and that foreign nationals (including OCI Cardholders) who visit India for commissioning surrogacy are not to be granted visas.
3. India, for instance, does not allow Indian citizenship for the commissioned child. In cases where the genetic parent is an overseas citizen of India (OCI), person of Indian origin (PIO), or a non-resident Indian (NRI), the same status will be transferred to the child resulting from a surrogacy arrangement.
4. For a discussion of this case see Smerdon (2013).
5. See the statements made by Justice Diane Bryant http://www.smh.com.au/national/chief-justice-diana-bryant-confident-commercial-surrogacy-will-be-legalised-in-australia-20150429-1mvzn1.html and Millbank (2015).
6. See, Mennesson and Others v. France (no. 65192/11), and Labassee v. France (no. 65941/11).
7. Form 1259, Information about DNA testing for visa and citizenship applicants, Department of Immigration and Border Protection, Commonwealth of Australia 2015, http://www.border.gov.au/forms/Documents/1259i.pdf, cited 16 August 2016.
8. Between 2001 and 2003 only 0.5% of the total family migration intake underwent kinship testing (Gesche 2006: 227).
9. ALRC 96: Essentially Yours: The Protection of Human Genetic Information in Australia, 2003, Sect. 37.52, available at http://www.alrc.gov.au/sites/default/files/pdfs/publications/ALRC96_vol2.pdf, cited 16 August 2016.
10. This would explain the much bigger amount mentioned in the story about Lovette which this chapter opened with.
11. See Strathern (2004) for a discussion on approaching comparison in this way.
12. See also Van Wichelen (2016) for an exploration of biolegality in Australian court cases.
13. See Lynch and McNally (2009).

14. This happened recently, in the so-called Baby Gammy case, where a boy with Down syndrome was left behind with his Thai surrogate in Thailand by his contractual parents in Australia. See http://www.smh.com.au/world/baby-gammy-granted-australian-citizenship-20150119-12tum2.html.
15. See also Jasanoff (2001).

Acknowledgments This research was funded by the Australian Research Council for Early Career Researchers (DECRA, project number DE140100348). I am very grateful to the editors Vincenzo Pavone and Joanna Goven for inviting me to contribute to this volume and for their extensive and thoughtful comments that have made it a much better piece. Much of the ideas for this paper came out of a workshop *Reproductive Biopolitics* organized by Catherine Waldby at the University of Sydney in December 2014. I would like to thank her and the other participants for a productive discussion to an earlier version of this chapter. I would also like to thank Marc de Leeuw for helping me develop the argument presented here.

References

Andreas, P., & Biersteker, T. J. (2003). *The rebordering North America: Integration and exclusion in a new security context.* New York: Routledge.

Banerjee, A. (2014). Race and a transnational reproductive caste system: Indian transnational surrogacy. *Hypatia, 29*(1), 113–128.

Btihaj, A. (2013). *Governing through biometrics.* New York: Palgrave Macmillan.

Castles, S., Hugo, G., & Vasta, E. (2013). Rethinking migration and diversity in Australia: Introduction. *Journal of Intercultural Studies, 34*(2), 115–121.

Cooper, M., & Waldby, C. (2014). *Clinical labor: Tissue donors and research subjects in the global bioeconomy.* Durham: Duke University Press.

De Leeuw, M., & Van Wichelen, S. (2012). Civilizing migrants: Integration, culture and citizenship. *European Journal of Cultural Studies, 15*(2), 195–210.

Feldman, G. (2011). *The migration apparatus: Security, labor, and policymaking in the European Union.* Stanford: Stanford University Press.

Gammeltoft-Hansen, T., & Sørensen, N. N. (Eds.). (2013). *The migration industry and the commercialization of international migration.* New York: Routledge.

Gesche, A. H. (2006). Protecting the vulnerable: Genetic testing and screening for parentage, immigration, and aboriginality. In M. Betta (Ed.), *The Moral, Social, and Commercial Imperatives of Genetic Testing and Screening* (pp. 221–236). Dordrecht: Springer.

Goven, J., & Pavone, V. (2015). The bioeconomy as political project a polanyian analysis. *Science, Technology and Human Values, 40*(3), 302–337.

Harland, A., & Limon, C. (2016). Recognition of parentage in surrogacy arrangements in Australia. In P. Gerber & K. O'Byrne (Eds.), *Surrogacy, law and human rights* (pp. 145–160). New York: Routledge.

Heinemann, T., & Lemke, T. (2013). Suspect families: DNA kinship testing in German immigration policy. *Sociology, 47*(4), 810–826.

Heinemann, T., & Lemke, T. (2014). Biological citizenship reconsidered: The use of DNA analysis by immigration authorities in Germany. *Science, Technology and Human Values, 39*(4), 488–510.

Heinemann, T., Helén, I., Lemke, T., & Naue, U. (2015). *Suspect families: DNA analysis, family reunification and immigration policies*. London: Ashgate.

Heinemann, T., Naue, U., & Tapaninen, A. M. (2013). Verifying the family? A comparison of DNA analysis for family reunification in three European countries (Austria, Finland and Germany). *European Journal of Migration and Law, 15*(2), 183–202.

Helen, I. (2014). Biological citizenship across the borders: Politics of DNA profiling for family reunification. *Distinktion: Scandinavian Journal of Social Theory, 15*(3), 343–360.

Jasanoff, S. (2001). Ordering life: Law and the normalization of biotechnology. *Notizie di Politeia, 17*(62), 34–46.

Jasanoff, S. (2011). *Reframing rights: Bioconstitutionalism in the genetic age*. Boston: MIT Press.

Keyes, M. (2013). Australia. In K. Trimmings & P. Beaumont (Eds.), *International surrogacy arrangements: Legal regulation at the international level* (pp. 25–48). Oxford and Portland: Hart Publishing.

Lynch, M., & McNally, R. (2009). Forensic DNA databases and biolegality. In P. Atkinson, P. Glasner, M. Lock, S. Gibbon, & S. Hilgartner (Eds.), *The handbook of genetics & society: Mapping the new genomic era* (pp. 283–301). London: Routledge.

Margalit, Y., Levy, O. A., & Loike, J. D. (2014). The new frontier of advanced reproductive technology: Reevaluating modern legal parenthood. *Harvard Journal of Law & Gender, 37*(2), 108–139.

Menz, G. (2009). The neoliberalized state and migration control: The rise of private actors in the enforcement and design of migration policy. *Journal of Contemporary Central and Eastern Europe, 17*(3), 315–332.

Menz, G. (2013). The neoliberalized state and the growth of the migration industry. In T. Gammeltoft-Hansen & N. N. Sørensen (Eds.), *The migration industry and the commercialization of international migration* (pp. 108–127). New York: Routledge.

Mezzadra, S., & Neilson, B. (2013). *Border as method, or, the multiplication of labor*. London: Duke University Press.

Millbank, J. (2015). Rethinking "commercial" surrogacy in Australia. *Journal of Bioethical Inquiry, 12*(3), 477–490.

Murdock, T. R. (2008). Whose child is this: Genetic analysis and family reunification immigration in France. *Vanderbilt Journal of Transnational Law, 41*, 1503–1534.

Pande, A. (2016). Global reproductive inequalities, neo-eugenics and commercial surrogacy in India. *Current Sociology, 64*(2), 244–258.

Pottage, A. (2007). The socio-legal implications of the new biotechnologies. *Annual Review of Law and Social Sciences, 3*, 321–344.

Pugliese, J. (2005). In silico race and the heteronomy of biometric proxies: Biometrics in the context of civilian life, border security and counter-terrorism laws. *Australian Feminist Law Journal, 23*(1), 1–32.

Pugliese, J. (2010). *Biometrics: Bodies, technologies, biopolitics*. New York: Routledge.

Pugliese, J. (2014). The alleged liveness of "Live": Legal visuality, biometric liveness testing and the metaphysics of presence. In A. Wagner & R. K. Sherwin (Eds.), *Law, culture and visual studies* (pp. 649–669). Dordrecht: Springer.

Rose, N. (2009). *The politics of life itself: Biomedicine, power, and subjectivity in the twenty-first century*. Princeton University Press.

Smerdon, U. R. (2013). India. In K. Trimmings & P. Beaumont (Eds.), *International surrogacy arrangements: Legal regulation at the international level* (pp. 190–207). Oxford and Portland: Hart Publishing.

Storrow, R. F. (2005). Quests for conception: Fertility tourists, globalization and feminist legal theory. *Hastings Law Journal, 57*, 295.

Strathern, M. (2004). *Partial connections*. Walnut Creek: Rowman & Littlefield.

Strathern, M. (2005). *Kinship, law and the unexpected: Relatives are always a surprise*. Cambridge: Cambridge University Press.

Tapaninen, A., & Helen, I. (2016). Finland: Securing human rights, suspecting fraud. In T. Heinemann, I. Helén, T. Lemke, & U. Naue (Eds.), *Suspect families: DNA analysis, family reunification and immigration policies* (pp. 33–54). London: Ashgate.

Van Beers, B. (2014). Is Europe giving into baby markets? Reproductive tourism in Europe and the gradual erosion of existing limits to reproductive markets. *Medical Law Review, 23*(1), 103–134.

Van Wichelen, S. (2014). Medicine as moral technology: Somatic economies and the making up of adoptees. *Medical Anthropology, 33*(2), 109–127.

Van Wichelen, S. (2015). Scales of grievability: On moving children and the geopolitics of precariousness. *Social & Cultural Geography, 16*(5), 552–566.

Van Wichelen, S. (2016). Changing rights to family life: Biolegalities in the globalization of reproduction. *Sociolegal Review, 12*(1), 34–54.

Vora, K. (2009). Indian transnational surrogacy and the commodification of vital energy. *Subjectivity, 28*(1), 266–278.

Weiss, M. G. (2011). Strange DNA: The rise of DNA analysis for family reunification and its ethical implications. *Genomics, Society and Policy, 7*, 1–19.
Whittaker, A., & Speier, A. (2010). "Cycling overseas": Care, commodification, and stratification in cross-border reproductive travel. *Medical anthropology, 29*(4), 363–383.
Žižek, S. (2002). *Welcome to the desert of the real!: Five essays on September 11 and related dates.* London: Verso.

Author Biography

Sonja van Wichelen is Senior Research Fellow with the Department of Sociology and Social Policy at the University of Sydney. Her research broadly engages with the body, law, and science in the age of globalization and the effects that changes in these areas have on our understanding of citizenship.

CHAPTER 10

Embedded Promissory Futures: The Rise of Networked Agribusiness in Argentina's Bioeconomy

Pierre Delvenne

"The bioeconomy is one of the things that in principle everyone agrees on." (Personal interview with a public official from the Science, Technology and Productive Innovation Ministry)

Introduction

In Argentina, as elsewhere, policy discourse on "the bioeconomy" addresses an abstraction as if it had a concrete existence. The policy plan *Bioeconomia Argentina 2016* states: "Argentina has features that offer multiple opportunities for the development of the local bioeconomy. The country has a vast territory, large variety of climates and biodiversity, an important area of native and planted forests, and highly competitive agricultural and livestock and agribusiness sectors".[1]

As argued in this book, the concept of the bioeconomy is an act of political imagination. According to Goven and Pavone (2015: 6), "[i]t is

P. Delvenne (✉)
SPIRAL Research Centre, University of Liège, Liege, Belgium

V. Pavone and J. Goven (eds.), *Bioeconomies*,
DOI 10.1007/978-3-319-55651-2_10

a promissory construct that is meant to induce and facilitate some actions while deterring others; most explicitly, it is meant to bring about a particular set of changes that will shape the parameters of possible future action." Argentinean policymakers today instrumentally use the concept to try to re-territorialize economic activities by promoting regional/provincial bioeconomies (emphasizing the plural) and, more importantly, to conclusively transcend a historically entrenched dichotomy between agriculture and industry.

Recent institutional embodiments of these political attempts can be observed, for instance, with the creation of a Secretary of State for Bioeconomy (rebranded in late 2015 as a "Secretary of State for Added Value") and its institutional lodging within the Ministry of Agroindustry (formerly "Ministry of Agriculture, Cattle Raising and Fishing") in the aftermath of the most recent presidential elections. Mauricio Macri, the newly elected President, stressed in his first speech that a key challenge of his administration was to "achieve the agroindustrialization of Argentina". This is probably where the political value of the bioeconomy concept lies: it is an attempt at definitively crossing a formerly settled boundary between agriculture and industry. As a political project, bioeconomy appears as the latest stage of a co-production process at the intersection of the political and technoscientific levels. At the technoscientific level, the global adoption of technological packages (biotechnologies + herbicides + new farming practices) started long before the emergence of the bioeconomy concept and provided a hospitable ground for its successful integration into national discourses. At the political level, the roots of the bioeconomy are to be found in the neoliberal 1980s but, interestingly, both progressive (so-called "anti-neoliberal") and conservative governments sit comfortably with the concept. Whereas Argentinean politics usually is extremely polarized and antagonistic, policymakers from opposed sides of the political spectrum have tried to enact a specific world with the help of similar bioeconomic narratives.

The success of the bioeconomy as a political project, however, lies in the possibility for this world to fit the collective capacities of imagination, which are deeply entrenched in the past. Effectively, the interactions between emerging imaginative capacities and already existing collective imaginations enable and constrain what actions will be possible in the future. When a new master narrative such as the bioeconomy circulates and embeds itself in specific countries (Delvenne and Hendrickx 2013), it sediments on previously existing histories, path dependencies and technological trajectories that constitute the strata of knowledge/social orders as they have unfolded up to the present day. In this chapter, I use the "idiom of co-production" (Jasanoff 2004) to track and discuss the regularities of this sedimentation process from the 1970s onwards, as well as the tensions that arose when disruptive events affected the normal course of sedimentation, and past social fractures and conflicts resurfaced. The idiom of co-production, briefly stated, "is shorthand for the proposition that the ways in which we know and represent the world (both nature and society) are inseparable from the ways in which we choose to live in it" (Jasanoff 2004). A co-production framework stresses the importance of contextualization to make sense of the emergence and stabilization of new technoscientific objects or the resolution of scientific and technological controversies (Bonneuil et al. 2014). Importantly for this chapter, co-production opens up "the possibility of seeing certain 'hegemonic' forces, not as given but as the (co-)products of contingent interactions and practices. These insights may, in turn, open up new opportunities for explanation, critique and social action" (Jasanoff 2004).

Whereas many authors have undertaken co-productionist analyses mainly focusing on certain epistemic claims (e.g. scientific) and certain social formations (e.g. the state), this chapter additionally makes use of it to analyze the development of a political economy as both epistemic and social order (Rajan 2012; Jasanoff 2012; Birch 2016). Inspired by Joly (2015), I undertake a "strong co-productionist analysis", meaning that I pay attention to the diversity of processes that operate at different scales (from local socio-technical arrangements to system level) and to their interactions. A strong co-productionist approach "considers power as relational and it is attentive to lasting asymmetries of power, the use of force, constraint, 'fait accompli', and the accumulation of resources and competences by some actors at the expense of others" (ibid).

This chapter analyzes the "bioeconomy" of genetically modified soy in Argentina, the world's third leading producer and exporter of GM crops. GM soy production is a central source of extraction of economic value, which has provided the economic oxygen to the country since it declared a partial default on its national debt in 2001. As of 2014, soy sector exports represented 28% of total Argentine exports and accounted for USD 20 billion in foreign income (INDEC 2015).

The chapter is based on fieldwork carried out between 2010 and 2016, as well as secondary sources (press articles, regulatory and legal documents). Fieldwork included 62 semi-structured interviews (8 in 2010,

31 in 2011 and 6 in 2015, 17 in 2016) carried out with individuals in Buenos Aires and Rosario who are involved in the GM soy complex or who have had a say in the public discussion around it (including members of academia, public administration, seed and agrochemical companies, agricultural production, regulation, distribution sectors, financial investors, managers of sowing pools, and related civil society organizations). Some of the key informants were interviewed up to three times, and the interviews were systematically analyzed, categorized and tagged with the qualitative software Mosaiqs.

The text is structured as follows. It begins with a discussion of the international and domestic political economy of Argentina during the past 30 years, to stress the co-production dynamics of the GM soy model with its political-economic contexts (neoliberal in the 1990s and national-populist in the 2000s/2010s). The next section focuses on the techno-scientific level, in particular how the adoption of a technological package (biotechnology + herbicide + new farming techniques) has created an opportunity structure supporting the emergence of a vanguard vision aimed at "revolutionizing" agriculture. Following that, I analyze a particularly consequential development within Argentina's GM soy bioeconomy: "*pooles de siembra*" (sowing pools), the name given to an agricultural production system characterized by the organization of a financial enterprise system that assumes temporary control of agricultural production.

The Political Economy of Genetically Modified Soy

In the 1970s, Argentina went through a series of transformations toward a new regime of capitalist accumulation. On March 24, 1976, a military coup brought to power a violent dictatorship that, until the recovery of democracy in 1983, devastated the country economically, politically and socially (Giarracca and Teubal 2004; Hernandez 2013). Contrary to what happened during the same period in neighboring countries, especially Brazil, where the dictatorship was supportive of industry, Argentina went through a time of deindustrialization and openings of its economy to financial liberalization. The "authoritarian bureaucratic state" (O'Donnell 1979) implemented a series of measures that profoundly affected the labor, agricultural, capital, and financial markets. Wage increases were frozen in the public sector and the dictatorship adopted provisions favoring large companies and financial and speculative activities. Consequently, the external debt dramatically increased and constituted a severe drag on policies and strategies of subsequent democratic governments. Total external debt increased by almost USD 36 billion between 1976 and 1983, the year of the democratic recovery (Teubal and Teubal 2011: 63). At the end of the 1980s, when the government temporarily suspended debt repayment, there was a massive and generalized capital flight, which accelerated the devaluation of the currency, led President Alfonsin to resign and provoked the hyper-inflationary outbreak period of 1989–1991. These conditions set the scene for the profound process of neoliberalization during the administrations of President Carlos Menem (1989–1995 and 1995–1999).

The agricultural sector was at the forefront of this neoliberalization process: production was re-oriented towards export markets; the institutions and mechanisms that regulated production for the internal market were eliminated; and global agribusiness corporations expanded their operations (Lapegna 2016: 6). In the early 1990s, the structural adjustment plan carried out by the first Menem government deregulated the markets for goods and capital, enacted the Convertibility Law, which pegged the value of the peso to that of the US dollar (Giarracca and Teubal 2004), and privatized public assets. These macroeconomic conditions stimulated the free movement of financial capital and, in return, undermined the little competitiveness that remained in the Argentine industrial sector. The concentration of land and capital in agriculture was facilitated by the enactment in 1992 of Law No. 24083 of mutual fund investments, under which the organization of sowing pools was designed, in order to turn agribusiness into a privileged niche for quick returns on financial capital investments (Hernandez 2013).

The neoliberal agrarian restructuring accelerated in the second half of the 1990s. The national government approved the commercialization of herbicide-tolerant GM soy in 1996, and the biotechnology regulatory regime was transformed on the basis of market principles benefiting large corporations (Otero 2012; Delvenne et al. 2013). Combined with new agricultural techniques well adapted to Argentinean conditions, particularly no-till farming, glyphosate-tolerant soy has been by far the most rapid adoption of any seed variety in Argentina, including those introduced in the Green Revolution (Penna and Lema 2003). Consequently, soy has replaced beef and wheat (the primary goods consumed by the urban working class) as the country's leading export.

As Gras and Hernandez (2014: 343) explain, "a key explanatory element of the fast adoption of GM soy was the local economic environment of the mid-1990s, which was shaped by a combination of volatile international commodity prices and the impact of Argentina's neoliberal economic policies on farmers". These impacts on the agricultural sector were indeed significant: Menem's government reduced farm subsidies and trade protections, and increased interest rates on agricultural loans. In these conditions, for many indebted farmers the only way forward was to embrace the new biotechnological wave:

> You have a problem of technological shock in the mid-90s. You have a problem of variation in international prices and profitability; farmers were indebted and the only way out was an upgrade to biotechnology and the new technological package, whereby at the end of the 2000s, you had 24–25 million cultivated hectares and 14 million mortgaged hectares. Yes, Schumpeter is expensive… (Agricultural economist 1)

After 1997, the economy experienced a sharp decline and unemployment soared (Caceres 2015: 120), eventually leading to the 2001 crash, when the neoliberal model of a decade of *Menemismo* collapsed and forced Argentina to declare a partial default on its national debt (Teubal and Teubal 2011). At that time, half of Argentines were living in poverty, amongst the highest rates of inequality in history, with immense despair in a context of institutional, political, and economic crisis (Leguizamon 2014: 155; Carranza 2005). Five different presidents succeeded each other in less than 10 days during the 2001 crisis. The fifth of them was Eduardo Duhalde, who decided to abandon convertibility with the USD, to devalue the *peso,* and to call for a general election (Caceres 2015: 121). The Peronist political party then came back to power: Néstor Kirchner was elected president (2003–2007), to be followed by his wife Cristina Fernández de Kirchner (2007–2011 and 2011–2015), what political analysts called the "Kirchnerist" period. Kirchnerism was however not a monolithic block. Argentinean political economist Mathias Kulfas (2016) considers that the country has known three types of *kirchnerism.*

The first kirchnerism started with Néstor Kirchner's presidency, which was a historical moment for soy export and profitability. During this time there was a sustained international demand for soy products (notably from emerging markets of China and India) coupled with a high price of commodities on global markets. Richardson (2009) coined the term "export-oriented populism" to argue that under Néstor Kirchner's government, changes in agricultural production (soy replacing wheat and meat as the country's leading export) and productivity (which dramatically increased as an effect of the adoption of the GM soy technological package, see next section) created the conditions of a new variant of populism, temporarily eliminating important sources of political and economic instability that had plagued previous incarnations of Argentine populism. "Because soy is not consumed by the working class, Kirchner could both promote and tax their export, generating fiscal revenue for populist programs while not harming the effective purchasing power of urban workers or provoking a balance-of-payments crisis" (Richardon 2009: 228). Export-oriented populism even expanded under the second kirchnerism, coinciding with Cristina Fernández de Kirchner's first presidency (2007–2011),[2] as she decided to increase the export taxes from 30 to 35%. "Increasing foreign income from agricultural exports and higher fiscal revenue explain how it was possible for the Kirchners to return to a model of state intervention, investing in infrastructure and social spending, without enacting major structural reform, thus keeping intact the agro-export model devised in the neoliberal 1990s" (Leguizamon 2014: 156).

The Kirchners thus conserved and reinforced the transgenic soy model they inherited. As Amalia Leguizamon explains:

> [T]he rise of a self-proclaimed anti-neoliberal and progressive government, the Kirchners', has in fact not dismantled the mode of production set in place by the previous neoliberal administration. To the contrary, the Kirchners' administrations have created favorable conditions for the expansion of GM soy. Driven by debt, both neoliberal and post-neoliberal governments have relied on state policy to intensify Argentina's comparative advantage. (Leguizamon 2014: 158)

It is important to note that although state intervention is key to neoliberal policy, in most cases it is often so only in the supply side of the economy. In a way, without dismantling the supply side, Nestor and later Cristina Kirchner have been able to use part of the surplus to reactivate, selectively, some demand-side oriented policies. However, this strategy was severely challenged by a series of crises. The global financial crisis of 2008–2009 eventually eroded Argentinean growth rates; a major drought in 2008 curtailed soy production; the slow erosion of international prices of agricultural commodities reduced returns on soy; and most importantly, in 2008 the countryside (the *campo*) rose up against Cristina Kirchner's government, after she announced her intention to increase the export tax on soy to 50%. Kirchner attempted to do this through executive measure ("*Resolucion 125*") to avoid having it voted on by Parliament (see Fraga and Baistrocchi 2011). That attempt failed and sparked a conflict that paralyzed the country for several months: roads and bridges were blocked and there were massive street demonstrations and public events against Cristina Kirchner. This conflict disrupted the ideological patterns of classical Peronism, historically marked by a dichotomous view of the bourgeois agrarian countryside (considered as the rich elite and the enemy of the Peronist cause) and the industrial workers' urban areas (considered as the poor people to protect and emancipate from the elite's domination). This was a real turning point:

> The powerful thing with the conflict with the 'campo' was that it joined factions that were previously impossible to unite [...] It gathered all spectrums of the rural producers: from the peasant to the agribusiness, and then many urban people who had never been supportive of the countryside; suddenly in this union we are all 'el campo', we are all with the 'campo' against the government. (Environmental sociologist)

That conflict, and the related socio-economic and political struggles that occurred in its aftermath, seriously plagued the third kirchnerism (corresponding to the second mandate of Cristina Kirchner, 2011–2015), which experienced a significant economic downturn: almost zero growth, a sharp decrease in private investment, and a sustained drop in commodity prices. It is in this light that we need to examine the successful emergence of "the bioeconomy" as a concept and political project in Argentina. Policymakers started to make explicit allusions to the bioeconomy from 2012 onwards, often explicitly referring to the OECD, the European Commission, International Energy Agency and FAO's definitions.

Argentina Innovadora 2020 was designed in 2012 as a Science, Technology and Innovation plan for the period 2012–2020. The Plan's objective was markedly similar to the Lisbon Agenda goal: "sustainable development with social inclusion by improving the country's economic competitiveness". The Ministry for Science, Technology and Productive Innovation (MINCYT) focused its efforts on specific areas where the country has or can have comparative and/or competitive advantages. One of them, perhaps the most important one because of the relevance of this sector for Argentina's economic performance, is the agro-industry sector (Martinez Demarco 2013). Interviewees share this relevance when they address the bioeconomy today:

> The issue of agro-industrialization is the main issue for us concerning the bioeconomy. (Public official from the Science, Technology and Productive Innovation Ministry)

For Kirchnerists, it was a way to accommodate to the moving political-economic landscape of a post-2008 Argentina, which threatened both the reelection of Cristina Kirchner and the legacy of one decade of recovered Peronism. Thus, Kirchnerist policymakers realized that the inherited historical dichotomies between agriculture and industry were falling apart, and considered it necessary to move beyond the idea of agriculture as a backward, elite rent-seeking sector.

> Of course, agroindustry is the concept. I think there appears the value, why it was so important for Argentina [...] I think in that sense the bioeconomy came to offer a smart way to rethink that and to go above and beyond the conflict [between the government and the *campo*] and to renew the agricultural sector that came to appear as a hub of innovation. (Agricultural economist 2)

> The issue is that we believed we had to get out of [the historical dilemma between agriculture and industry] and that the word bioeconomy helped us to raise the issue of agro-industrial development. So that also had the virtue of putting value creation at the territorial level, bringing producers closer to more stable markets, industrial contracts et cetera. [...] Bioeconomy helps retain the population, which is important because we have to somehow stop internal migration. (Public official from the Science and Innovation Ministry)

The bioeconomy concept proved to be of particularly high political value because it allowed policy-makers to return to discussing previously existing issues (especially, the political construction of a boundary between agriculture and industry) and to overcome the struggles over the distribution of surplus in the aftermath of the 2008 conflict. Interestingly, bioeconomy today is one of the few things that reconciles Kirchnerism and Macrism. Both Peronist and conservative governments could indeed endorse this vision of rural development as an advanced way to integrate the territory and to seize the historic competitiveness of Argentina.

The Technoscientific Transformation of Agriculture

The durable transformation of agriculture and the joint progression of ideas favoring the bioeconomy as a political project did not solely happen as a consequence of political-economic measures taken "from above". Crucial developments also emerged "from below", with farmers' grass-roots activities as a crucial reservoir of power and action toward a new vision of agriculture. In this section, I discuss the emergence of that sociotechnical vanguard vision (Hilgartner 2015), and I show how it became possible for one small group equipped with this vanguard vision to shape the future of agriculture by mobilizing resources and linking them up with their own technoscientific projects. Vanguard visions emanate from "relatively small collectives that formulate and act intentionally to realize particular sociotechnical visions of the future that have yet to be accepted by wider collectives" (Hilgartner 2015: 34). In doing so, these vanguard actors "typically assume a visionary role, performing the identity of one who possesses superior knowledge of emerging technologies and aspires to realize their more desirable potential" (Hilgartner 2015: 34).

At the end of the 1980s, a small group of farmers who were both concerned by the environmental costs of their farming techniques and seeking productivity gains initiated what was going to become a major breakthrough in farming practices. These pioneers, who were inclined to make alterations to their agricultural machinery to accommodate desired changes in practice, started by replicating some machines they had seen in Brazil in order to avoid tilling the soil. With the support of technical advisors from Monsanto, they were encouraged to establish a non-till farming association, AAPRESID, which was created in 1989. AAPRESID members were concerned with articulating a new vision, for which they "[took] action on the ground actually to prototype, build and configure practices that should—literally—realize their ideas" (Hilgartner 2015: 35). By doing so, their vision set the scene for and anticipated the rapid and massive diffusion of no-till farming techniques. Today, Argentina is the world's leading country in no-till farming techniques,[3] with a rate of 92% of adopters, and AAPRESID has become a very influential actor on the national and international scenes (Goulet and Hernandez 2011; Goulet 2013; Hernandez 2013).

Declarations of a key AAPRESID official about a new agricultural revolution reflect her important ambitions for the country and the world. She claims:

> We take care of the soil for food security, which is essential; and we also want to achieve energy security because we are cultivating energy as well […] We are working hard toward our 2020 strategy and we are a strategic ally for the state partners, because we believe we can guarantee food, energy and environmental security. (AAPRESID official)

Interestingly, as the vision of a desirable future attainable through an "agricultural revolution" was taking shape, the dichotomies agriculture/industry and rural/urban came to be increasingly challenged in the technoscientific sphere, as they were in the political sphere. Crucially, new biotechnologies' encounter with the vanguard vision promoted by AAPRESID linked that vision with an imaginary of national greatness, a technology-intensive agro-industry that would give Argentina back its rightful place in the economic geography of the world.

> Argentina has this thing that even though we are a society that is export-oriented with agriculture and the granary of the world and so on, […] we are very urban and at the cultural level actually we have this thing we want to be Europe, then… uh… there was a big disconnection that turned out into something else, this super-massive agrarian transformation that included adoption of biotech. (Environmental sociologist)

The cultural and organizational change that began with changes in farming practices has indeed been deepened and accelerated half a decade later, in 1996, when Roundup Ready (RR) soy was introduced into the country. Together with glyphosate-based herbicides and no-till farming, RR soy formed a homogenous "technological package" (i.e. actors use these terms to refer to the combination of these three elements to stress their joint and mutually reinforcing diffusion and commercialization), which transformed the paradigm of agriculture well beyond a reorganization of production and existing business models. The technological package revolutionized production and dramatically increased soy's profitability, particularly when cultivated on a large scale (Trigo et al. 2002; Pena and Lema 2003). Because of its homogeneity and inherent simplicity, and in spite of numerous socio-technical controversies (see Delvenne et al. 2013 for an overview), the package could easily travel across the country and be incorporated by farmers seeking higher yields and productivity. In turn, this further promoted the invention of a sustainable agro-industry based on the constant incorporation of new technologies

(GPS, smartphones, drones, precision agriculture), in which agricultural innovation is portrayed as desirable and most likely to succeed if undertaken by "knowledge workers". These workers interact across and even beyond the traditional sectors of agricultural value chains. Under this conception, which AAPRESID both represents and reproduces, the farmer is no longer portrayed as the one who cultivates his own lands but rather as a flexible, mobile and innovation-friendly entrepreneur.

Sowing Pools and the Rise of Networked Agribusiness

With the expansion of the technological package, the figure of the farmer has been re-framed into an innovator-entrepreneur who never shies away from taking risks, incorporating new technological features into his collection of sophisticated machines, making his knowledge (technical, financial, agronomic, managerial) available to others through his active involvement in farmers' associations,[4] and being perfectly able to integrate networks of producers, investors and input suppliers in a win-win capital accumulation operation.[5] This emerging figure embodies new relationships and a redistribution of roles in agricultural production. While the Green Revolution was based on the production of knowledge by public institutions that provided farmers in need with "top-down" agricultural-technology solutions, farmers themselves now play a central role in the development and testing of "bottom-up" innovative solutions. Public institutions are deemed "backward-looking" and incapable of knowing the real needs of farmers. Under this new configuration, private players are seen as best equipped to meet the need for expertise and technical support for farmers, particularly through professional organizations such as AAPRESID or AACREA (Goulet and Hernandez 2011). Public institutions' epistemic authority over agricultural innovation is thus challenged by the emergence of a farmers' identity empowered by farmers' personal networks and knowledge, oriented toward entrepreneurship, and supported by technological innovations, such as biotechnologies, no-till farming and precision agriculture. This co-produced worldview does not seem to leave any room for family farmers, who do not have the mind-sets of innovators-entrepreneurs and who are incapable of keeping up with ever-growing demands of capitalization, technologization and land concentration. Their continued existence as farmers is clearly in peril, now that even public institutions defending family farming evoke their programmed disappearance:

> Small producers, people who have very little, five or four hectares, and finally end up getting subsidies through various channels - that is what they do - are perpetuating inefficient production. Actually what they ought to do is to quickly adjust to change and to strive for a better access to infrastructure, education, roads. All those people have to leave the rural sector. (Economist 3)

Epistemic claims by and about farmers becoming innovators-entrepreneurs are reciprocally conditioned by the rise of new social formations such as sowing pools (*pooles de siembra)*, which are taken here as a central instance of this agricultural bioeconomy. Sowing pools can be seen as instantiating a new hegemonic organizational model of agribusiness (Grosso 2010; Murmis and Murmis 2012; Gras and Hernandez 2013, 2014). Sowing pools are defined as "agricultural trusts consisting of farmers seeking to extend their scale of production, who gather temporarily (usually one planting season) to lease tracts of land as well as services for the main farming operations (planting, spraying and harvesting) and sometimes for transport. They also look for national or international investors, who may come from outside of the primary sector (banks, finance companies) as well as inside (agro-industrial firms, providers of agricultural inputs) to finance soybean production" (Choumert and Phelinas 2015: 134).

Some sowing pools operate with their own capital, including land ownership, but most pool capital from external investors and operate under short-term contracts (Leguizamon 2014: 153). The managers of the pools as well as those who do the farming are paid a salary, and investors receive profits on their investment depending on the export prices and volumes at the end of the season.

In these new social formations, small and medium-sized farmers have become "unable to catch up with capitalization and land scale demands, [so they] have opted out of production by leasing their land to larger farmers or investors, thus becoming rentiers" (Gras and Hernandez 2014: 344).

Interviewer (I):	These rentiers what do they do then?
Agronomist 3:	Usually they are sitting in the pub of a village with a cell phone with which they can communicate with the contractor; if any decision needs to be taken he handles it very well at a distance.

Interviewer (I):	The rentiers those who stopped farming who own lands … Some became contractors and others not, what do they do?
Agricultural economist 2 (AE2):	These others became nothing. Nothing.
I:	They do nothing they live well with just renting the land?
AE2:	Well, it depends on how much land you have and the quality of land. If you have 50 ha and live in a village, with current rent levels you live phenomenally well.

Sowing pools' managers, often working from afar in their Buenos Aires offices, are therefore interacting with landowners to lease the lands they need, while landowners themselves have also left the fields to live in villages and small towns. In this configuration, the closest producers to the land—and the bearers of the new farmer identity—are the so-called *contratistas* (rural contractors), a group of about 12,000 people duly trained and equipped with the best possible machinery to offer contracted services across the country, such as, for example planting seeds, fertilization, herbicides and insecticides spraying, packaging the harvest, et cetera.

> Contractors emerge as small producers who have excess machinery and begin to provide services to the neighbors. Now, why did these small producers have excess machinery in the first place? That is the question, why had they bought more tractors? Because in Argentina I would say historically that there was no mortgage or credit to purchase fields. But you could access from public banks to buy machinery. So there have always been plans to purchase more machinery than what was needed… (Agricultural economist 3)

Most of the time, there is no contract, strictly speaking. This is a verbal agreement between a service provider (the contractor) and its temporary employer (the manager of the pool or the landowner). Contractors are very vulnerable to weather conditions, as one bad season means the collapse of their business if they do not obtain sufficient income to cover the leasing of their machines. Even though the first contractors were small producers with their own machinery, most contractors today lease the machinery they use in order to benefit from the most recent technological advances and to be more competitive on the market of *contratistas*. The machinery industry incentivizes contractors by providing them free training sessions and by offering them credit facilities. However, getting the most recent machines is not always financially possible for every contractor. Those who cannot afford it are thus locked in inferior technological possibilities (compared to their competitors) and they are paid with inferior salaries. Contractors have become a central actor in the networked agribusiness of sowing pools.

> The sowing pool needs contrastistas like water, because it will go 500 km away, rent land and seek some service contractors. If he can't find any service contractor, the guy [from the sowing pool] is dead. Therefore the system is a network. (Agricultural economist 1)

Contractors became the vector of innovation, information and knowledge. The quality of the services they offer and the productivity that can be expected from a sowing season depend on contractors' capacity to keep up to date with cutting-edge innovations:

> Each seller of agricultural machines has half its people training contratistas to use the new machines [...] Because the seller is obviously interested in that the John Deere machine has the best possible performance in the field, and to achieve that John Deere trains each type who buys the machine and offers him an ad hoc course, with yearly updates. Why? Because it is the form of economic capture when the guy changes the machine. Have you understood how much the system is innovative? [...] It is a network of interests. The one leasing lands can pay high rents and capture more land if he manages to balance his budget. He does not have to manage the Chicago market, he manages productivity, and productivity is obtained with contractors. (Agricultural economist 2)

The technological package not only transfigured agricultural techniques, upset traditional roles of farmers, and challenged the epistemic authority of scientific public institutions; it also transformed agricultural management into network management.

> [Previously], informal networks began almost by word of mouth and information flowed through more direct contacts; it was not private circuits at that time, this came much later, as now the networks are completely managing the distribution of inputs, large multinationals working in seeds that

> have established their own distribution centers, etc. And you, as a producer you're in a network. (Agricultural economist 1)

Among the paradigmatic examples of agribusiness "made in Argentina" is the company Los Grobo which, although it is demonized in the media because it is often accused of abusing its market power, is "cited in almost all the manuals, courses and reports on agribusinesses as having successfully changed their mentality" (Gras and Hernandez 2014: 346). Los Grobo developed an innovative vertical integration strategy named "associative network company".[6] The complexity of Los Grobo's networks has served as a model for agro-industrial specialists and academic programs like Harvard Business School (Bell and Scott 2010). The model is based around building a network of input and service providers, including landowners, agronomists, contractors and agricultural production branch managers, so that instead of directly hiring employees or owning land or machinery, the company operates through land leases and third-party contracting (Leguizamon 2016).[7] The charismatic Director of the company, Gustavo Grobocopatel, envisioned the transformation of agriculture into a knowledge-based industry and rather than opting for the "heaviness" of lands and machinery, he preferred "an asset-light company: investing in soft assets—top talent, training, and state-of-the-art technology, for example—was a better investment for future growth in the twenty-first century" (Bell and Scott 2010: 5). Of course, not every sowing pool can be Los Grobo. From the basic idea of generating economies of scale by pooling resources have stemmed very different models of sowing pools: informal local pools (50–300 ha), large sowing pools (1000–50,000 ha) and network companies (up to 350,000 ha) (Grosso 2010).

> A pool is a strategic alliance between people, where the management is centralized. It is the minimum expression of a pool, in my case I have a motorcycle and it is the entire heritage that my pool has (laughs), that and nothing else. Then, you have got that pool with 50 thousand hectares planted, with offices in Buenos Aires, trained economists, own contratistas, 10 trucks and 15 agronomists… (Sowing pool manager 4)

However, in recent years, and particularly since 2012, producers have been seeing slimmer profit margins for soy (soy was until recently worth 600 USD/ton, and it's now around 220 USD) and rising tensions with the government over export taxation. Consequently, many sowing pools have broken up, some have significantly decreased the magnitude of their operations, and still others have diversified their businesses or expanded abroad to neighboring countries. Los Grobo, for instance, moved part of its agricultural business to Brazil and also to Uruguay and Paraguay, where the lack of good infrastructure is compensated for by the absence of soy export taxes. The *contratistas*, however, did not cross the borders with the asset-light companies. They remained attached to the lands they usually cultivate with their own machines. Under networked agribusiness, greater power and success seem to correlate with greater mobility. Those who remain attached to particular pieces of land or particular points in the agribusiness network have fewer options. The mobility of powerful actors is not limited to geographic mobility: because the cost of coordination of its network had become too expensive in relation to soy's decreasing profitability, Los Grobo (and other mega-network companies) expanded its network into other agribusiness activities, such as the production of fertilizers and other agroindustrial inputs and even digital agriculture.[8] This seems to be a general trend, which has been highlighted by Murmis and Murmis (2012: 496): "once a certain scale is reached, other avenues of business expansion start to dominate as the companies receive foreign investments, initiate projects in neighboring countries, move along or integrate additional parts of the supply chain, and create or extend networks with different levels of inter-company integration". They add that network companies "seem to progress through an evolutionary path to a form where other non-farming avenues of business expansion start to dominate" (Murmis and Murmis 2012: 496).

These observations resonate with the findings of Carla Gras and Valeria Hernandez who studied the evolution of the heterogeneous large-scale farming sector in Argentina from the neoliberal 1990s onwards (Gras and Hernandez 2014). They found that the impact of neoliberalization has differently impinged on the category of large-scale farmers. According to them, network companies adopt inter-sectorial business models, are more flexible, and are less anchored in specific territories, and therefore are much more likely to adapt to changing circumstances than the traditional capitalist actors (agrarian bourgeoisie or large-scale farmers below the scale of network companies). Gras and Hernandez (2014) conclude that the remarkable growth of Argentinean agricultural production was accompanied by a transformation of the agrarian capitalist class that weakened the traditional capitalist firms and favored the networked agribusinesses.

Conclusion

The bioeconomy is relatively new in Argentina as a concept and a political project. To understand its emergence and to anticipate its future as a political project, this chapter has examined four decades of co-produced transformations at the political and technoscientific levels. 'Agribusiness' looks like the habitat of today's bioeconomy in the country. It encapsulates "a new worldview [not limited to] specific technological or organizational innovations, but rather a systematic change that involves material, ideological and symbolic elements" (Gras and Hernandez 2014: 345). Agribusiness gained a hegemonic position; it contributed to regularizing and stabilizing the course of capital accumulation, including the "displacement of conflicts, and crisis tendencies elsewhere and/or into the future" (Jessop 2008: 27). These include, most notably, the denial of the numerous socio-technical controversies on social and environmental impacts attributed to the "soyization of Argentina" (see Delvenne et al. 2013 for an overview), such as deforestation, the displacement of indigenous populations, tensions with other productive activities like cattle raising, soil degradation because of proliferating monocultures, and a rise of cancers and malformations due the increasing use of agrochemicals (on collective actions challenging the bioeconomy, see Arancibia 2013).

To some extent, for the Argentinean agricultural bioeconomy, macro-trends were clearly over-determined in important ways by the features of the local situation. We have seen that international developments, such as globalization, financialization, or corporatization have as much shaped new social relations and political configurations as local specificities and imaginations, national history, and the internal dynamics of the agricultural production system. First, to deal with the aftermath of the structural neoliberal dismantlement of the country, to face the 2001 bankruptcy, and to feed the financial appetite of vulture funds, the Kirchners' governments transformed a neoliberal agricultural model into the motor of an export-oriented national-populist system. Second, as part of a global phase of consolidation and transnationalization of the agriculture market, which was dominated by multinational companies, new technological packages were introduced in Argentina in the mid-1990s. Their adoption was far from mechanical, as the imaginative resources of farmers subsequently adapted the technological package in order to propagate their vision of sustainable agro-industry throughout the country. Third, the expansion of the GM soy model was driven by global trends toward capitalization, land concentration, and "salarization"; while these trends severely impacted small-scale and traditional capitalist farmers, they also created a favorable environment for "asset-light companies" to flourish.

I have characterized this form of agribusiness in terms of networks, in order to stress its engagement with actors through both their (im) material possessions (machinery, know-how, technologies, lands, expert knowledge) and their moral commitment to networks, social capital, diversity and information-sharing. The powerful actor in this imagined world privileges mobility and autonomy over security. S/he should not be attached to a family heritage, such as land, because it equates to fixity and immobility. As I have shown, to be successful under networked agribusiness, the farmer—whoever this category today corresponds to—should favor leasing over ownership. This is simply the result of the need to travel light—from one province to another, from one country to another, from one sector to another—to move more easily when circumstances are changing (Boltanski and Chiappello 2005).

Lastly, unlike many other examples of hegemonic policy concepts circulating from OECD countries to Latin America (for instance, the concept of national innovation systems; Delvenne and Thoreau 2017), the bioeconomy has not been promoted in Argentina as part of a normative agenda of what had to happen in order to "catch up" with Northern countries. Rather, it results from an explicit choice of policymakers to refer to the concept in an instrumental way in order to achieve a national political project that seems to transcend the well-entrenched divisions between self-proclaimed "anti-neoliberal" and conservative governments. As a preferred route for reaping the benefits of the country's competitiveness, the bioeconomy extends the agribusiness logic that has dominated for decades and, more importantly, it allows a discursive *industrializing* of agriculture. In bioeconomic terms, at least in the way "bioeconomy" is currently used in Argentinean policy arenas, agriculture is thus "diluted" into industry, revamped as just another (Pehlivan et al.) industry that can thus be supported by Peronist and conservative governments alike, without generating internal contradictions.

In Argentina, the bioeconomy is at the center of intense imaginative activities to re-think agriculture as-we-knew-it, and to re-name it as "agro-industry". These developments are evidence of an attempt to reject the idea of agriculture as the reactionary stronghold of a backward bourgeoisie and instead embrace agriculture as generative of an industrial avant-garde that promises political-economic transcendence.

Notes

1. http://www.bioeconomia.mincyt.gob.ar/bioeconomia-argentina/.
2. Nestor Kirchner remained very influential until his sudden death in October 2010.
3. The most technically advanced option of no-tillage is direct seeding: no ploughing is done before planting, which is done directly in the mulch (remains of vegetation) on the ground (hence the term "direct-seeding"). Two sine qua none elements are necessary for the realization of these particular seedings: (i) the use of drills designed specifically for this purpose, able to plant the seed using a disk system into unploughed soil, and (ii) the use of chemical herbicides offsetting the absence of ploughing, one of whose functions is precisely to improve the soil structure, and to mechanically destroy weeds (Goulet 2013: 441–442).
4. In addition to participating in AAPRESID congresses and meetings, medium and large-scale farmers are very often also members of AACREA, a farmers' association of over 200 local "CREA groups", distributed in 18 different regions all over the country and focussing on all kinds of agricultural production. In exchange for a fee (between 100 and 500 US$/month), each member can access the so-called community of AACREA farmers, participate in monthly discussions (gathering 10–15 producers from the same region) and benefit from dedicated R&D support. Each of these groups has a technical coordinator and one of the producers is appointed president of the group for a limited period of time, to voice and represent the group's interests at regional or national meetings.
5. The formal training of knowledge workers to act like innovators is also an important asset for the functioning of the system. A press article reporting on the findings of a comparison of the profile of medium-scale farmers in Argentina and the United States revealed interesting findings: the Argentinean average farmer is 7 years younger than his American counterpart (47 years old in Argentina, 54 in the US), almost twice as much educated in universities (46% of farmers in Argentina have a university degree, for only 25% of US farmers) and twice as many Argentinean farmers (10, versus 5% in the US) have postgraduate degrees. See http://www.lanacion.com.ar/1317397-ventaja-argentina.
6. See the company's website. http://www.losgrobo.com.ar.
7. See also the article published in *The Economist*, "Farming without Fields" (2014), http://www.economist.com/news/business/21592662-argentine-farming-group-heavy-science-and-light-assets-farming-without-fields, accessed on 24 of October 2016.
8. Los Grobo recently partnered with another company to create Frontec, a company that offers farmers "a technological platform that combines the

latest in geospatial and agricultural science with ICT, to offer innovative and sustainable solutions to the agribusiness value chain". See http://www.frontec.net/eng/, accessed on 24 of October 2016.

References

Arancibia, F. (2013). Challenging the bioeconomy: The dynamics of collective action in Argentina. *Technology in Society, 35*(2), 79–92. http://doi.org/10.1016/j.techsoc.2013.01.008.

Bell, D., & Scott, C. (2010). *Los Grobo: Farming's future?* Boston: Harvard Business School Publishing.

Birch, K. (2016). Market vs. contract? The implications of contractual theories of corporate governance to the analysis of neoliberalism. *Ephemera, 16*(1), 107.

Boltanski, L., & Chiapello, E. (2005). The new spirit of capitalism. *International Journal of Politics, Culture, and Society, 18*(3), 161–188.

Bonneuil, C., Foyer, J., & Wynne, B. (2014). Genetic fallout in bio-cultural landscapes: Molecular imperialism and the cultural politics of (Herbrand and Paternotte) seeing transgenes in Mexico. *Social Studies of Science, 44*(6), 901–929.

Cáceres, D. M. (2015). Accumulation by dispossession and socio-environmental conflicts caused by the expansion of agribusiness in Argentina. *Journal of Agrarian Change, 15*(1), 116–147.

Carranza, M. E. (2005). Poster child or victim of imperialist globalization? Explaining Argentina's December 2001 political crisis and economic collapse. *Latin American Perspectives, 32*(6), 65–89.

Choumert, J., & Phélinas, P. (2015). Determinants of agricultural land values in Argentina. *Ecological Economics, 110,* 134–140.

Delvenne, P., & Hendrickx, K. (2013). The multifaceted struggle for power in the bioeconomy: Introduction to the special issue. *Technology in Society, 35*(2), 75–78.

Delvenne, P., & Thoreau, F. (2017). Dancing without listening to the music: Learning from some failures of the national innovation systems in Latin America. In S. Kuhlmann & G. Ordoñez-Matamoros (Eds.), Research Handbook for Innovation Governance in Emerging Economies (pp. 37–59). Cheltenham: Edward Elgar.

Delvenne, P., Vasen, F., & Vara, A. M. (2013). The "soy-ization" of Argentina: The dynamics of the "globalized" privatization regime in a peripheral context. *Technology in Society, 35*(2), 153–162.

Fraga, A. J., & Baistrocchi, G. (2011). *Campo de batalla: crónicas de la resolución* (p. 125). Argentina: Ediciones B.

Giarracca, N., & Teubal, M. (2004). Que se vayan todos. In *Good governance in the Era of global neoliberalism: Conflict and depolitization in Latin America, Eastern Europe, Asia and Africa* (p. 56). London: Routledge.

Goulet, F. (2013). Narratives of experience and production of knowledge within farmers' groups. *Journal of Rural Studies, 32*, 439–447.

Goulet, F., & Hernández, V. (2011). Vers un modèle de développement et d'identités professionnelles agricoles globalisés? *Revue tiers monde, 3*, 115–132.

Goven, J., & Pavone, V. (2015). The bioeconomy as political project a polanyian analysis. *Science, Technology and Human Values, 40*(3), 302–337.

Gras, C., & Hernández, V. (2013). El agro como negocio. *Producción, sociedad y territorios en la globalización*. Buenos Aires: Editorial Biblos.

Gras, C., & Hernández, V. (2014). Agribusiness and large-scale farming: Capitalist globalisation in Argentine agriculture. *Canadian Journal of Development Studies/Revue canadienne d'études du développement, 35*(3), 339–357.

Grosso, S. (2010). Les pools de culture: diversité des combinaisons financières et productives. *Dossier «Argentine», Revue DÉMÉTER*: (pp. 223–254).

Hernández, V. (2013). Genealogía de una elite rural: elucidación antropológica de una práctica de poder. *Mundo agrario, 13*(26). http://www.redalyc.org/pdf/845/84527468004.pdf

Hilgartner, S. (2015). Capturing the imaginary: Vanguards, visions, and the synthetic biology revolution. In S. Hilgartner, C. Miller, & R. Hagendijk. (Eds.), *Science & democracy: Knowledge as wealth and power in the biosciences and beyond*. New York: Routledge.

INDEC (Instituto Nacional de Estadísticas y Censos). (2015). Complejos exportadores: evolución año 2014. Informe de prensa 08/05/15. Retrieved from http://www.indec.gov.ar.

Jasanoff, S. (Ed.). (2004). *States of knowledge: The co-production of science and the social order*. London: Routledge.

Jasanoff, S. (2012). Taking life: Private rights in public nature. In K. S. Rajan (Ed.), *Lively capital: Biotechnologies, ethics, and governance in global markets* (pp. 155–183). Durham: Duke University Press

Jessop, B. (2008). A cultural political economy of competitiveness and its implications for higher education. In B. Jessop, N. Fairclough, & R. Wodak (Eds.), *Education and the knowledge based economy in Europe* (pp. 13–41). Rotterdam, NL: Sense.

Joly, P. B. (2015). Governing Emerging Technologies? The need to think outside the black-box. Science and democracy: Making knowledge and making power in the biosciences and beyond. In S. Hilgartner, C. Miller, & R. Hagendijk (Eds.), *Science & democracy: Knowledge as wealth and power in the biosciences and beyond*. New York: Routledge.

Kulfas, M. (2016). *Los tres kirchnerismos. Una historia de la economia argentina 2003–2015*. Buenos Aires: Siglo Veintiuno.

Lapegna, P. (2016). Genetically modified soybeans, agrochemical exposure, and everyday forms of peasant collaboration in Argentina. *The Journal of Peasant Studies, 43*(2), 517–536.

Leguizamón, A. (2014). Modifying Argentina: GM soy and socio-environmental change. *Geoforum, 53,* 149–160.

Leguizamón, A. (2016). Disappearing nature? Agribusiness, biotechnology and distance in Argentine soybean production. *The Journal of Peasant Studies, 43*(2), 313–330.

Martinez Demarco, S. (2013). *European biofuels technology platform: A science, technology and innovation policy instrumentation.* Unpublished master thesis, European Studies on Science, Technology and Society, Maastricht University.

Murmis, M., & Murmis, M. R. (2012). Land concentration and foreign land ownership in Argentina in the context of global land grabbing. *Canadian Journal of Development Studies/Revue canadienne d'études du développement, 33*(4), 490–508.

O'Donnell, G. (1979). Tensions in the bureaucratic-authoritarian state and the question of democracy. In *The New Authoritarianism in Latin America,* (pp. 285–318). Princeton: Princeton University Press.

Otero, G. (2012). The neoliberal food regime in Latin America: State, agribusiness transnational corporations and biotechnology. *Canadian Journal of Development Studies/Revue canadienne d'études du développement, 33*(3), 282–294.

Penna, J. A., & Lema, D. (2003). Adoption of herbicide tolerant soybeans in Argentina: An economic analysis. In N. Kalitzandonakes (Ed.), *The economic and environmental impacts of agbiotech. A global perspective* (pp. 203–221). New York: Kluwer Academic.

Rajan, K. S. (2012). *Lively capital: Biotechnologies, ethics, and governance in global markets.* Durham: Duke University Press.

Richardson, N. P. (2009). Export-oriented populism: Commodities and coalitions in Argentina. *Studies in Comparative International Development, 44*(3), 228–255.

Teubal, M., & Teubal, H. M. (2011). La crisis de 2001–2002 y el colapso del neoliberalismo en la Argentina. *Realidad económica, 261,* 58–84.

Trigo, E., Chudnovsky, D., Cap, E., & López, A. (2002). *Los transgénicos en la agricultura argentina: Una historia con final abierto.* Buenos Aires: Libros del Zorzal.

Author Biography

Pierre Delvenne holds a PhD in Political and Social Sciences. He is Research Associate of the Fund for Scientific Research (FNRS) and the Co-Director of the SPIRAL Research Centre at the University of Liège, Belgium, where he coordinates the Research Unit in Science, Technology and Society.

PART IV

Capitalizing on Altruism: Donation, Care, and Unequal Relations in Global Bioeconomies

CHAPTER 11

Egg Donation in the Making: Gender, Selection and (In)Visibilities in the Spanish Bioeconomy of Reproduction

Sara Lafuente Funes

Introduction

Since in vitro fertilization (IVF) proved successful in 1978, reproductive material, its development, and its economic value, entered ongoing processes of re-signification, which modify what they were and are, what is to be expected from them and the actual practices in which they participate and that they generate. The rapid generalization of assisted reproductive technologies (ARTs) was quickly followed by the development of what has been named Selective Reproductive Technologies (Gammeltoft and Wahlberg 2014) and by an intensive use of reproductive material and capacity of third parties, something that we will here refer to as socio-technical practices of the Transference of Reproductive Capacity (TRCs). All these took place in the context of new bioeconomies (Goven and Pavone 2015) characterized by particular ways of framing life and the living. Indeed, the expansion of ARTs led to what has been named bioeconomies of reproduction (Pavone

S. Lafuente Funes (✉)
Institute of Public Goods and Policies (IPP), Consejo Superior Investigaciones Científicas (CSIC), Madrid, Spain

V. Pavone and J. Goven (eds.), *Bioeconomies*,
DOI 10.1007/978-3-319-55651-2_11

2012; Thompson 2005) within which the role of eggs, and egg donation in particular, are of growing interest. This is the case particularly in Spain due to the increasing use of donated eggs both for national and international patients. Spain functions as a destination country of what tends to be called reproductive tourism (Pennings 2002), a trend that makes its reproductive bioeconomy increasingly dependent on donors.

This chapter looks at the reproductive bioeconomies in Spain through the lens of egg donation. The Spanish bioeconomy of assisted reproduction has undergone a rapid growth since the first baby conceived through IVF was born in 1984. Many of these "first steps" along with this growth have been led by the private sector (Coroleu Lletget 2011), though the public health system also provides for a (limited) number of attempts for both artificial insemination (AI) and IVF for couples (mainly, heterosexual, but with some exceptions) which meet certain criteria (such as women being under 40). Nevertheless, and related to both restrictions and long waiting lists in the public system, most of the reproductive treatments take place within the private sector.[1] This sector is also the one in which the activities referred to as reproductive tourism or exile occur (Matorras 2005; Pennings 2002, 2005). According to our informants, many of the people coming to Spain for reproductive treatments do so in the search for donated oocytes, mainly due to restrictions on access in their countries of origin. IVF with donated eggs represents a high percentage of the total number of cycles done in the clinics, making oocytes highly valuable biological material in both the emotional and monetary senses. Previous work on egg donation in Spain has focused either on donors' motivations and experiences (Orobitg et al. 2013; Orobitg and Salazar 2005), on patients' experiences in relation to phenotypical matching while traveling for egg donation (Bergmann 2012), or on kinship reconfigurations (Álvarez Plaza 2006, 2014). Here we focus on the roles of egg donation in IVF clinical practices and in professional's narratives.

Treatments involving donated eggs were presented to us by clinics as their "most successful technique", as they have higher success rates than traditional IVF. In that sense, I argue that certain donated eggs–that is, those from (particular) young(er) women—might be playing a complex role within clinics around the resolution of low success rates of ARTs. Most IVF procedures are done with the patient's own gametes or embryos, though IVF cycles relying on donated material are increasing. Moreover, the use of donated eggs is higher than that of donated sperm[2] or embryos. In only four years (2009–2013) the use of donated material has rapidly increased. Thus, 34% of embryo transfers came from donated eggs in 2013 (compared to 29% in 2009), 7.5% from donated sperm (6.5% in 2009), and 1.7% from donated embryos (showing a sharp increase from the 0.6% of 2009[3]) (SEF 2013, 2009).

Taking as a starting point Sarah Franklin's question regarding what else we are reproducing, in addition to babies, through ARTs (Franklin 2013), we aim at bringing to the core of the debate the role that particular economic arrangements have in the making of reproductive material-semiotic entanglements (Haraway 1997) that make eggs and their transference meaningful today. We will do so taking into account both feminist insights on what the economy is (Hewitson 2014; Pérez Orozco 2012; Pérez Orozco 2006, 2014; Picchio 1992) and STS approaches towards bioeconomies (Birch 2006; Birch and Tyfield 2013; Goven and Pavone 2015) and assisted reproduction (Franklin 2002, 2013; Friese et al. 2006; Thompson 2005). With these theoretical inputs we aim at better understanding the ways in which identities, markets, materialities and subjectivities are made and entangled within current reproductive bioeconomies.

Even though the term "bioeconomics" was first used by those critical of the way dominant conceptions of development ignored biological and physical limitations of the sources linked to economic growth (Bonaiuti 2014), the concept of the bioeconomy has instead been used to argue for new strategies of, again, economic growth, which, while labeled *sustainable,* leave the dominant model of development and the idea of possible permanent growth untouched. The bioeconomy has been described as a political project aimed at framing both problems and solutions in particular (neoliberal) ways (Goven and Pavone 2015). Though very much signified by this political project and framed within it, what we name reproductive bioeconomies here have internal logics making them distinct from other bioeconomies. This is mainly due to how they bring together commercialization and production with the world of intimacy and reproduction, making business out of what have tended to be feminized areas of the economy, such as those linked to care work and social reproduction. These areas and tasks have tended to be privatized and feminized (in both households and gender roles) as studies of global care chains have shown (Orozco and Gil 2011). The notion of care, key to the development of these areas and the logics embedded in them, is linked to "a signifier of devalued ordinary labors that are crucial for getting us through the day" as well as to the "material and affective task related to communication, the production of sociability and capacity of affect" (Puig de la Bellacasa 2011). The separation of the productive sphere (thought to follow rational self-interest logics) from the reproductive one (thought to follow altruistic logics) has been gendered, with masculinity linked to autonomy, self-sufficiency, and self-interest and femininity to altruism, love, and care (Picchio 2005). Reproductive bioeconomies, thus, link the worlds of care, reproductive work, and intimacy with those of economic negotiations, commercialization of health, and market development. Here I argue that if bioeconomics was once used to point out the limits of the planet from an ecological or a degrowth perspective, feminism is nowadays well-placed to tackle the limits linked to power, bodies, and subjectivities. In particular, a feminist analysis can reveal the ways in which these bioeconomies draw on gender imperatives, such as those that construct femininity as linked to unpaid and undervalued care work.

To do so, I draw on interviews[4] carried out with 20 staff members, most of them working directly with egg donation, at 10 Spanish reproductive clinics. We focused on private clinics (9 out of 10) because the vast majority of egg donation procedures occur there; this is at least in part due to the fact that the public sector does not compensate donors whilst the private sector is used to compensate economically each donation.[5] The clinics' web pages were also analyzed and valuable information derived from them to match the explanations given to us on the way in which techniques and procedures are normally explained to donors and patients. Thus, I studied the role of donated eggs within clinics through the ways in which staff members talked about reproduction failure and success, donors, patients, egg donation and eggs themselves. Interviews lasted between 30 and 70 min and were a combination of individual interviews (13) and group interviews involving two (2) or three (1) people. They were transcribed and analyzed using thematic analysis, but focused on critical discourse analysis (Jager and Maier 2009; Martín Criado 2014) to better understand information, narratives and stories told by professionals.

From Egg Donation to Ovodon: The Making of a Successful Treatment Within Particular Socio-economic Arrangements

What "egg donation" means is in a constant process of (re)construction: through fieldwork I was able to grasp several ways in which the meanings of it are constantly negotiated and stabilized. This process involves, on one side, women[6] who enroll in programs to develop and give away their oocytes and, on the other, IVF patients (either women on their own or couples) wanting to receive them. In between those, the eggs themselves, their accompanying biological material, reproductive professionals, donation programs and protocols, techniques, matching categories, regulations and so on, mediate and work to generate what we currently name "*donación de óvulos*" (egg donation)[7] and/or "*ovodon*" in the Spanish context. Those categories are used nearly as synonyms though they tend to appear in different moments. In the web pages of the clinics, they normally talk about "egg donation" in sections that discuss the donation process and are oriented toward potential donors, while they use *ovodon* or *ovodonación* in sections that discuss the treatments available for getting pregnant and are oriented towards potential patients. This distinction reflects a particular way of *doing* egg donation, which highlights the medical and technical side of this transference of biological material rather than the relational or biological side. These gestures render the agency of donors and of eggs less visible to patients than that of embryologists and clinics. Thus, egg donation within the clinics studied tends to be presented more as a reproductive medical treatment than as a transference of reproductive potentiality between women.

The final treatment, offered as a package, involves particular sets of eggs aimed at ensuring both high success rates and the presence of certain characteristics in the potential resulting children. Thus, *ovodon* as a treatment is built to offer not only the possibility of a new child but also a certain amount of confidence regarding the health of the potential baby as well as its suitability for a particular family (the meaning of suitability here is explained below). Eggs within these processes are enrolled in particular networks which change their meanings and which, in turn, are affected by them. Donated eggs, in particular, come to the clinics framed under narratives of altruism (seen as *bio-gifts*) but are then assigned very different meanings (as stock, as sum of their potentialities, etc.).

Here I will first look at how the process of obtaining eggs is framed as *donation* by the staff, through an emphasis on altruism, the non-exploitation of donors and the non-commercialization of eggs. Then I examine how the eggs are portrayed in the way in which the treatment is explained: eggs are made to be not just any egg capable of producing a baby, but rather particular eggs with the potential of producing healthy, "suitable" and situated babies. Finally, I will argue that precisely the strict selection of donors needed for that distinction between "any egg" and "a particular egg" is key to building competitive reproductive treatment within these types of economies. A careful and tailored selection, on the other hand, depends on having a big pool of potential donors from which to choose, something seen as dependent on the existence of monetary compensation for the "altruistic" collaboration of donors.

Egg Donation as a Donation: The Construction of a Narrative of Altruism

Egg donation, a key treatment within the Spanish reproductive economy, increases the range of patients that can be treated in any given clinic (including those unable to use their own eggs to reproduce) and raises their overall success rates, due to the much higher success rates of IVF with donated eggs. Thus, donated eggs are a very valuable resource within the reproductive bioeconomies of countries like Spain, in which IVF treatments with donated eggs constitute a high percentage of all IVF treatments. In Spain, egg donation is compensated with up to 1,000 €, but the law establishes that the transaction "will never have a lucrative or commercial character" as the compensation "will only be allowed to strictly compensate for the physical discomfort and the displacement and working expenses that derive from the donation and it shall not be an economic incentive for it" (Law 14/2006, p. 19954). In practice, this compensation is only offered in the private healthcare system, as the public one lacks funding for this.[8] The law reflects a problematization of the character of the act of giving away the oocytes and the role of compensation, setting women's motivations to donate as the defining criterion for nature of the transaction.

Here I focus not on donors' motivations in themselves, but rather on the narrative of donors' motivations constructed by the clinics' professionals. The "narrative" analyzed here is the one produced in the course of the interviews. This is complemented by information provided for prospective donors and patients on the clinic websites. Clinic staff builds a narrative of this practice as a *donation* through linking it with logics of altruism as opposed to those of exploitation and commercialization. But the emphasis on, and positive valuation of, altruism sits alongside an acknowledgement of the role of economic motivation in the process. This acknowledgement surfaces particularly when issues of donor recruitment and characteristics are discussed.

The discourse of altruism is presented mainly as solidarity between women, building on the cultural ideal of womanhood as linked to maternity, care and altruism, but also, even if less often, through highlighting the propensity to donate (blood, organs) within Spain in general, as somehow a *national culture of donation.* They, nonetheless, do not reflect over the fact that these other donations are not monetarily compensated while egg donation is.

> I think that also when they have been mothers themselves they will see even more the interest of helping. When we ask them they tell you that they do it to help. But well, really I think it is for both things [referring to monetary incentives]. (F2)

> We are in a society that understands donation very well, to be able to help others; if we have an exceptional problem that we need blood or something like that... and so with this... it is indeed different but takes place over a learned basis of, well, that, that donation exists and that you should help others, etc. So most of the women, basically the ones we do have, they do it in a way... in fact many times other people said to us, 'won't they do it for the money?' Well, if it were for the money they would do it heaps of times and most of them don't, they do it twice, two or three times and they stop. (BBI)

As this last quote showed, the emphasis on altruism is often a reaction to previous problematizations of economic compensation and motivations. As one doctor said:

> There are some very pragmatic people that when they talk about this they think they have the absolute truth and that donors only do it to have the money and I am more idealistic and I do not want to believe that. (F)

This *idealism* points to the preference for altruistic narratives, as she continues: "Why not an act of solidarity [in reference to donors]?? Who are we to put a weight into that act that might have nothing to do with it?" (F) In other words, why assume that donors are motivated entirely or primarily by the money, when we can equally assume that are motivated by altruism? Some interviewees explained that they tried to make sure that the donors were not in great monetary need; that is, because giving one's eggs in exchange for a *needed* (as opposed to relatively superfluous) sum of money might suggest exploitation, those in need were excluded. Some professionals took pride in adopting an ethical rhetoric in which they see themselves as safeguards or carers of these women, adopting at times a rather patronizing approach. They, nevertheless, say that *other clinics* might act differently:

> In one of the clinics of which they informed me, I am not going to say the name (...) they told me they had a [donors] social profile quite *in the limit* [as a way to suggest that they were in clear need for money, linked to poverty] It is not our perception [of our donors]. That does not mean that maybe some girl well, yes, that she might have a profile that you wonder that maybe she is doing it because she needs it [money] a lot. In that case, if we see in the test that she has an urgent need, economic, normally the psychologist will look at the case closely because you could be doing almost an exploitation, wouldn't you? (F1)

Some professionals admit, with concern, facing blurred situations regarding women's willingness to donate, but those are pointed out as exceptions. The following quote shows this happening in an interview with clinic staff members, the egg donation coordinator (A1) and the head in chief of the whole unit (A2):

A_1: We have had some weird cases, very few, where we wondered what to do. I do not know how to say it. The typical case in which the girl donates but you know that she does not want to donate [Interviewer: How is that? That she is forced to do it?] I think so, yes... that in some cases... I mean, we do not know for sure but at the end, especially in the day to day, when you are with a person, you kind of get to feel it [...] especially when you are directly dealing with the patient often you have the feeling that this girl signs everything, she tells you that is ok, she tells you yes to everything, but you...

A_2: But maybe it is only that they have certain immatureness that they are not fully aware if they want or do not want to do it.

A_1: Exactly, or maybe there is someone behind it, a relative, their partner, whatever.

A_2: But these are the fewer. [...] These are things... you can suspect that there might be problems but you cannot dig any further...

The issue of exploitation or of unclear willingness on the side of donors, nonetheless, was not often brought up within the interviews. More common was the definition of donation as contrary to selling. This was mainly done through presenting the amount of compensation as appropriate: enough to secure the availability of sufficient eggs to meet demand, but not so much that the transaction constitutes a clear commercialization of eggs. Avoiding commercialization, it was argued, was necessary not only as a fulfillment of the law, but also as a taken-for-granted moral value in itself.

> There might be a motivation... an economic compensation that is more or less attractive, but if we look at egg provision in other countries we are talking here of nothing, right? Here donation is donation, because really the cost that eggs have is not what is done, it is just compensation, as the law says, for the inconveniences caused. (BB2)

> Compensation is not an amazing sum as it can be in United States. So, a component of altruism is there. (BB1)

> Well, the economic motivation is there, it is there, but also... I mean, you cannot live only off of donating eggs. (A)

The narrative here links quantity with incentive, as if a higher amount would mean commercialization but lower amounts, even if known to be playing a key role in the donors' participation, would not mean the same. Interestingly, the same rhetoric of egg donation as gifts (as opposed to sperm donation that is as well monetarily compensated but more frequently seen as an economic of working activity) is also to be found in countries in which the donation is compensated with much higher sums, such as the USA (Almeling 2011). However, this narrative is not able to actually explain the key role that economic compensation plays in the Spanish surplus of potential donors and the fact that donors virtually donate only to the private healthcare system (where compensation is paid). It also does not provide a convincing explanation for some significant shifts in the profile of the donors, which interviewees in several clinics related to the worsening of living conditions of the general population caused by post-financial crisis austerity policies. These shifts affected not so much the number of donors but rather their ethnic profile[9], with an increase of those defined as Spanish donors and a decrease of those from Latin America. As this shift occurred, clinics became more likely to reject potential donors from Latin America, describing them as "less desired" from the perspective of phenotypic coordination.

> We used to have more foreign women, especially from South America, now there are many Spanish ones. (BB2)

> The nationality is Spanish, right? Because now we turn down the ones from South America. (F2)

There was sometimes a near-acknowledgement of increased economic need as a driver of the increased willingness of so-called Spanish women to donate eggs: "They have seen an easy way… an easy way of earning money". (F2)

Making Donors to Make Babies

The success of *ovodon* depends on the active construction of donors, both through the application of selection criteria and through disciplining interactions. Behavioral, medical-genetic, and phenotypical-esthetic factors all come into play in making donors, and through them, potential and particular babies.

Behavioral Expectations

Even though the altruistic motivation of donors in general is insisted upon in all the interviews, we can also see that altruism is actively shaped and enforced through behavioral expectations placed on donors. For example, staff talked disapprovingly of potential donors trying to negotiate for higher compensation or noted that some potential donors ask directly about the amount of money to be *paid* to them. Moreover, even if some monetary incentive is assumed, when the donor is too explicit about it, clinicians say that might result in rejection:

> Many [donors], obviously, even if they do not say it openly, you know that many of them come because there is an economic factor that is going to help them. But as I said, if we know that there is one that comes for the money, clearly she is rejected. That does not take away the fact that they all want their economic compensation, it does not take that away. But if there is someone that [says] "I come for the 900 euros", that one already has the cross […] I am not going to kick her out of the talk, but I am not going to call [her] […] "How much are you going to pay me?" They ask you openly without any will to cover it up... "How much are you going to pay me?" OK, discarded. (G)

This last quote indicates that altruistic narratives are welcome by professionals as a sort of necessary complement of economic motivations. In other words, when a potential donor is able to maintain a façade of being motivated by altruism, the staff feels more relieved in relation to their otherwise assumed economic motivations. Economic motivation is accepted, but only if it remains implicit, while explicitly the potential donors talk of "helping others". This, however, varied among clinics and professionals: in general terms, male doctors who had less contact with donors (that is, those working with patients) tended to insist on altruistic motivations and showed a more judgmental attitude towards monetary incentives, while women (both doctors and program coordinators) working directly with donors normally told a more complex story, in which they both acknowledged more mixed motivations and took a less antagonistic attitude toward economic needs.

To ensure that egg procurement is performed within this scheme, those who are motivated to offer their eggs need to be constructed, or need to construct themselves, as (proper) donors in several ways. Staff members, thus, engage in disciplining gestures in order to nudge donors' behavior towards the explicit altruistic model. Indeed, the term "egg donation" itself may be seen as part of this process.

First of all, to be acceptable as donors, women must exhibit an appropriate motivation. Motivation is read as embodied in donors, and the staff showed enthusiasm when donors' behaviors showed a desire to help others. Occasionally, when the desired script is not followed, the staff appears to be attempting to reshape women's motivation (or at least their account of their motivation) in the "right" direction.

> There is something we always tell the donor when she comes to the echography, when I [the doctor] talk with her, 'cause the first visits are a filter, to explain the programs and so on. But when I have the donor in front of me in the box right before doing the echography I always tell them: "This is about helping someone. (...) I know there is an economic compensation that is more or less attractive but you need to remember that this is about helping somebody." (BB2)

Second, donors are expected to be invisible (as donors) to patients in the clinics. This is enacted through clinics not only devoting different times of day to donors and patients and meeting them in different buildings, but also including in the selection criteria a preference towards women able to *pass* as patients. The latter seems to translate into a preference towards those with higher socio-cultural as well as economic capital, as a way of trying to make the class differences that are acknowledged to exist between donors and patients less visible.

> When they get to the clinic we have an interview where we ask them questions of psychological type, about their habits, their social environment,

familiar environment, we check what type of environment is the one in which these women move and the habits they have. (E)

These donors afterwards will have to come here, they sit in a waiting room full of patients, and so they need to go unnoticed. They need to be up to a standard... to come here and not stand out. (G_2)

Third, the clinics seek donors with certain personality characteristics that will both make them more suitable for the patients' waiting room (calmness, patience) and more likely to complete the process successfully (reliability).

"[Donors are accepted if professionals see that] they are trustworthy girls, committed, that they do not come clearly for the money... If we like them and we see that they are engaged..." (G)

Clinics, summing up, show a clear preference towards having donors that behave in particular ways: altruistic, calm, responsible, and discreet. This might affect both selection procedures and their later interactions in the clinics.

Medical-Genetic Criteria: Managing Risk

Whereas behavioral criteria relate to the needs of the clinics' performance of *ovodon*, medical-genetic criteria (and phenotypical-esthetic criteria, discussed below) relate to what eggs are thought to embody, as traits to be potentially transmitted to prospective children. Medico-genetic selection aims to maximize the chances of a successful pregnancy and a so-called healthy baby. However, unlike sperm, eggs themselves cannot be measured in quality terms within the labs (either they are "ready" to be fertilized, in metaphase II, or they are not, but all those reaching that phase are ready to be used[10]). Therefore, the *quality selection* comes not within the eggs themselves but through selecting the women carrying them.

ARTs generally try to achieve pregnancies with the genetic contribution of patients, either through artificial insemination, IVF with or without ICSI (direct injection of a single sperm into an egg, or intracytoplasmic sperm injection) or using PGD and PGS to choose particular embryos. In the case of *ovodon*, however, the genetic contribution of, at least, the woman, is not included. Where donated eggs are involved, staff showed a higher level of concern over the health of babies who may be born through this technique, engaging in a fairly extensive evading-risks narrative.

This translates into trying to reduce risks of disease transmission by finding healthy women who test negative for all contagious diseases, but particularly by ensuring that the risk of genetic transmission of disease is low. Subsequently, potential donors must also respond well to hormone therapy and have good ovarian reserves. Thus, "the doctor does a complete exploration, a complete interview with gynecological examination, echography" (BB1) or, as the woman in charge of human reproduction at a hospital said: "we make a clinical history of them, a gynecological review with cytology and echography, a general analytic, we do a serological study to discard infectious diseases, HIV, hepatitis, syphilis, chlamydia, well... sexually transmitted diseases and infectious [diseases] in general and then we do a genetic study of them" (E) This "genetic study" looks both for potentially hereditary diseases and for genetic markers of increased risk of disease. The genetic study is carried out through collecting information about family history and through genetic tests which either look at karyotypes and/or screen the genetic compatibility of donors with the men whose semen is going to be used:

> Actually we are doing a test of genetic compatibility that screens more than five hundred genes to cover up to six hundred diseases of this type [...] One can cover all the genome if he wants [...] the important bit is to know the mutations of the two, cross them and make it so that the kid will not have the 25% of chance of having the disease. (BB1)

> We are currently applying the latest genetic test that serves to detect recessive autosomic diseases. That means that we are all carriers of diseases that do not manifest, we are not affected by them. The recessive ones are not the ones making a person ill, but they make you a carrier. Then, this test helps you to detect two hundred recessive diseases, and so we do this test on the donor and on the husband, the partner of the receiver, or if it is double donation, of the egg donor and the sperm donor, to both of them and you make sure that the mutations are not the same. (F)

Some clinics are stricter than others when it comes to disease risk; this is reflected in the type of information they seek and the ways in which they manage it. Stricter clinics claimed that:

> If we have any doubts we do not proceed. This is not a problem for you [referring to the donor], nor for your health, neither for your offspring, but, with gamete donation, it needs to be clear that there is not even the slightest risk [...]. If we do have even the smallest suspicion that there can be a transmission, we do not accept [the donor]. (BS)

Thus, medical-genetic criteria are meant to ensure three main things: that donors have a good reproductive capacity, that they are healthy (in the sense of the absence of existing or transmissible diseases), and that the risk factor for genetic diseases is low. This selection of donors takes place under a principle of precaution framed through logics of potential risk control (both known and suspected). These readings of healthiness and risk control can involve conditions whose character is contested (such as differences in karyotypes, contested psychological or psychiatric diagnoses, etc.) as well as better established (less contested) conditions.

Phenotypical-esthetic Criteria: Selective Similarity

Phenotypical-esthetic selection occurs through a discourse of potentiality linking the eggs to images of future children who will be suitable for particular households. Staff displays a concern with assuring patients that the babies who will be born through egg donation will fit well within the patient's family. In so doing, prospective mothers' (and fathers) views are said to be taken into account. In particular, staff recognize that using donated eggs, and thus "giving up" on the woman's genetic material, is an emotionally (as well as monetarily) costly process, which is ameliorated not only through the birth of a healthy baby, but also through the connectedness and resemblance of the baby to the family.

This suiting of prospective children to parents is accomplished through a re-interpretation of affiliation as traceability and a building of relatedness through the reproduction of appearance. The staff's effort to match donors with patients is partly about the ability to mask the fact of donation, if this is desired (as can be noted by the importance of matching blood types).

Creating children who look like full biological children of their parents is key in the selection of donors, but this is not a straightforward process. The matching process is performed by the staff but involves a peculiar negotiation with patients over donor characteristics. What these negotiations mean varies: some staff see themselves as ethical barriers to what is seen as patients' desire for *too much* selection; others see their mediation as a moral and contractual obligation to fulfill, as far as possible, the wishes of their clients-patients. In the latter case the relationship is framed in terms of free choice from autonomous consumers (with limits as established in the law: the final choice is the responsibility of doctors). In the former case, two logics appear: a more traditional paternalistic doctor-patient perspective and a perspective within which choosing too much is seen as negative:

> She will never be the one making the decision, the decision is made by the medical team, ok? But she sets her preferences and if in that moment we have it, well then, why not? (A)

> [Some] think that this is like in the movies, that they pick up a book 'I want this donor, one that likes music, who is an athlete, who goes to university' And I am really sorry but... I, as a center, need to be sure to give her a healthy donor. I do not care if this woman works in El Corte Inglés [a large department store] or if she is in the University (...) We are based on blood type, in race, skin color, eye color, hair, type of hair and physical complexion. I mean, I always tell the patient 'look, we are going to look for a donor, not that she looks like you, but indeed with similar general physical characteristics. (G)

> If you ask for a specific profile or that she likes, I do not know, music or photos, well then yes, we can look at it, yeah, why not, but you will need to wait more. (BB1)

> I do try to adapt to what the patient wants. Legally I am not obliged to do it, just with complying with the minimum data is enough. I indeed do not want to be a traitor of the trust that a patient put in me. (E)

Larger IVF providers share their egg banks among their different clinics, having more capacity of adjusting their supply to specific demands. They see their banks as enabling them to make "very perfect matching, a very good one" (BB3), as "[i]f I have a donor here that has similar characteristics to this receiver, but above that I have some eggs that are more exact, you pick the ones that are more exact" (BB3). Here, matching processes seem to be playing a key part in the marketization of the egg donation treatments, turning them into quasi-products embedded in making treatments ready to compete in the reproductive bioeconomy. Some smaller clinics, on the other hand, take pride in keeping their selection criteria at a more general level, separating themselves from logics of (too much) selection.

All these selection processes render some eggs more desirable than others. Two criteria were regarded as basic to desirability: matching blood type and race. From there negotiation starts. Issues open to negotiation seem to get amplified in the core of normativity, narrowing down in the margins of it. This might be linked both to availability (more normative potential donors available) and to the biased frame through which professionals look at both patients and donors. In other words, patients seen as white Spanish, and in general white European, have more opportunity to specify additional characteristics than do other patients.

Thus, if a particular patient is regarded as part of the general population (understood as a non-explicit "us" built through European notions of whiteness marked by a situated Spanish reading[11]), the number of particular characteristics (type of hair, eye color, and so on) seen as possible to find in donors will be greater, as donors in those categories are said to be more common (and thus, a variety of characteristics easier to find). This situated reading of race places the idea of *caucasian* as the most common profile, used interchangeably with categories such as "normal", "like me and you"[12], or "from here"—the latter being either in reference to the city or province (in the case of Málaga for instance) or in more general terms meaning either "Spanish" or "Basque" (in Basque clinics).[13] The term caucasian is also used, though in this case it refers to "particular types of caucasian"—that is, women from northern Europe and generally meaning blonde hair and pale eyes. This broad category (caucasian) was often taken for granted (unnamed) and used as a starting point from which to take many aspects into consideration: hair color and type, eye color, complexion, specific requests by patients as seen above, etc.

When patients are perceived to be outside that *norm*, race stands as the basic criterion to be fulfilled, and all other aspects almost disappear, at least according to the way the staff describe the process:

> The Spanish, the Italians, they want Spanish donors, Italian, or... And this is normal, and the Asian wants an Asian donor and the black woman wants a black donor. (F2)

> Maybe a blonde with blue eyes which is [blood type] B, that is a very complicated blood type, because we can have blonde with blue eyes but that on top of that she has a B blood type well, you reach a moment in which

> you tell her 'hey, do you mind..., because there is this really good donor but maybe [instead of] blue eyes she has green eyes or that instead of being pure blonde she is pale brunette. (A)

> For example if they are black, well really black race there is not much, and so of what we have, we have some, we have some that are Dominican or Brasilian mulattas or Cuban mulattas or Dominicans. And so then there, there we... [laughs] we talk with them, which is because really of course, if they come here from a country in Africa in which they are of a very intense black race, we tell them that well, the black women we have in Spain, that talk Spanish and have emigrated here well, they are from the Caribbean. And so well, there are some people who struggle more with accepting this... (A)

Variation seems to be better understood in relation to the "caucasian us" in which staff include themselves. Thus, (imagined) national identities are seen as equally relevant to these patients as broader categories such as "Asian" or "black" are to others. What comes to be negotiated with "caucasian" patients seems to be much more detailed than what is negotiated with "black" or "Asian" patients, at least in some clinics.

Within this "caucasian" norm, there is a further expectation that donors will meet normative standards of beauty as well. In the context of private Spanish clinics, this means white, tall and slim and without particular characteristics that could lead to them being regarded as ugly. While compatibility with patients is emphasized in relation to some characteristics, other characteristics seem to be chosen differently, prioritizing a socially normative standard[14]:

> All of them are beautiful... all the ones that come here at least... If they have, let's see, a [potential] donor that has a very pronounced characteristic well, then, you are always going to try not to. This is a reality. You have to try to assign a donor that you... that could work for you. I think this is the way to work. That if you are seeing that you would not like her for yourself... [you would not take her]. (G2)

Thus, (some) professionals make explicit that not only physical resemblance to the patient matters. Indeed, a high degree of homogeneity among donors is often assumed, though this is not the case for patients. As another doctor said "we have to take into account that all donors are tall and slim" (E).

In these senses, so-called phenotypical matching of donors and patients might tend to semiotic-material reproductions of the same and normativization of potential offspring. That is, women who fit in normative standards are chosen as donors; by selecting for normative donors, clinics are also in effect selecting for normative babies. This takes place first, through the ways in which donors' physical diversity is mainly acknowledged (or seen as deserving acknowledgment) in what comes to be seen as "normal" and through the reduction of just racial categories to those situated as "other". Second, all those bodily aspects that are not given importance in itself as to be *matched* might be considered and chosen under logics of adjusting to normative beauty standards, preferring donors described as slim, medium-to-tall, cute young girls.

Discussion & Conclusions: Creating Ovodon *as the Most Successful Technique*

Donated eggs have been key to building Spanish bioeconomies of reproduction by enhancing general ART success rates and making it possible for many women and couples (from both Spain and abroad) to gain access to eggs of younger women. *Ovodon* has been mainly developed within the private sector, due to the fact that the public sector places age limitations on IVF treatments and does not offer compensation to donors. The way in which patient-donor matching works within these private clinics, particularly in the bigger ones, builds patients as consumers within certain logics of choice, limited by a rather flexible law which gives final choices to doctors.

Ovodon is currently flourishing in the Spanish private healthcare system by not only offering high(er) success rates (than regular IVF), but also deploying strategies to render the use of donated gametes invisible. First, by portraying *ovodon* as a technique, power and agency are attributed to the biomedical side rather than to eggs or donors' reproductive capacity. Second, donors themselves are made less physically apparent through confining them outside the regular areas of clinics or selecting and advising them to be indistinguishable from patients. Third, the use of donated gametes is made less evident through phenotypic coordination and the search for matching blood types between parents and children. Familial relatedness is re-interpreted as physical similarity, both in terms of readily visible characteristics of appearance and the less visible but potentially undermining characteristic of blood type. This concurs with the compulsory anonymity of donors.

I argue here that an alternative framing of this practice might be needed, given the dynamics, power relations, economic configurations, and reproductive possibilities that it entails. Therefore, I propose to talk about Transference of Reproductive Capacity (TRCs) instead of assisted reproductive technologies (ARTs) to refer to treatments involving third parties' bodies and processes in their development. Moreover, the way in which *ovodon* currently works in the context studied, which is said to be rejecting the majority of potential donors, might as well be functioning as an SRT or selective reproductive technology "used to prevent or allow the birth of certain kinds of children" (Gammeltoft and Wahlberg 2014) through enhancing biological reproduction of certain women rather than others.

These treatments are presented as the clinic's most successful technique and key to the future development of the reproductive market, to be expanded and re-configured by the development of egg banks, both within clinics and independent of them. *Ovodon* has a key role within current functioning of reproductive economies but relies on a key practice, the donation of eggs from young women, that is not acknowledged as a form of economic transaction or work but rather framed as a form of altruism that coincides with gendered and patronizing views of these women and their motivations. This process of understanding female donation as moved by logics of care and gendered imperatives (linked to caring, motherhood and feminity) are neither new nor exclusive to Spain, as they share many elements with those found by Rene Almeling in the USA when comparing egg and sperm donation experiences (Almeling 2011). Here I argue that making donors' motivation the key element in understanding egg donation enables the construction of donors as (only, mainly) moral agents, without acknowledging that the practice has been actively manufactured in these terms by the previous readings of healthcare business professionals, doctors, bioethicists and/or law-makers.

Reproductive laws and clinics insist that egg procurement is based on altruistic donation, which, however, needs a little monetary compensation to function (that is, it is assumed that without compensation, donors would not exist and thus, compensation is allowed). Economic compensation is set to a maximum of €1000, as that sum is considered enough to avoid commercializing the transaction. That is, it is seen as not paying for the value (however interpreted) of the eggs, but rather compensating donors for the costs and inconveniences involved. Comparisons with higher sums of money offered in other countries are used to support the view that, in Spain, the compensation payment constitutes neither inducement nor commercialization.

However, this claim ignores the actual economic conditions of the country in general and of the potential donors (young women), in particular. In fact, the actual meaning of this sum of money for potential donors in Spain can be better appreciated if we consider these data (2013): the unemployment rate for women was 76.26% for those aged 16–19; 50.74% for those aged 20–24, and 31.99% for those aged 25–29; the average annual wage for women under 25 was € 9287.90; and the minimum monthly wage, on which many rely, stands at 646 euros. It is plausible that the incentivizing effects of the compensation would vary depending on the economic situation of potential donors; indeed, this could explain the reported change in donors' profiles during the so-called economic crisis.

In theory, the setting of compensation at a level that is allegedly insufficient to function as an incentive—that is, as a payment for which women would be willing to transfer eggs to the clinic—means that "donation"-with-compensation can be considered an altruistic activity rather than an economic transaction. Staff, however, recognize that the payment does play a role in donors' decisions, even if it is mixed with other, more altruistic, motivations. This focus on donors' motivation—not, for example, the motivation of the professionals involved or the profits derived from the donated eggs—as the determinant of whether the process has been commercialized, subjects potential donors to disciplining processes of moral scrutiny and judgment. These judgments appear to reflect an expectation, or ideal, of women located in a domestic or non-economic domain governed by altruism, solidarity, and love, while the clinics and their staff are located in societal or market domain that is legitimately governed by economic rationality. In this way, *ovodon* builds not only on a gendered division of what is (and what is not) economic but also on the construction of femininity as linked to caring, discretion and helping others, echoing similar narratives found in care work

(Pérez Orozco and López Gil 2011). Current readings of egg donation that do not fully recognize the involved agents as economic subjects (and as subjects deserving of rights as workers) reinforce the idea that the economy is divided into two different spheres in which two different logics are at play. In turn, this dichotomy renders invisible the actual complexity of all agents, agencies, motivations, desires, and needs at play in the reproductive bioeconomies.

Both orthodox and most heterodox economic perspectives tend to limit what we see in particular directions, removing from the view the scope and importance of the reproductive sphere in current economies and the interdependency of the different spheres. This makes it difficult to grasp the ways in which different spheres conjoin within current reproductive bioeconomies. One of the first starting points for feminist economics was to stress the economic condition of non-monetized economic spheres (social reproduction, in rather Marxist terms, or households), and the economic rationalities of women and altruism (Peterson and Lewis 2001; Picchio 1992, 2005). Consequently, the very pervasiveness of the dichotomies through which the economy was thought of and portrayed (market/non-market, production/social reproduction, selfishness/altruism, etc.) have been questioned, and a rather different view, which situates interdependence within the sustainability of life, has been proposed (Pérez Orozco 2006, 2014). The research discussed in this chapter, thus, can help to question the rigidity of these boundaries (particularly those between production and social reproduction, egoism and altruism, and market/non-market)- and to highlight how the reading of certain agents and activities as non-economic (and moved by altruism) plays a key role in bioeconomies of human care and reproduction.

Notes

1. We have been unable to get solid data on the exact percentage of treatments done within the public and private clinics. We do, nonetheless, know that around 75% of registered centers are private as it is said in the webpage of the Comisión Nacional de Reproducción Humana Asistida (Asisted reproduction national committee). Futhermore, an article from El Pais stated that 80% of treatments took place in the private clinic, and we take that as a somehow illustrative figure (see here: http://sociedad.elpais.com/sociedad/2013/07/23/actualidad/1374611147_796311.html).

2. The use of donated sperm is also to be seen in artificial inseminations, that are less than half of the overall treatments studied (29.550 de 72953) 25.5% of those were done with donated sperm (but there is no record how many of those were due to male fertility issues or simple lack of men) (SEF 2013).
3. This might be related to the fact that embryo donation is cheaper than egg donation and these years were the ones in which economic crisis affected more directly the general population, with a particular focus on middle classes.
4. This paper draws on fieldwork done as part of a broader project nationally funded to study the "The Bioeconomy of reproduction: the mutually constitutive interaction between assisted reproduction and regenerative medicine in Spain and the UK" led by Vincenzo Pavone (CSO-2011-26019).
5. Previous work with a focus on the public sector (Álvarez 2006) shows that, as far as we found during fieldwork, availability of egg donation in the public sector seems to have decreased since the economic crisis.
6. Even though I am aware that trans men can be egg donors and receivers as well, we will be referring to women within this chapter as those are the ones towards which the donation programs are designed and, as far as we have found in the clinics, with whom those are implemented. A further study focusing on the ways in which ARTs are implemented within trans people in Spain would be of much interest, though it is not the focus here.
7. Donación de óvulos translates as oocyte donation; "ovodon" or "ovodonación" is just a game with the words "óvulo" (egg) and "donación" (donation) making it one.
8. The only program of egg donation in the public sector we had the chance of getting to know by first hand used crossed donation: each patient needing donated eggs had to bring a woman willing to donate whose eggs would be given to another woman bringing a donor herself so as to maintain the compulsory anonymity. The donors were described as either family-related, friends or colleagues, whose profiles were "the same as the one of the patient taking them" and their motivation said to be "clear: to help your cousin going through an egg donation" (D).
9. Ethnic profile was differently dealt with in different clinics. Generally the staff talked about "race" and referred to either nationalities or the categories black, caucasian or Asian. People from South America were described by nationalities except for one interviewee that talked about "indian origins" in reference to women coming from Ecuador, Bolivia and Perú. In many clinics the online contact form prospective donors have to fill in has already some categories from which to choose, some of which make reference to skin color rather to those categories later used.

10. Even if this was explicitly explained during fieldwork, we came to see that this is not always exactly the case, as in some clinics, if the donor produces fewer eggs than expected (a rare case), the procedure stops, as patients are assured a minimum number of fertilized eggs. Thus, if lesser eggs than expected are retrieved, some clinics cryopreserve them and others will consider them waste.
11. This is not a general Spanish reading of race, but rather a particular reading of race as done within these (private) clinics.
12. In reference to the interviewee and interviewer (also author of this paper). As said by a doctor during an interview.
13. This distinctions showed a strict division between a non-mentioned "us" and "others" that reminds to that previously pointed out while studying whiteness and difference within the European context (Griffin and Braidotti 2002).
14. A rather interesting research question would be the way in which these clinics deal with both gender and sexual diversity of potential donors. However, as this implies observation of clinical practice or interviews with potential donors, it remains out of the scope of both this research and this particular paper.

References

Almeling, R. (2011). *Sex cells: The medical market for eggs and sperm.* Berkeley: University of California Press.

Álvarez Plaza, C. (2006). Múltiples maternidades y la insoportable levedad de la paternidad en reproducción humana asistida [Multiple maternities and the unbearable lightness of paternity in assisted human reproduction]. *Revista de antropología social, 15,* 411.

Álvarez Plaza, C. (2014). La diversidad familiar y la divulgación de los orígenes genéticos a los niños nacidos a partir de donantes y/o gestación subrogada. *IM-Pertinente, 2,* 17–43.

Bergmann, S. (2012). Cambiar el óvulo manteniendo el fenotipo: la búsqueda de semejanza en la FIV con donación de óvulos transnacional. In Perez Sedeño and Ibañez Martín - Cuerpos Y Diferencias. Plaza y Valdéz—Calíope, Madrid.

Birch, K. (2006). The neoliberal underpinnings of the bioeconomy: The ideological discourses and practices of economic competitiveness. *Genomics Society and Policy, 2,* 1–15.

Birch, K., & Tyfield, D. (2013). Theorizing the bioeconomy biovalue, biocapital, bioeconomics or... What? *Science Technology Human Values, 38,* 299–327.

Bonaiuti, M. (2014). Bioeconomics. In *Degrowth: A vocabulary for a new era.* Abington: Routledge.

Coroleu Lletget, B. (2011). Orígenes, Antecedentes e Hítos más Importantes de la Especialidad en España. In *Libro Blanco Sociosanitario. La Infertilidad En España: Situación Actual Y Perspectivas* pp. (71–85). Imago Concept: Madrid.

Franklin, S. (2002). *Embodied progress: A cultural account of assisted conception*. Abingdon: Routledge.

Franklin, S. (2013). Re-thinking nature—Culture anthropology and the new genetics. *Anthropological Theory, 3,* 65–85.

Franklin, S. (2013). *Biological relatives-IVF,* stem cells and the future of kinship. Durham: Duke University Press.

Friese, C., Becker, G., & Nachtigall, R. D. (2006). Rethinking the biological clock: Eleventh-hour moms, miracle moms and meanings of age-related infertility. *Social Science & Medicine, 63,* 1550–1560.

Gammeltoft, T. M., & Wahlberg, A. (2014). Selective reproductive technologies. *Annual Review of Anthropology, 43,* 201–216.

Goven, J., & Pavone, V. (2015). The bioeconomy as political project a polanyian analysis. *Science Technology & Human Values, 40,* 302–337.

Griffin, G. & Braidotti, R. (2002). Whiteness and European situatedness. In *A reader in European women's studies* (pp. 221–236). London: Zed Books.

Haraway, D. J. (1997). *Modest- witness@ second- millennium. FemaleMan- Meets-OncoMouse: Feminism and Technoscience.* London-New York: Routledge.

Hewitson, G. (2014). The commodified womb and neoliberal families. *Review of Radical Political Economics, 46,* 489–495.

Jager, S. & Maier, F. (2009). Theoretical and Methodological Aspects of Foucauldian Critical Discourse Analysis and Dispositive Analysis. In Methods of Critical Discourse Analysis (pp. 34–61). Thousand Oaks: Sage.

Martín Criado, E. (2014). Mentiras, inconsistencias y ambivalencias. Teoría de la acción y análisis de discurso. *Revista Internacional de Sociología, 72,* 115–138.

Matorras, R. (2005). ¿Turismo reproductivo o exilio reproductivo? *Revista Iberoamericana de Fertilidad, 22,* 85.

Orobitg, G., Bestard, J., Salazar, C., (2013). El cuerpo (re) productivo. Interés económico y altruismo social en las experiencias de un grupo de mujeres donantes de óvulos. *Revista Andaluza de Antropologia* 2013, *5,* 91–104.

Orobitg, G., & Salazar, C. (2005). The gift of motherhood: Egg donation in a Barcelona infertility clinic. *Ethnos, 70,* 31–52.

Pavone, V. (2012). Ciencia, neoliberalismo y bioeconomía. *CTS Revista Iberoamericana de Ciencia Tecnología y Sociedad, 7,* 145–161.

Pennings, G. (2002). Reproductive tourism as moral pluralism in motion. *Journal of Medical Ethics, 28,* 337–341.

Pennings, G. (2005). Reply: Reproductive exile versus reproductive tourism. *Human Reproduction, 20,* 3571–3572.

Pérez Orozco, A. (2006). *Perspectivas feministas en torno a la economía: el caso de los cuidados.* Madrid: Consejo Económico y Social.

Pérez Orozco, A. (2012). Crisis multidimensional y sostenibilidad de la vida. *Investigaciones Feministas, 2*, 29–53.

Pérez Orozco, A. (2014). *Subversión feminista de la economía.* Mapas-Traficantes Sueños Madr: Aportes para un debate sobre el conflicto capital-vida.

Pérez Orozco, A., & López Gil, S. (2011). *Desigualdades a flor de piel: cadenas globales de cuidados.* ONU Mujeres, Madrid: Concreciones en el empleo de hogar y políticas públicas.

Peterson, J. & Lewis, M. (2001). *The Elgar companion to feminist economics.* Cheltenham: Edward Elgar.

Picchio, A. (1992). *Social reproduction: The political economy of the labour market.* Cambridge: Cambridge University Press.

Picchio, A. (2005). *Unpaid work and the economy: A gender analysis of the standards of living.* Abingdon: Routledge.

Puig de la Bellacasa, M. (2011). Matters of care in technoscience: Assembling neglected things. *Social Studies of Science, 41*, 85–106.

SEF. (2009). Registro de la Sociedad Española de Fertilidad: Técnicas de reproducción asistida (IA y FIV/ICSI).

SEF. (2013). Registro de la Sociedad Española de Fertilidad: Técnicas de reproducción asistida (IA y FIV/ICSI).

Thompson, C. (2005). Making parents: The ontological choreography of reproductive technologies. Cambridge, MA: MIT Press.

Author Biography

Sara Lafuente Sociologist by Universidad Complutense of Madrid and MSc. on Biomedicine, Bioscience and Society by the London School of Economics, Sara Lafuente has recently completed her PhD on Bioeconomies of Reproduction, focusing on the meanings and materialities of eggs in the Post-IVF Biology and their associated clinical practices. Her work aims to stablish a dialogue between STS, feminist economics and queer theory.

CHAPTER 12

Thinking (Bioeconomies) Through Care: Patients' Engagement with the Bioeconomies of Parenting

Pablo Santoro and Carmen Romero-Bachiller

Introduction

At that time, in my husband's family, there was a condition… *one of these so-called "rare diseases". Fatal familial insomnia.* [Pauses]. It is really hard. You stop sleeping and 6 months later you're dead. And we didn't know…

Research on which this essay is based was funded by the Spanish Ministry of Education through the 2013–2016 project led by S. García Dauder "Análisis de la producción y circulación de saberes expertos/legos en prácticas biomédicas" [*Analysis of the production and circulation of expert/lay knowledges in biomedical practices*, Ref: FFI2012-38912-C02-02], integrated in the coordinated research "Visiones y versiones de las tecnologías biomédicas" (VIVERTEC) (*Visions and Versions of Biomedical Technologies*, Ref. FFI2012-38912-C02-01), with Eulalia Pérez Sedeño as Main Researcher.

P. Santoro (✉) · C. Romero-Bachiller
Complutense University of Madrid, Madrid, Spain

V. Pavone and J. Goven (eds.), *Bioeconomies*,
DOI 10.1007/978-3-319-55651-2_12

> We'd just learned that it was something that ran in his mother's family. They started collecting clinical records from all the family and there were a lot of cases. Its causes are unknown. And there are some people who carry it, but it never activates. But others… When they turn 45 or so, it starts. And they don't sleep. They don't sleep at all. They start losing cognition, they stop walking. It is a horrible death, in 6 months time. And it is a familial condition, as the name says. So it was like: *let's do that cord thing.* With my first and second children we couldn't do it, but with the third… just in case something happened to him, or to one of his siblings. [Pauses again]. Just remembering it gives me the shivers. [The moderator of the group, knowing there is no current application for a cord blood transplant in that case, asks if she looked for more specific information.] The thing is that, back then, *I was in utter panic.* [Pauses]. And I was being physically abused by my husband—now he's my ex-husband. My children didn't sleep because they saw and heard many things. So my children didn't sleep. And I said to myself: they have it. Everything was… They couldn't sleep, not because they had it, but because… they were anxious and… It was this man's fault. I have to say, nonetheless, that my mother-in-law dared to go and take the test. Her mother had died from it, her younger sister had died from it. And a nephew, a 25 year old, had died from it too. They make you a blood test and search for a gene, the "p" something… I don't remember precisely now, but *back then I read a lot, it was like doing a master's degree on the disease. I spent the whole day on the internet, on patient forums and… I even became friends with a man in Galicia*, he told me in that area there are a lot of cases. I was so anxious and fearful but *I couldn't stop reading.* I can't remember exactly, it's a specific gene, twenty-something… Today I have lost all contact with my husband's family, but back then, I was pregnant, and I already had two children, so I thought: my god, if my husband has it, my children… they can have it or not… And my mother-in-law was brave, because, the disease being so cruel, some people do not dare to take the analysis. Just imagine it is a positive: there are some people who have the gene but it never develops. *So imagine living the rest of your life with that uncertainty.* But this woman went and took the test; she didn't say a word to anyone. We just learned about it when it was a negative. And since she doesn't have it, nor does my husband, and nor do my children. So I cried… I have never cried that much since or before. But even with that, *we thought banking the cord was a wise decision. And it was just 2,000 euros, we could afford it.* (Focus Group 1, June 2015. Emphasis added.)

How should we account for a story such as this, which we heard during a focus group from a woman who had hired the services of a private cord blood bank? How should we interpret this woman's reasoning for contracting a private cord blood account? Is she a selfish and imprudent "hoarder" (Fannin 2013), who preserves a bodily tissue just for a potential and uncertain use inside her own family, even knowing that there is currently no clinical application for it and that it could have been donated for treating other people's diseases? Is she a "victim", abused by a neoliberal regime of health and the endless greed of commercial services, as well as by her husband? Or maybe one or another kind of "biological citizen"—a "sufferer" in Petryna's version (Petryna 2004), a "biological consumer" in Rose and Novas' (Rose and Novas 2007). Was she—as her mention of a "master's degree" evokes—on the way to becoming a "lay expert" (Prior 2003), a potential collaborator with medical research on rare diseases, alongside those novel "epistemic communities" (Akrich 2010) that are carrying out biomedical research "in the wild", beyond its former limitations to professional clinicians and medical researchers (Callon and Rabeharisoa 2003)? Or is she all of these things at once? And what to make of other elements of the story she told: allocations of guilt and responsibility; the severing of family ties, along with the building of new relationships through virtual communities; fears and anxieties about her children's future health; stories about blood tests and genetic markers, about the uncertainty of facing the diagnosis of a life-threatening disease; economic valuations on the price of a medical service. All these elements, as heterogeneous as they are, become entangled into what is called the new "bioeconomy", but, how to account for its complexities and ambiguities? How to understand, in short, the subjectivities of patients as a nodal point of current bioeconomies?

This chapter explores some of the ambivalences and tensions of patients' engagement with bioeconomies in advanced capitalist societies. We draw on empirical materials collected through two research projects carried out in Madrid (Spain), which dealt with health decisions related to childbirth and parenting. The first one studied practices and discourses around umbilical cord blood (UCB) banking, especially the dichotomy between donating the cord stem cells to the public healthcare system or hiring a private familial account in one of the many commercial cord blood banks operating in Spain and globally. The second project investigates a quite common situation during breastfeeding, the painful breast inflammation known as mastitis, its different scenarios of treatment and the different social networks formed around them.

In both cases, as in many other areas of the reproductive and biomedical economies, the relationship of new mothers with biomedical companies could turn them not only into *passive* recipients of healthcare, treatments, and drugs, but also into *active* providers of blood, breast milk, stem cells, human microbiota or other samples of biological materials from their families, which will be preserved, processed, and employed in research, or sometimes even used to commercially generate other bioproducts. Some recent critiques of the bioeconomy are indeed emphasizing the previously neglected role of patients, donors, and participants in clinical trials as necessary providers of tissue, cells, reproductive biomaterials, biodata or health information (Cooper and Waldby 2014; Lupton 2014). The ambivalence of this dual position of patients—both consumers and "producers"—traverses most of the scholarly literature on the role of patients in biomedicine. We argue that, in order to better understand the position that patients occupy in bioeconomies, we should focus on this ambivalence. For ambivalence and complexity are crucial in our two studies and in the current organization of bioeconomic companies focused on parenting, childbirth, and child health.

The chapter starts by describing two common readings in STS of patients' engagement in contemporary health regimes. We argue that these perspectives will benefit from a more nuanced view of the subjectivity and agency of individuals. We then introduce the concept of *care* developed in some recent STS and feminist writing, suggesting that this perspective can help us to develop an alternative understanding of new parents' involvement with biomedical research and bioeconomic markets. We then present our argument by using the concept of care in analyses of our UCB banking and mastitis case studies.

The Role of Patients in Health Bioeconomies: Empowered Patients or Bio-exploited Individuals?

However we define "bioeconomy", it is evident that a medical bioeconomy requires the participation of patients. As citizens and "consumers" of healthcare services, patients are the end-users of (human) health bioeconomies and whatever products they may bring to the clinic or the market. As providers of biomaterials—such as stem cells and genetic material or data; ova, sperm, and embryos for assisted reproductive technologies; breast milk; or the very body, in the case of clinical trial subjects and surrogate mothers (Cooper and Waldby 2014)—patients are the beginning and, in some way, the raw material of the "circuits of vitality" (Rose 2007) that organize these sectors. Patient representatives and social movements also intervene in diverse forms in many processes related to the bioeconomic sectors that deal with health and healthcare. The involvement of patients and individuals, in any of these senses, is therefore essential to the maintenance and growth of bioeconomies. But how should we understand the socio-political significance of this involvement?

This question has generally been approached in STS and social studies of health in two different ways. The first one focuses on the ideal of "participation", stressing its democratizing potential and the redistribution of knowledge and agency it provides. Patient participation—along with other related values such as empowerment, autonomy or choice—has indeed become relevant in a wide variety of social scenarios related to health, from policy documents to healthcare clinics, from activists' groups to medical ethics committees, and could thus be judged as one of the central regulatory ideals in contemporary health and medicine (Del Savio et al. 2016). Even in one of the most technocratic fields of healthcare policy, Health Technology Assessment (HTA), the ideal of patient involvement is gaining momentum (Gauvin et al. 2011).

STS scholars have been particularly interested in patients' organizations, given their growing relevance as legitimate actors to be consulted in health policy and medical research (Akrich et al. 2008; Wehling et al. 2014). Moreover, this new position of patients' organizations provides an access point to the changing relations between expert and lay knowledge in current technoscience. Since Steven Epstein's pioneer study on HIV activism (Epstein 1996), new forms of patients' organizations and social movements on health have been variously conceptualized as "embodied health movements" (Brown et al. 2004), "epistemic communities" (Akrich 2010) or "evidence-based activism" (Rabeharisoa et al. 2014), which go beyond the role of mutual support traditionally ascribed to groups of patients; they get involved in practices of production and dissemination of biomedical knowledge—what Vololona Rabeharisoa and Michel Callon, termed "research in the wild" (Callon and Rabeharisoa 2003). By focusing on organized, public, self-conscious, and (usually) successful forms of patient-groups' involvement with biomedical research, many of these studies seem to partake to a certain degree in the widespread enthusiasm for the ideal of public participation, projecting onto patients' organizations a strong, active, and empowered subject position. A similar view can be seen in celebrations of the "expert patient" (UK Department of Health 2001; Fox and Ward 2006) or in popular accounts of the Do-It-Yourself genomics movement (Maher 2013).

A second, alternative, approach can be found in critical analyses of the bioeconomy. Whereas literature on patients' organizations has tended to understand participation in a positive light, most critiques of the bioeconomy interpret the same set of processes—e.g., calls for patient engagement, patient choice, autonomy, self-management, and self-entrepreneurship of health—through a theoretical framework built on notions of biological materiality, commodification, and neoliberal forms of governmentality. In the tissue economies (Waldby and Mitchel 2006), individuals are mobilized as consumers and "self-entrepreneurs", but also as providers of biomaterials, biodata, or "experiential knowledge" and "patient experiences" (Caron-Flinterman et al. 2005; Lupton 2014) that can later be transformed into economic value and biocapital (Sunder Rajan 2006). Following Foucault's famous description of modern biopower, the mantra of participation in contemporary healthcare would then correlate with novel processes of exploitation and creation of biovalue from individual bodies (Waldby 2002; Rose 2007; Cooper 2008; Cooper and Waldby 2014), paradoxically underpinned by ideas of individual choice and informed consent. Melinda Cooper and Catherine Waldby, for instance, stress how the bioethical principle of informed consent many times conceals the production of an exploited bioeconomic labor force, and how this framework redistributes risks and potential bodily harm in terms of transnational, economic, gendered and racialized lines, thus working to reinforce global and local inequalities (Cooper and Waldby 2014: 14). And while STS literature on patients' organizations focuses on new forms of patients' empowered subjectivities and forms of agency, bioeconomy theorists tend to portray patients as disempowered: as isolated and individualized bio-consumers, or as bio-laborers who are subjected to new processes of objectification and alienation.

While this may be something of a simplification of a diverse field of scholarship, nonetheless these two constructions of patients as subjects can readily be found in social studies of biomedicine during the last decade or so. Though both literatures have provided crucial insights into some of the transformations that are shaping our current global health regimes, their opposing constructions of patients' subjectivity and agency are partial and limited, and miss the complexity and ambivalence in patients' relationships with health bioeconomies. We suggest that a different literature may help us here: recent STS and feminist writing on care practices (Mol 2008; Mol et al. 2010; Puig de la Bellacasa 2011; Pérez Orozco 2014; Murphy 2015). Here we can find a more nuanced conceptualization of the patient as subject: neither the strong, empowered subject position of expert patients involved in collective formal action, nor the alienated subjectivity of the bio-exploited individual, but an entangled subject, unavoidably involved in socio-material networks mediated by vulnerability, interdependence, and care (Butler 2004, 2009; Winance 2010; Ruiz Marcos 2015).

Thinking Through Care

"Care is a slippery word" (Martin, Myers and Viesu 2015: 1). Originating in feminist ethics, the focus on care directs our attention towards what seems mundane and unimportant, yet is in fact vital in a remarkable way: for care is about the sustaining of life (Puig de la Bellacasa 2011; Pérez Orozco 2014; Ruiz Marcos 2015). Joan Tronto's often quoted definition of care includes "everything that we do to maintain, continue and repair 'our world' so that we can live in it as well as possible. That world includes our bodies, our selves and our environment, all of which we seek to interweave in a complex, life sustaining web" (Tronto 1993: 103). Housework, child and elderly care, professional healthcare, care for the sick and the disabled; affection, intimacy, emotional care, bodily work for others and for the self; mothers caring for their children, nurses caring for their patients, farmers caring for their animals, artisans or lab technicians caring for their objects: making visible and valuing all these activities—historically overlooked, devalued, feminized, and relegated to the private sphere—is central to care. But for feminist politics, care is also a critical approach, an ethical and political stance that foregrounds the relational character of our existence, emphasizing connectedness, interdependency, and vulnerability as basic characteristics of human life. And caring is also "a mode, a style, a way of working" (Mol et al. 2010: 7) that appeals to us as scholars: studying care should not be set apart from *thinking with* care—that is, a critical reflection on our politics of knowledge and our relationship with the people we study. From this perspective, then, care is simultaneously "a signifier of necessary yet mostly dismissed labors of everyday maintenance of life, an ethico-political commitment to neglected things, and the affective remaking of relationships with our objects" (Puig de la Bellacasa 2011: 100).

In STS literature there has been an increasing interest in multiple dimensions of care: for example, in the role of objects and materiality in care; in the local and contextual character of caring; or in the ethnographic account of the recurrent, dull, and sometimes painful labor that accompanies care work (Mol 2008; Pols 2010; Taylor 2010). STS ethnographic approaches have insisted that care is a *practice*, something that is done, bringing to the fore many apparently banal but crucial elements in science, technology, and medicine and showing how care is a matter of tinkering with technologies (Mol et al. 2010). However, sometimes the detailed ethnographic attention to tinkering practices leaves little space to consider how long-term, historically sedimented patterns of inequality and logics of power are recurrently inscribed in care (Collins 2004; Haraway 1997). But there is a critical edge that is inherent to thinking about care: as María Puig de la Bellacasa states, we need "to ask critical questions about *who* will do the work of care, as well as *how* to do it and for *whom*" (Puig de la Bellacasa 2011: 92).

It is impossible to ignore that care practices are still far from being equally distributed. Women have historically borne the burden of care work, and feminism has long criticized the unpaid and unrecognized workload of women in the domestic sphere. Marxist feminists talked of "care labor" as a way of highlighting the need for its redistribution and the socialization of the responsibilities it entails, while simultaneously rejecting any kind of naturalization of care to a woman's "essence" (Federici 2013; Pérez Orozco 2014). Yet, progress made in gender equality has not ended stratification in care work. In western countries like Spain women still devote much more time than men to housework and child care, and becoming a mother is still a turning point in a woman's professional career, when time dedicated to care and domestic work tends to increase and professional expectations tend to freeze or to be heavily reduced (Prieto and Pérez de Guzmán 2013). Care is also stratified along lines of race and class: in advanced capitalist societies, much care work, both formal and informal, is performed by a low-wage and precarious workforce, usually female and from a migrant and/or working-class background (Gutiérrez-Rodríguez 2010). New economic and racial divisions under conditions of globalization are re-articulating these "global chains of care", but do not end their gendered character: many professional women and couples in western countries shift their "burden" of care to transnational migrant women, who in turn have to leave their children under the caring of other women (Pérez Orozco and López Gil 2011; Vega 2009). Care also has a normative dimension: some—for example, mothers—are expected to care, while others may find it easier not to care. Obligations and expectations surrounding care are equally gendered, classed, and racialized.

Engaging with care as an object as well as a perspective of analysis—*thinking through* care—implies, then, paying "attention to the ambivalent rhetorics and practices taken up in its name" (Martin et al. 2015: 6). Thinking through care emphasizes the collective maintenance of life, the affective labor involved in sustaining it, the interdependency and the material relationality it mobilizes. But it also insists in making visible inequalities and relations of power, the gendered, economic, and symbolic obligations irreducibly entangled in everyday caring practices. Biomedicine, healthcare, and parenting are privileged sites to apply this perspective, since our dependency on dense networks of caring relations becomes salient in situations where the Western ideal of a disembodied, independent, and rational subject clearly cannot be applied: chronic diseases (Mol 2008), situations of disability (Winance 2010; Sánchez Criado et al. 2015), dementia and Alzheimer's disease (Taylor 2010; Ruiz Marcos 2015); and, of course, caring for a newborn baby.

Thinking the bioeconomy through care also implies understanding patient engagement in a more complex way, highlighting the ambivalent character of practices of bio-consumption, bio-data sharing, and forms of clinical labor. In the case of childbirth and parenting, the mobilization of parental expectations is fuelled by deeply affective bonds as well as by social norms and imperatives: children embody hope, vulnerability, and obligation, compelling us to do everything we can for them. This not only makes reproduction, parenting, and childbirth a direct target of many bioeconomic companies, but it also requires studies of the bioeconomy to attend to these affective and caring dimensions. In what follows, we apply the perspective of care to our two case studies.

Private Cord Blood Banking

Since the first cord blood transplant was performed in Paris in 1988 (Gluckman et al. 1989), UCB has been used as a substitute for bone marrow transplantation in the treatment of certain child blood malignancies. As a promising therapy and an alternative source of hematopoietic stem cells, it was immediately incorporated into existing systems of tissue and organ donation organized around public national banks. But almost at the same time, a new biotech business was invented: the private cord blood bank. Still following the original model established by the first UCB private bank—the Cord Blood Registry, founded in California in 1992—these companies encourage new parents and expecting couples to open a private UCB account for their child. This is not to take advantage of existing therapies, since the vast majority of such therapies use allogeneic transplantation: that is, transfusion between unrelated but immunocompatible individuals. Instead, private UCB banks emphasize the potential—and, for the moment, still quite uncertain—personal future uses that biomedical research could find for those stem cells. After paying a fee of around 2000 euros, blood is collected at birth and cryopreserved for a period of at least 20–25 years—some banks also charge an annual fee for maintenance—being available in case the child or other family member acquires a condition that may be treatable with UCB in the future. Some scholars suggest that, while donation to *public* UCB banks operates in a "regime of truth", in which UCB is stored for its current applications in oncology and hematology and its use on present-oriented and evidence-based research in those fields, *private* UCB banks function in a "regime of hope" aimed at the future prospect of regenerative medicine (Brown and Kraft 2006; Martin et al. 2008; Beltrame 2014).

Private UCB banks are a burgeoning bioeconomy: according to Paul Martin and colleagues, cord blood companies amounted to 35% of the stem cell market in 2006 (Martin et al. 2006), and their business model seems to have survived the impact of the global financial crisis on the stem cell industries. Private banks are present today in most countries: as of July 2016, the Parent's Guide to Cord Blood Foundation website listed a total of 389 private banks worldwide, most of them concentrated in North and Latin America, Europe, and East Asia, but also in some countries outside those regions—e.g., Kenya, Saudi Arabia, and Iran.[1] Even in countries where their business model is not authorized or is subjected to strict regulations, such as Italy, France or Spain,[2] a vast majority of pregnant mothers get to hear about "family" UCB banks from the mass media, medical professionals or personal acquaintances (Katz et al. 2011).

One of us (Pablo) started following the debates over private UCB banks in 2007. The press news in 2006 about the Spanish prince and princess having stored stem cells from the cord of their first daughter in a US private bank spawned a very heated public debate, influenced the implementation of the 2004 EU Tissue and Cells Directive into Spanish law, and became a fortunate piece of advertising for the private banks that were just starting to appear in the country (Santoro 2009, 2011). In 2008 and 2009, Pablo conducted interviews with key stakeholders in the Spanish public and private UCB banking sectors and with users of private banks, and analyzed media coverage as well as European and national regulatory frameworks and advertising materials from private and public banks. From 2013, in the context of a joint national project on biomedicine, he has carried out a second phase of research, surveying discourses about UCB in Spanish e-forums on maternity, interviewing several more health professionals and, jointly with Carmen, the other author of this chapter, conducting three focus groups with recent mothers.

Mastitis

Mastitis is described as"an inflammation of the women's breast caused by an infection that occurs commonly during lactation, with a prevalence of 3–33% of lactating mothers, and constitutes one of the main reasons to cease breastfeeding" (Arroyo et al. 2010). According to medical pathology, this is due to a decompensation of mammary microbiota by the overpopulation of a commensal bacteria colony, or by the presence of a pathogenic agent—most commonly *Staphylococcus aureus* (Fernández and Rodríguez 2013: 34). Its symptoms include fever, pain, abscesses, and difficulties with breastfeeding. Starting in 2014, Carmen has been investigating different treatments, reactions, and discourses around mastitis in various settings: monthly meetings of a breastfeeding group of *La Leche League*; a breast milk analysis laboratory; online forums; and medical literature. Methods include formal and informal interviews with women and professionals dealing with mastitis, as well as an auto-ethnographic account.

Mastitis occupies a rather peculiar position, as it is quite common in breastfeeding mothers, yet it seems relatively invisible, an "open secret". This lack of visibility is even more striking since bovine mastitis is a disease widely researched in veterinary sciences, as it decreases the productivity and profitability of the dairy industry. This aligns with other developments in reproductive sciences during the twentieth century, like artificial insemination, whose origins are in techniques for improving livestock reproductive rates and traits (Cooper and Waldby 2014: 38–40).

Whereas mastitis may not seem at first to be as concerned with the bioeconomy as UCB, an examination of trends in treatment suggests otherwise. According to current medical practice in Spain, there are two main options—not necessarily mutually exclusive—for treating mastitis. The first option focuses on medical procedures: that is, the administration of antibiotics and, if infection leads to an abscess, surgical intervention. Even though taking antibiotics does not always require abandonment of breastfeeding, the persistence of pain and discomfort lead in many cases to its termination. The second option involves "natural" procedures not requiring biomedical intervention, including massages, alternative application of hot and cold showers, breast pumps and, particularly, frequent breastfeeding in specific mother-baby positions, so the sucking of the baby helps in getting rid of the mastitis pathogens. In that case, as one participant in our focus groups put it, "it's the child that takes the mastitis from you." (Focus Group 2, June 2015). Compared to medical procedures, nonetheless, these treatments—which tend to be recommended by midwives and breastfeeding support groups—are slower in healing the mastitis and, if not complemented with pain relievers, in taking away the pain, besides being more demanding for the mother.

However, in recent years a third option has emerged: new treatments with probiotics, produced out of various strains of Lactobacilli that have been isolated from donated human breast milk and patented in 2004 (Patent WO2004003235 A2) by *Biosearch Life*. *Biosearch Life*—previously *Puleva Biotech*—is a Spanish biopharmaceutical company that trades on the stock market.[3] After a successful clinical trial in human subjects performed by a Spanish research team (Arroyo et al. 2010), it released a product in Spain in January 2014, marketed as *Lactanza Hereditium*, in the form of 28 capsules containing *Lactobacillus fermentum LC40 (CECT5716)*, with an approximate cost of €28. At the beginning of 2016, two more probiotics were commercialized in Australia under the license of *Biosearch*'s patent: *Quiara*, by Puremedic Pty Ltd, containing 21 sachets of *Lactobacillus fermentum LC40 (CECT5716)*, and *Profutura Mastitis Relieved* including *Lactobacillus salivarius (CECT 5713)* and commercialized by *Danone Nutricia*, one of the largest international dairy companies.[4] Also, different clinical tri als can be found in National Health System registries, including Australia, the US, and the Netherlands, which inform us of an emergent international expanding business of probiotics in the treatment and prevention of mastitis.

Although its commercialization is quite recent, information about a researcher, "the veterinarian of mastitis",[5] who offered probiotics for mastitis has been widely disseminated on Spanish breast feeding online forums and blogs, as well as by health professionals. There is an increasing awareness of the probiotics option, as one of our focus-group participants indicates:

> I have around me many people [with mastitis]. I know cases of all types of mastitis. People now treat it with probiotics and they do great. It is being

> recommended by their midwives, and even by the same doctors that used to prescribe amoxicillin. Actually mastitis is bacteria, bacteria that sprout out of your nipple and so… With probiotics, that is very natural and totally innocuous both for you and for the baby… and people are delighted. Of course, what happens? Amoxicillin, you go to the pharmacy and it costs you 1.50. Probiotics cost you 36 euros. (Focus Group 3, June 2015)

This informant, who participated in a breastfeeding support group, engages in a discourse that opposes allopathic medicine as artificial, to probiotics presented as "natural" and "innocuous". They are "natural" as they proceed directly from several bacterial strains of human milk microbiota provided by anonymous donors. Human milk donation is in fact an easy and quite well standardized procedure. Breastfeeding mothers are regularly informed by midwives of milk banks and milk donation procedures—there is currently one milk bank in Madrid at the *Hospital 12 de Octubre.* Yet, human milk donated to milk banks is carefully pasteurized to prevent dangerous bacteria being ingested by babies at neonatology units. Milk donated for probiotics research and samples from mothers with mastitis for lab analysis, on the contrary, preserve mammalian microbiota. Notably, milk donated for probiotics research is converted into products that may be beyond the means of the donor: as our participant noted, while antibiotics are cheap, probiotics are expensive, and therefore not available to all.

Activating Care in UCB Banking and Mastitis

For those of us who research parenting and childbirth—especially if we, too, are mothers or fathers—there are inevitably "big" questions lurking in the background: what is a "*good*" mother (or father)? What is "good care" *here*—and *when*, and *how much*, and *in which form*, is it needed? Or more specific ones: what should be done in *this* situation, or *this* other one? Should I have banked the cord of my baby? Would probiotics allow me to continue breastfeeding, as I want? Will they take away the pain? Even interviewees request one's judgments or seek one's approbation. For parenting has to do with constantly taking decisions, many of them difficult and messy. All of them engaging care.

Decisions concerning UCB and reactions to mastitis cannot be separated from care, in all its complexity. For example, private UCB banking is repeatedly justified in terms of affection and concern for the child, as "wanting to do for my child everything, anything that I can" (Focus

Group, 2, June 2015). In the case of familial or genetic conditions like the one that opened our chapter, it is impossible not to see how fear and concern for the child enters the decision. Some of our interviewees made a significant economic effort to pay the banking fee, even in the knowledge that the usefulness of UCB is, at best, very uncertain. If there are other siblings, their well-being also comes to mind when privately banking the cord of a new brother or sister, and some participants in focus groups expressed their anxiety if they had preserved UCB from only one of their children: "I had these things in my head. When they grow up, imagine the elder becomes ill… And what if his brother needs it later and can't have it?" (Focus Group 3, June 2015). In this sense, the hypothetical futures of potential diseases and therapies evoked by UCB are always "futures with emotional resonance" (Brown and Kraft 2006)—or, we could say, "care futures".

In mastitis, concern about the child and treatments for her mother's pain run in parallel, mobilizing networks of care and interdependence that can include a multiplicity of actors, objects, technologies, and social relations:

> Of course I was scared that the doctors could tell me to stop breastfeeding. Because I wanted to. I have breastfed both of my children until the age of 2. But they said: "Don't worry, in fact it's the child that will cure your mastitis"…They said: "We're going to prescribe you some pills. But you can continue breastfeeding. In fact, what you have to do is breastfeed him often, as much as you can. He is the one that will take it from you". So I did that, but each time my husband said the baby was hungry, I started crying. I had a towel and I bit it each time I put the baby on my breast. And one evening my mother came to visit and said: "Let's see". And she started massaging my breast. She took the breast pump, a manual one, not electric, that a friend had given to me. And she started… I was biting my towel crying, my mother pumping and pumping… And finally there was a noise and something poured out of my breast…. (Focus Group 2, June 2015)

Doctors and nurses, pills and probiotics, breast pumps, towels, and also partners, friends, mothers—even the child herself as part of the healing process. It is easy to see that parental care involves heterogeneous networks of people and objects relating to each other, and that any bioeconomic product or process also enters these networks. In relation to mastitis and UCB, indeed, personal relationships are reinforced, re-experienced or re-configured. For instance, it has been suggested that the private UCB bank, alongside other trends towards the geneticization of kinship, participates in a symbolic and novel re-shaping of family ties among parents and siblings (Brown and Kraft 2006). But, as happens with mastitis, many times it also involves the actual participation of the extended family: as some of the health professionals interviewed explained, in Spain the private UCB account is a relatively popular childbirth gift from new grandparents in affluent families (Private gynecologist, October 2008; Midwives team in public hospital, March 2016). Most of the interviewees referred to talking with sisters, cousins, or any other family member who had banked the cord before them, or discussing the decision with other relatives and acquaintances. Deciding to bank or to donate the cord can lead as well to the creation of new friendships and symbolic ties, for example with workmates who had done the same thing (Interview with couple, October 2009).

Digital networks also play a prominent role in our two cases: websites, online discussion forums, blogs, social media, messaging services, all of these digital media have been used by parents since the beginnings of the Web to look for information and advice, but also to give and receive support and to connect with other parents (Lupton et al. 2016). Breastfeeding and its difficulties, in particular, are one of the main subjects of discussion in parenting websites, blogs, and online forums, and several mothers told us about finding support and advice for their mastitis and making friends in these digital environments, as well as getting to know more about its different treatments (including probiotics). Information and advice on UCB is also sought on the Internet and online forums, and most, if not all, private UCB banks are active on Facebook and other social media, encouraging their customers to become followers. In this sense, promises, expectations, and new therapies surrounding bioeconomies are leading not only to new professional alliances (involving, e.g., clinicians, commercial actors, and regulators) and new logistical, manufacturing, and clinical delivery infrastructures (see Gardner, Webster, and Mittra, this volume), but also to new configurations of care and kin relationships and new ways of associating with others.

Apart from these dimensions of affection and relationality, care entails obligations, normativities, inequalities, and impositions which are unevenly distributed in terms of gender, class, ethnic origin, and so on. These logics of stratification also play a part in the rationales that lead parents to engage with bioeconomies. In the case of mastitis, the gendered nature of parenting is clearly shown: current stereotypical representations of the "good mother" include, according to all our interviewees, successful breastfeeding during at least the first 6 months of life. Since mastitis can interfere with this aspiration, it induces in many women feelings of guilt and frustration, of not providing "enough" care, that add to the physical pain and suffering. New treatments with probiotics promise to allow breastfeeding to continue, so they serve to perform those ideals of "good motherhood". Many stories on mastitis we heard also speak about a lack of care that is also strongly gendered—about emotions of loneliness often experienced by new mothers, or about an extended disregard of women's pain, which is often unattended, trivialized, or underestimated. Here, the combination of the traditional ideal of motherhood as an absolute devotion to her children and the lack of support and care that many mothers themselves face produces a physically and emotionally punishing dilemma. Success in breastfeeding when mastitis appears often becomes an exercise in individual heroism, reinforcing traditional stereotypes of the mother who sacrifices even her life and her health for her children, and where pain is only thought as a by-product, a collateral effect, of care.

UCB banking has less to do with gender and more with class and economic inequalities. There is a logic of distinction and "exclusiveness" associated with the private UCB account that, in the Spanish case, was reinforced when it became known that the royal couple had banked the cord of their first child (Santoro 2009), something still remembered in our focus groups after 9 years. This class dimension is also exhibited in the process of choosing between public and private cord banks: in a 2009 European survey of pregnant mothers, 53% of the respondents said they would not store in a private bank because the price is too high (Katz et al. 2011), and some of our interviewees said they donated the cord because they did not have the money to pay for a private account, even though they would have wanted to (Focus Group 3, June 2015). Being able to pay for a private UCB account—and therefore, to "activate" the care that the advertising of private banks ascribe to that decision—entails a logic of class differentiation, which adds to the more commonly noted opposition between the "altruism" of donation and the "selfishness" of private banking. The advertising motto for a Spanish private UCB bank, which states: "let your baby be born with an advantage" (www.safetycord.com), shows this dimension of class reproduction quite bluntly.

Rather than thinking of mothers with mastitis and mothers or couples who privately bank the cord of her/their child as either empowered and active patients or as powerless and exploited victims, we argue that a more nuanced understanding comes from thinking about them from the perspective of care—as people involved in logics and networks of interdependence and support, in affective and embodied relationships which require care work, love, friendship, and support to and from others, but which also entail obligations, impositions, oppressions, and gendered and economic normativities. Such networks encompass a multitude of relationships with other actors, individual or collective, both human and not human, "natural" and "technical". From this perspective, engagement with health bioeconomies could be interpreted as an *activation of care*.

Conclusion

In this chapter we have explored the involvement of patients with health bioeconomies, focusing on two bioeconomic sectors that target childbirth, breastfeeding, and the first moments of parental life: private UCB banking and treatments for mastitis (particularly new promising treatments employing probiotics). Using the notion of care as a theoretical alternative to the opposing constructions of patients' subjectivities as either empowered and strong actors or as powerless and exploited victims, we have tried to emphasize the ambivalent positions and forms of agency of patients, as well as the diverse, and sometimes conflicting, networks of care—familial, medical, social, digital—in which they become entangled. Care, as an overarching term for the multiplicity of practices and relationships that sustain life, organize affections, and mobilize mutual support, has also shown its gendered, stratified, and unequal character.

Thinking the bioeconomies through care has a number of implications for studies of the bioeconomy. First, and foremost, it urges us not to forget that patients and citizens are central actors in the bioeconomic "circuits of vitality" and in the everyday practices of valuation of the life sciences and biotech sectors. We can agree with Kean Birch that, in current bioeconomies, "value is not inherent in biological materials" provided by patients and individuals, but rather constructed through the variety of "political-economic actors, knowledges, and practices involved in the creation and management of value" (Birch 2016: 1). But without those materials, even as fetishes or mere objects for speculative financial operations, the promises and economic value of the bioeconomy would not be possible. The bioeconomy largely relies on a series of procurement and care practices that need to be framed and understood as altruistic or "moral" practices in order to conceal their economic value and to prevent patients, carers, and other subjects of these practices from becoming formal economic subjects.[6] Therefore, individuals and patients are encouraged to donate, sell, preserve, use, or buy biomaterials and bio-objects (Vermeulen et al. 2012) and to invest and participate in the bioeconomy through a variety of practices that would not be understandable if excised from the networks of care of which they partake. Health bioeconomies depend on an activation of care.

Thinking the bioeconomies through care also impels us to preserve ambivalence and complexity. Care is "both necessary to the fabric of biological and social existence and notorious for the problems that it raises when it is defined, legislated, measured, and evaluated" (Martin et al. 2015: 625). The notion of care changes our perspective, directing it to practices and relations that are not just economic, or generative of financial value, but that are configured and displayed inside networks of affect and relationality: decisions like hiring a private UCB account, donating breast milk to biomedical research, or opting for probiotics instead of antibiotics to treat mastitis would not be understandable without these dimensions. But normativities, obligations, and social imperatives must also be taken into account, as well as the dynamics of gender and class reproduction that are involved in care. As Michelle Murphy writes, "the exercise of power operates through care in many divergent ways" (Murphy 2015: 720). By thinking our cases through care we wished to avoid "flat" readings of parents and patients as either empowered or exploited, and to show how forms of engagement in bioeconomic circuits are more complex and ambivalent than these perspectives can account for. And this complexity is irreducible, not least because practices of care—e.g. caring for a newborn, planning for her future healthcare, healing a sick breast—face urgent demands and requests that cannot be ignored. Care—besides being strongly gendered and unequally distributed—is not an individual, rational, altruistic "choice" but a powerful social imperative, something that obliges us, something we find ourselves involved in. Preserving its complexity, ambivalence and multivocality is then a way of, in Donna Haraway's terms, "staying with the trouble" (Haraway 2010) instead of resolving it too quickly.

Finally, the perspective of care forces us to reflect on how to engage with the worlds and the people we study. María Puig de la Bellacasa's proposal to *think with care in STS* suggests that caring, as simultaneously "an affective state, a material vital doing and an ethico-political obligation" (2011: 90), has implications for our politics of knowledge: "Understanding caring as something we do extends a vision of care as an ethically and politically charged practice…care connotes attention and worry for those who can be harmed by an assemblage but whose voices are less valued, as are their concerns and need for care" (2011: 90–92). Patients, carers, and citizens, and particularly those more vulnerable or less privileged, are the targets of the bioeconomy, both as bio-consumers and bio-laborers. Their disposition for care is activated in bioeconomic processes—but their voices are not always listened to, their needs for care not always valued. How could we, when studying the bioeconomies, recognize the effects of power without denying people any capacity of agency and resistance? How to communicate our analyses to, and share our analytical tools with, citizens and collectives concerned or affected by the bioeconomic project? Can we, indeed, collaborate with patients, citizens, and activists to imagine and co-construct a "more livable" bioeconomy? These political and ethical questions have a methodological counterpart: if the two studies presented in this chapter resorted mainly to "traditional" methodologies (focus groups, interviews, ethnographic observation), a more caring approach to researching the bioeconomies, we suggest, has another task: imagining other forms of research and engagement with the public; other, more collaborative, methodologies that could, at the same time, make care visible as a central element of bioeconomies and make research of the bioeconomy more caring.

Notes

1. http://parentsguidecordblood.org/en/family-banking-world-map (last accessed 28 July 2016).
2. Private UCB banking has in Spain an ambivalent situation: companies are allowed to operate in the national territory and pursue their clients, yet legally the samples of private banks have to be available to the public healthcare system in case it reclaims them. Given this peculiar regulation, most private banks preserve UCB in other countries: already in 2009, more than 30,000 "Spanish" UCB units were stored in foreign banks (Santoro 2009). This kind of transnational arrangement of the international cord blood market (Brown et al. 2011) is not strange in a sector

that, like many other bioeconomies, operates globally, in the sort of post-statist political geography of what Andrew Barry called a "socio-technical zone" (Barry 2001).

3. Biosearch Annual Financial Report 2015, http://www.biosearchlife.es/wp-content/themes/x-child/descargas/Informe%20Anual%20Gobierno%20Corp/IAGobiernoCorporativo2015.pdf (last accessed 27 July 2016).
4. A paper critical of the use of probiotics for mastitis treatment has appeared in the *International Breastfeeding Review* in June 2016. The authors question the methodological rigor of the clinical trial in which the probiotics' efficacy was established. They also question the terminology used in the study (acute mastitis, subacute mastitis and sub-clinical mastitis), pertinent to bovine mastitis but not to human mastitis; in women, only acute mastitis is described as mastitis, while the other conditions are regarded as minor forms of blocked duct (Amir et al. 2016). The preventive use of probiotics for mastitis, which is increasingly an advertising strategy for probiotic drugs, is also uncertain.
5. This is the nickname of Miguel Ángel Rodríguez, professor at the Faculty of Veterinary at the University Complutense in Madrid, and one of the signers of the referred patent. As I (Carmen) was informed in various interviews, prior to their commercialization he used to prepare the probiotics himself and hand them freely to women searching for relief for their mastitis–something that also deserves to be termed care.
6. We thank the editors for this point.

References

Akrich, M. (2010). From communities of practice to epistemic communities: Health mobilizations on the internet. *Sociological Research Online, 15*(2), 10.

Akrich, M., Nunes, J., Paterson, F., & Rabeharisoa, V. (Eds.). (2008). *The dynamics of patient organizations in Europe*, Paris: Presses de l'École des mines.

Amir, L., Griffin, Cullinane, M., & Garland, S. M. (2016). Probiotics and mastitis: Evidence-based marketing? *International Breastfeeding Review, 11*, 19. Retrieved from https://internationalbreastfeedingjournal.biomedcentral.com/articles/10.1186/s13006-016-0078-5.

Arroyo, R., Martín, V., Maldonado, A., Jiménez, E., Fernández, L., & Rodríguez, J. M. (2010). Treatment of infectious mastitis during lactation: Antibiotics versus oral administration of lactobacilli isolated from Breast Milk. *Clinical Infectious Diseases, 50*(12), 1551–1558.

Barry, A. (2001). *Political machines*. London: The Athlon Press.

Beltrame, L. (2014). The bio-objectification of umbilical Cord Blood: Socio-economic and epistemic implications of biobanking. *Tecnoscienza, 5*(1), 67–90.

Birch, K. (2016, August 10). Rethinking value in the Bio-economy: Finance, assetization, and the management of value. *Science, Technology and Human Values*. doi:10.1177/0162243916661633.

Brown, N., & Kraft, A. (2006). Blood ties: Banking the stem cell promise. *Technology Analysis & Strategic Management, 18*(3/4), 313–327.

Brown, P., Zavestoski, S., McCormick, S., Mayer, B., Morello-Frosch, R., & Altman, R. (2004). Embodied health movements: new approaches to social movements in health. *Sociology of Health & Illness, 26*(1), 50–80.

Brown, N., Machin, L., & McLeod, D. (2011). Immunitary bioeconomy: The economization of life in the international cord blood market. *Social Science and Medicine, 72*, 1115–1122.

Butler, J. (2004). *Precarious life: The powers of mourning and violence*. London: Verso.

Butler, J. (2009). *Frames of war. When is life grievable?* London: Verso.

Callon, M., & Rabeharisoa, V. (2003). Research 'in the wild' and the shaping of new social identities. *Technology in Society, 25*, 193–204.

Caron-Flinterman, F., Broerse, J. E. W., & Bunders, J. F. G. (2005). The experiential knowledge of patients: A new resource for biomedical research? *Social Science & Medicine, 60*(11), 2575–2584.

Collins, P. H. (2004). Comment on Hekman's "Truth and method: Feminist standpoint theory revisited": Where's the power? In S. Harding (Ed.), *The feminist standpoint theory reader. Intellectual and political controversies*. New York: Routledge.

Cooper, M. (2008). *Life as surplus. Biotechnology and capitalism in the neoliberal era*. Washington: University of Washington Press.

Cooper, M., & Waldby, C. (2014). *Clinical labor: Tissue donors and research subjects in the global bioeconomy*. Durham: Duke University Press.

de la Bellacasa, M. P. (2011). Matters of care in technoscience: Assembling neglected things. *History & Philosophy Of Science, 41*(1), 85–106.

Del Savio, L., Buyx, A., & Prainsack, B. (2016). Opening the black box of participation in medicine and healthcare, *Institute of Technology Assesment Manuscript ITA-16-01*, Austrian Academy of Sciences. epub.oeaw.ac.at/ita/ita-manuscript/ita_16_01.pdf.

Epstein, S. (1996). *Impure science. AIDS, activism and the politics of knowledge*. Berkeley: California University Press.

Fannin, M. (2013). The hoarding economy of endometrial stem cell storage. *Body and Society, 19*(4), 32–60.

Federici, S. (2013). *Revolución en punto cero. Trabajo doméstico, reproducción y luchas feministas*. Madrid: Traficantes de Sueños.

Fernández, L. & Rodríguez, J. M. (2013). *Mastitis. El lado oscuro de la lactancia*. Madrid: Probisearch.

Fox, N., & Ward, K. (2006). Health identities: from expert patient to resisting consumer. *Health, 10*(4), 461–479.

Gauvin, F. P., Abelson, J., Giacomini, M., Eyles, J., & Lavis, J. N. (2011). Moving cautiously: public involvement and the health technology assessment community. *International Journal of Technology Assessment in Health Care, 27*(1), 43–49.

Gluckman, E., Broxmeyer, H. E., & Auerbach, A. D. (1989). Hematopoietic reconstitution in a patient with Fanconi anemia by means of umbilical cord blood from HLA identical sibling. *New England Journal of Medicine, 321,* 1174–1178.

Gutiérrez-Rodríguez, E. (2010). *Migration, domestic work and affect. A decolonial approach on value and the feminization of labor.* New York: Routledge.

Haraway, D. (1997). *Modest_Witness@Second_Millenium. Femaleman©_Meets_Oncomouse™.* London: Routledge.

Haraway, D. (2010). When species meet: Staying with the trouble. *Environment and Planning D: Society and Space, 281*(1), 53–55.

Katz, G. et al. (2011, March). Banking cord blood stem cells: attitude and knowledge of pregnant women in five European countries. *Transfusion, 51,* 578–586.

Lupton, D. (2014). The commodification of patient opinion: The digital patient experience economy in the age of big data. *Sociology of Health & Illness, 36*(6), 856–869.

Lupton, D., Pedersen, S., & Thomas, G. M. (2016). Parenting and digital media: From the early web to contemporary digital society. *Sociology Compass, 10*(8), 730–743.

Maher, B. (2013, June 27). Father's genetic quest pays off. *Nature, 498,* 418–419.

Martin, P., Coveney, C., Kraft, A., Brown, N., & Bath, P. (2006). Commercial development of stem cell technology. *Regenerative Medicine, 1*(6), 801–807.

Martin, P., Brown, N., & Turner, A. (2008). Capitalizing hope: The commercial development of umbilical cord blood stem cell banking. *New Genetics and Society, 27*(2), 127–143.

Martin, A., Myers, N., & Viseu, A. (2015). The politics of care in technoscience. *Social Studies of Science, 45*(5), 625–641.

Mol, A. (2008). *The logic of care. Health and the problem of patient choice.* London: Routledge.

Mol, A., Moser, I., & Pols, J. (Eds.). (2010). *Care in practice. On tinkering in clinics, homes and farms.* London: Transcript.

Murphy, M. (2015). Unsettling care: Troubling transnational itineraries of care in feminist health practices. *Social Studies of Science, 45*(5), 717–737.

Pérez Orozco, A. (2014). *Subversión feminista de la economía. Apuntes para un debate sobre el conflicto capital-vida.* Madrid: Traficantes de Sueños.

Pérez Orozco, A., & López Gil, S. (2011). *Desigualdades a flor de piel: cadenas globales de cuidados.* Madrid: ONU mujeres.

Petryna, A. (2004). Biological citizenship: The science and politics of Chernobyl-exposed populations. *Osiris* 2nd Series, *19,* 250–265.

Pols, J. (2010). Telecare. What patients care about? In A. Mol, I. Moser, & J. Pols (Eds.), *Care in practice. On tinkering in clinics, homes and farms* (pp. 171–194). Bielefeld: Transcript.

Prieto, C., & Guzman, S. P. (2013). Desigualdades laborales de género, disponibilidad temporal y normatividad social. *Revista Española de Investigaciones Sociológicas, 141,* 113–132.

Prior, L. (2003). Belief. *Knowledge and Expertise: The Emergence of the Lay Expert in Medical Sociology, Sociology of Health and Illness, 25*(3), 41–57.

Rabeharisoa, V., Moreira, T., & Akrich, M. (2014). Evidence-based activism: Patients', users' and activists' groups in knowledge society. *Biosocieties, 9*(2), 111–128.

Rajan, K. S. (2006). *Biocapital: The constitution of postgenomic life.* Durham: Duke University Press.

Rose, N. (2007). *The politics of life itself. Biomedicine, power, and subjectivity in the twenty-first century.* Princeton: Princeton University Press.

Rose, N., & Novas, C. (2007). Biological citizenship. In A. Ong & S. J. Collier (Eds.), *Global assemblages: Technology, politics and ethics as anthropological problems.* Oxford, UK: Blackwell Publishing Ltd.

Ruiz Marcos, L. (2015). *Cuando la memoria pasa por la piel. Escenarios del cuidado en la enfermedad de Alzheimer.* Unpublished PhD thesis. Universidad Complutense de Madrid.

Sánchez Criado, T., Rodríguez-Giralt, I., & Mencaroni, A. (2015). Care in the (critical) making. Open prototyping, or the radicalisation of independent-living politics. *ALTER–European Journal of Disability Research, 10*(1), 24–39.

Santoro, P. (2009). From (private?) waste to (public?) value. The regulation of private cord blood banking in Spain. *Science Studies, 22*(1), 3–23.

Santoro, P. (2011). Liminal biopolitics: Towards a Political Anthropology of the Umbilical Cord and the Placenta. *Body and Society, 17*(1), 73–93.

Solís, C. V. (2009). *Culturas del cuidado en transición. Espacios, sujetos e imaginarios en una sociedad de migración.* Barcelona: University of Catalonia.

Taylor, J. (2010). On recognition, caring and dementia. In A. Mol, I. Moser & J. Pols (Eds.), *Care in practice. On tinkering in clinics, homes and farms,* (pp. 27–56). Bielefeld: Transcript.

Tronto, J. (1993). *Moral boundaries: A political argument for an ethic of care.* New York: Routledge.

UK Department of Health. (2001). *The expert patient: A new approach to chronic disease management for the twenty-first century.* London: Department of Health.

Vermeulen, N., Tamminen, S., & Webster, A. (Eds.). (2012). *Bio-objects. Life in the 21st century.* Surrey: Ashgate.

Waldby, C. (2002). Stem cells, tissue cultures and the production of biovalue. *Health, 6* (3): 305–323.

Waldby, C., & Mitchell, R. (2006). *Tissue economies. Blood, organs, and cell lines in late capitalism.* Durham: Duke University Press.

Wehling, P., Viehöver, W., & Koenen, S. (Eds.). (2014).*The public shaping of medical research: patient associations, health movements and biomedicine.* London: Routledge

Winance, M. (2010). Care and disability. Practices of experimenting, tinkering with, and arranging people and technical aid. In A. Mol, I. Moser and J. Pols (Eds.), *Care in practice. On tinkering in clinics, homes and farms* (pp. 93–118). Bielefeld: Transcript.

Authors' Biography

Pablo Santoro is assistant professor at the Department of Social Theory, Faculty of Political Sciences and Sociology, Complutense University (Madrid, Spain).

Carmen Romero-Bachiller is associate professor at the Department of Social Theory, Faculty of Political Sciences and Sociology, Complutense University (Madrid, Spain).

CHAPTER 13

Who Is My Donor? A New Bioeconomy of Blood and Its Changing Ontology

Emma King

Introduction

The production of cultured red blood cells (cRBCs) from stem cells has the potential to reduce or eliminate the need for blood donors. This paper draws on interviews and focus groups carried out with Professor Catherine Lyall (Lyall and King 2016) as part of the Novosang project. The Novosang project is developing a process for growing cRBCs in the laboratory from stem cells and is seeking to scale up the process to meet the demand for clinical transfusion. The research discussed here explores the attitudes of various publics towards the introduction of cRBCs. In particular, I discuss public attitudes in relation to the changing bioeconomy around blood products, especially in light of the changing ontology of blood as it moves from an altruistic donation model to a commercialized one.

The development of cRBCs is designed to address a number of problems facing current blood-donation systems. These include an insufficiency and inconsistency of supply in many developing countries, an expected increase in demand in developed countries as a result of ageing populations, difficulties in meeting the needs of those requiring frequent

E. King (✉)
NMAHP-RU, University of Stirling, Stirling, UK

V. Pavone and J. Goven (eds.), *Bioeconomies*,
DOI 10.1007/978-3-319-55651-2_13

transfusions, and the possibility of transmitting infections through transfused blood. Transfusion Transmitted Infections (TTIs) are a risk in any donation system, and supplying safe transfusions requires many levels of safety procedures. Safety measures taken to prevent TTIs can exacerbate shortages by requiring some potential donors to be excluded on safety grounds (e.g., travel history).

Titmuss (1970) attributed an increase in TTIs, particularly in the US, to the way in which blood was *no longer* seen as a gift. Blood "donated" for compensation became a commercialized product in which it benefited the giver to conceal truths about their risk of TTIs in order to gain monetary rewards. Starr (1999) also attributes the slowness of authorities to inform patients or withdraw infected batches to the reverence given to blood within an altruistic donation system, and its symbolic nature as a social gift rather than a pharmaceutical product. The Penrose Report stated that before 1983 it was, incorrectly, believed that products from altruistic UK donors were less likely to be infected with hepatitis than commercial products from the USA (Penrose 2015). Titmuss, Starr and Penrose all recognized the socially embedded (rather than commercialized) nature of altruistic donation as positioning blood as a "social gift". This imbued blood with a specialness that made it difficult to recall and destroy batches of donated blood products, or to exclude categories of people (posited as "high-risk") from the practice of donation.

Given this link between the source of blood (through altruistic donation, and commercial transaction), on the one hand, and how blood is regarded (with reverence, appreciation, or distrust), on the other, how are cRBCs likely to be regarded, treated, and used? What kinds of new relations, commercial and otherwise, do they create? What are the implications of these relations for the way in which cRBCs are produced, regulated, and accepted by patient users? This chapter examines the differing attitudes to cRBCs shown through data collection with a wide variety of publics. It begins with an overview of the Novosang project, and goes on to describe the methodology that underpinned this data collection. I will then present and discuss three themes raised by our respondents—the commercialization and altruism associated with cRBCs and donated blood, how cRBCs change our concept of kinship within the donation system, and finally how participants reacted to the potential global production of cRBCs.

This is the first study of public attitudes to cRBCs, although others have examined public attitudes in the related fields of blood donation, tissue use, and stem cell research. One of the largest projects on alternatives to blood donation found that the public viewed blood substitutes as a "substandard replacement" for actual blood, an unnatural and synthetic alternative (Fleming et al. 2007; Ferguson et al. 2008). While others have concentrated on the regulatory implications of cRBCs (Mittra et al. 2015), I discuss the views of wider publics towards this form of "commercialization" of blood.

Promoters of the health bioeconomy point to major social and economic benefits to be gained through the pursuit of commercial and technological solutions to current healthcare and economic challenges (Goven and Pavone 2015; Wield et al. 2013), although research and development challenges have seen many potential areas fall short of their anticipated benefits. Innovative new health technologies are being developed at such a rate that they no longer fit within existing regulatory frameworks, reimbursement models, and operational structures (Mittra 2016), and tissue-products are challenging an approvals process more suited to pharmaceuticals (Webster et al. 2011).

As this project concerns the UK it is necessary to see blood in the context of the UK and its National Health Service (NHS), where blood transfusions are currently based on an altruistic, not-for profit donation model. Data from King (2013) showed that within the Blood Transfusion Services, the "bioeconomy" of blood was focused on the transfer of goods and services (procurement, processing, and testing). Parallels can be seen with the economy of other human tissues described in Waldby's (2002) "tissue economy". However, blood is unique in that both donated RBCs, and the expected Novosang product, do not contain nuclei and therefore have no method of reproducing. Donated RBCs do not therefore have the regeneration potential (or "biovalue amplification") discussed in relation to the tissue economy by Kent et al. (2006).

The Novosang Project

Blood transfusions allow surgeons to overcome the inevitable blood loss during surgery, and blood fractionation products are used to treat a range of blood-based diseases, such as hemophilia (Schneider 2003). A number of innovations have been proposed to overcome the dual challenges of donor recruitment and possible infection risk. These include reducing the requirement for blood by using autologous donation (that is, from the patient him- or herself) or clotting agents (Ferguson et al. 2008) or by boosting the RBC count of patients before surgery (Martyn et al. 2002). Hemoglobin-based or synthetic oxygen carriers (Cohn and Cushing 2009; Silverman et al. 2009) were also heralded as an alternative to RBC transfusion. Unfortunately neither product has performed well in clinical trials and neither is licensed for use in the UK or USA (Henkel-Honke and Oleck 2007; Grethlein and Rajan 2012).

The Novosang project (formerly known as the BloodPharma project) is a multi-partner research project, involving many universities and organizations. It represents a multidisciplinary approach that aims to provide a safe, affordable, and publicly acceptable alternative to blood transfusion (Mountford and Turner 2011). Whole blood transfusions are very rare in the UK, with most "blood" transfusions being composed of RBCs re-suspended in an additive solution for preserving cell viability and function. The biggest competitor of the Novosang partnership has, to date, been the team of Luc Doauy (2009) in Paris, who have published some of the seminal studies in this area.

The Novosang project has to date received around £12 million from the Wellcome Trust, Scottish Funding Council (SFC), Scottish Enterprise, Scottish National Blood Transfusion Service (SNBTS), and NHS Blood and Transplant (NHSBT).[1] The Novosang partnership regards the project as a demonstration of the world-leading research that is being developed within the UK, with many of the technologies developed having an impact beyond cRBCs. While not actively seeking to produce large future profits, the partnership is securing intellectual property rights and is seeking additional investors to support scale-up and clinical trials of the cRBC product. The long-term goal of the Novosang project is to produce a patient-ready cRBC product, of the O RhD Negative donor type, which could be used for the majority of transfusions in the UK as well as, eventually, meeting the global requirement for RBCs. While embryonic stem cells have been used for earlier research (Mountford 2008), a switch to Induced Pluripotent Stem cell (iPS cell) technology was required in order to provide the team with a source of O RhD Negative tissue. The immediate research aim of Novosang is to produce a small quantity of these cRBCs suitable for animal and human trials.

The requirement for red cell transfusions is expected to overtake the ability of donation systems to meet this demand. The future visions employed by the Novosang project see the cRBC product as potentially filling this gap. The Novosang product could also provide an improved therapy for patients who require regular transfusions due to conditions such as beta thalassaemia or sickle cell anaemia—reducing immune reactions and iron loading.

The potential market for cultured red blood cells in the UK is around 2 million red cell units per year. Even in jurisdictions like the UK where blood is altruistically donated, the need for a considerable infrastructure to collect, process, test, store, and distribute blood means that blood is not "free". In England, blood is supplied by NHSBT, which charges a standard tariff set yearly (NHS England and Monitor 2016), and hospitals pay for every unit of blood used. (In 2014/15, the cost was £121.85 per unit; in 2015/16, it was £120 per unit.[2]) In Scotland the SNBTS and the blood it produces are centrally funded from the core NHS budget, but the cost is similar to that in England and Wales. Thus, even within altruistic donation systems, with no recompense to the donor, charging within the health system places a monetary value on a unit of blood. The view of the team involved in Novosang is that the cRBC product, to be commercially successful, would have to be comparable in price to the current cost of a unit of donated RBCs or provide an improved therapy for multiply transfused patients that would justify a higher cost.

Funding received from the SFC came with the requirement for social science input into the project. In discussion with the scientific team it was decided that focusing on the views of the wider public towards the use of the Novosang product would be the most appropriate aim for this work. This was due to the attachment to current blood transfusion, and the knowledge that public reactions could play a crucial role in the uptake of a future clinical product. Seeking the views of various publics was also a way of understanding the awareness that people currently have around blood transfusion, which was used to design more effective outreach and educational material. The Novosang project has also commissioned a detailed report on the commercial opportunities for the cRBC product.

Methods

Focus groups and interviews were used to elicit the views of a wide variety of publics towards the development and use of cRBCs for transfusion. The main four target groups of our data collection were identified through discussions with the wider Novosang team, and through an exploration of the relevant literature. These were:

- Patients who undergo regular blood transfusions, for example for sickle cell disease, thalassemia, or myelodysplasia.
- Representatives of religious or moral groups who could contribute to the discussion on cultured red blood cells from an ethical or religious standpoint.
- Community groups chosen to represent a spread of publics, who may not have a background in science or medicine. These included sports clubs, arts groups, et cetera.
- Clinical groups, such as doctors and nurses, who use blood transfusions in the course of their work.

In total there were 27 data collection sessions, including interviews, full focus groups, and discussions (explained below). Most of these took place in Scotland, with the exception of two focus groups and one interview that took place in England. We attempted to obtain a representation of both rural and urban populations, a variety of ages, and variation in educational attainment.

Focus groups (Krueger 1988; Kitzinger 1995) were used to bring together those who might not have previously considered the issues involved in cRBCs and to allow the generation of ideas within the group setting. A number of existing groups were targeted, including community, patient, and sports groups, and there was a wide spread of demographics within these existing groups. Interviews were conducted with those judged "experts" in their field, including doctors, ethicists, and patients. Semi-structured interview and focus-group schedules (Fontana and Frey 1994; Bryman 2004) were designed following preliminary scoping conversations with the wider project team, an initial review of the literature, and a small number of pilot interviews. Two of our data collection events were termed "discussions", due to the situation which made an unscripted discussion more appropriate. This included the parent and toddler group, where the noise and the fact that parents were engaged in supervising their young children meant that the group was quite fractured. In the analysis of interview and focus-group transcripts, we used a grounded theory methodology, to draw information from the data rather than starting with preconceived theories (Strauss and Corbin 1994). Grounded theory also allowed for data collection to take place iteratively, with findings from earlier data collection sessions contributing to the conduct of later sessions. Transcripts (and notes from unrecorded sessions) were coded using NVivo software. It is worth noting here that, despite our targeting and recruitment efforts, a number of groups were underrepresented in the research: blue-collar workers, older men, and ethnic minority groups.

Findings

Here I focus on the findings from empirical data collection pertinent to the discussion of the bio-economy and commercialization of the cultured blood product. I will begin by discussing how our respondents imagined the potential commercialization of the cRBC product in light of the current altruistic donation system in the UK. Secondly, I show how the respondents' concepts of kinship changed with the potential introduction of a commercialized product. Finally, I discuss the respondents' views of how the cRBCs might be produced on a global scale and the implications of this.

Commercialization and Altruism

A key emerging theme was the potential commercialization of this new product. Currently blood in the UK is given altruistically, and was regarded by participants as a way for donors to "give something back" to their community. Both the literature and our data show an attachment to the blood transfusion services within the UK, which continued despite an awareness of the potential infection risks surrounding donated blood (as in Fleming et al. 2007; Ferguson et al. 2008). In contrast there was a marked unease regarding the potential commercialization of cRBC production, with commercial companies considered to be "out to make money". Participants felt this was likely to result in unsafe products developed in countries without strict safety regulations, as will be discussed further in the final theme on potential global production. There were doubts about commercial-sector scientists in the UK, who were regarded as potentially willing to cut corners in pursuit of profit. There was also a fear that commercial involvement would lead to the NHS being forced to rely on buying blood from profit-oriented companies, potentially resulting in patients becoming priced out of access to blood.

M1: ... somebody asked whether it would be a pharmaceutical company or not. For the UK as you mentioned it's irrelevant because in theory, for the moment we are on free NHS. Nobody knows what's going to happen tomorrow but we as the UK, and as [patient group], are interested in making sure that whatever is developed is developed in a manner that other countries can benefit, and if it is pharma that is going to be producing it, things become a lot more expensive than they should be. Because apart from [...], whether this thing is going to be called a drug or not.

M4: But it's a maximum of 12 [years, for patent protection], you can't go beyond that. The thing is, what's going to make this expensive is going to be the requirements of the quality standards of the labs that would produce the blood, or do the whole processing. (Patient group)

Throughout the focus-group discussions, participants drew parallels between cRBCs and the current UK transfusion system. As blood is given altruistically and then returned to the patient for free at the end of processing in the UK, it was regarded as a "free" product. Most of the participants were unaware of the unseen cost of producing blood products; they therefore saw the idea of a commercialized blood system as a departure from current practice. After focus-group participants were introduced to the true cost of blood within the UK, this appeared to raise awareness that blood already imposes a monetary burden on the NHS (NHSBT 2010). The cRBCs would not necessarily increase the cost that the NHS currently pays for blood transfusions, and may even reduce it, depending on the cost of the eventual cRBCs and the reduction in the requirements for donors, advertising campaigns, collection services, and clinical administration. Participants seemed to then be more receptive to the idea of commercial involvement, although they described a very black-and-white choice between a public system and a commercial system. The reality is more nuanced, with care practices having a cost but not necessarily being commercialized. There is a difference between a non-profit blood processing system and a for-profit system, however the participants did not appear to make this distinction, with attention focused instead on the trust placed in the current transfusion services, versus the potential for a for-profit private company.

Participants conversely were more accepting of commercial involvement in *pharmaceutical* production, believing that this was necessary for the huge development and manufacturing costs necessary for drug products. In reality a lot of the cost of developing pharmaceuticals is already borne by the public sector, for example through public research funds. Participants did not expect the NHS to pay for testing and development of pharmaceuticals and saw pharma companies as fulfilling a crucial role within the development of new medications, although public and NHS money is again often involved in these stages (see, for example, the chapter by Gardner et al. in this volume). As the quote below illustrates, they were also reflexive about how living in a country with a national health service has influenced their views on the involvement of private companies.

> Can I suggest, we're talking from a slightly naïve point. We in the UK have this perception of everything being free on the NHS, albeit we're paying for it through other methods, this is not something that is going to be exclusively done in the UK, this is going to be taken up in America or wherever else, either now or at some point in the future. ... Who's going to pay for that? Is it going to be private companies who see the long-term game plan that would make billions out of it, and without that private money going in, is it ever actually to get to the point of making it to production? I've no idea how much the Wellcome Trust has, and if they're the only body who are involved in this, but sooner or later surely they're going to say, 'wait a minute we've spent it all', and this is going to run for 20 years until it comes to production. (Sports group)

Views towards the commercialization of cRBCs varied even within groups and individuals. On one hand, there was an attachment to altruistically donated RBCs, and to the blood transfusion services within the UK, with participants feeling that blood was a commodity which should not be commercialized. On the other hand, respondents felt that organizations such as the NHS should not be paying for trials for experimental therapies, and saw the place of commercial companies in bringing a product into clinical use. Although in this case public and charitable funding has financed a large part of the early stage laboratory development, the many levels of clinical trials that will be necessary were seen by respondents as being a large financial burden.

It would appear that whilst having the NHS or blood transfusion services as the developer and distributer would be the choice of participants in an ideal world, the reality is accepted that such organizations do not have unlimited money. This is likely to be partly to do with participants not being familiar with research done within the NHS and therefore believing that any money spent on research would reduce the spending on healthcare. Participants saw funding from organizations such as the Wellcome Trust as an acceptable middle ground, but again thought at some point this pot of money would be spent. To participants, commercial companies were seen as useful in bridging this gap between the end of the public/charitable funding and bringing the cRBC product into clinical use. The Novosang consortium is more open to the idea of commercial involvement and investment, and are actively seeking investment in the developmental processes—many of which would also be applicable to other cell therapies and therapeutics.

Kinship, Cleanliness, and Ancient Concepts

When participants discussed current RBC transfusion they highlighted their appreciation for blood donors, with many participants having received donations themselves or knowing of a close family member or friend who had. The risk of potential TTIs in donated blood transfusions was raised, particularly by multiply transfused patients, but despite the potential risk of TTIs (albeit very small in the UK) there was still great attachment shown to the UK transfusion services and to donated RBC transfusions.

> Having such a low immune system as it was, I would always constantly worry in my head and it was always brought up at meetings with my consultant and stuff. Because I knew for a fact, I think from my school days that 50 percent of hepatitis is contracted through transfusion, so that was always in the back of my mind. Yes, it was a constant worry but there again if I didn't get the transfusions I had no quality of life … so I think the benefits outweighed the risks. (Patient)

Within the context of altruistic donation, participants were happy to receive blood from unknown volunteer donors without knowledge of who the donor was or where the blood had come from (this was despite the fact that in the UK there is a direct link between the donor and recipient, as pooled donations are not used). This, however, changed when the option for commercialization was raised, with participants discussing the possibility of having choice over the donor, or of wanting to know the identity of the donor. Here we see the ontology of blood changing as it moves from an altruistic “gift” to a commercialized product, as one respondent commented:

> That’s what happens when money comes into play, you want more choice, you think you’re entitled to something. If you’ve paid something you can bring it back and say this is unacceptable I want something better. But when it’s free you accept it more. (Bioethicist)

Although this quote was from the bioethicist who was interviewed, similar thoughts were expressed by other groups.

Participants were (not surprisingly) unable to suggest a satisfactory funding model to cover the costs associated with bringing the cRBC product to clinical use, which did not involve commercial involvement. This led to a feeling of inevitability surrounding the production of cRBC, and that if commercial involvement did become inevitable they would “want something for their money”. Having the option to choose the donor was also seen to lead to greater control for recipients of blood transfusions, especially multiply transfused patients.

> I think it’ll give the patient a bit more control, instead of them just turning up at the clinic and being told, ‘here’s your manufactured blood and I hope you feel better’. It might not make a difference at all to the medical procedure itself but I think it might give the patient a bit more control to go, he’s got blond hair I’ll have him. On a psychological note for the patient that could be a good idea. ... I can remember having my first ever transfusion and the next day my skin was very crazy, I broke out in so many spots and I thought that must have been a 17 year old teenager with acne or something, just joking to myself. (Patient)

Two methods were put forward by participants for increasing control or choice over the origins of the cRBCs. One is that, should the cells come from a single donor, then there was a desire to know who that donor was. Producing all the blood from a single donor is, however, seen by the public to be risky (King 2015). Another option raised was for the Novosang team to pursue, for example, six tissue lines from different donors. Recipients could then be presented with a list of characteristics and given the ability to choose which donor line their blood transfusion came from. It was not clear how the commercialization of cRBCs might then react to this aspect of choice, for example, whether blood from certain individuals would be more costly than others, and whether this would reflect the choices made by patients or by scientists about which donors have the "best" blood. This commodification changes the ontology of cRBCs from that of donated RBCs, as they move from a "gift" to a commodity, even though the commercial price for these two types of blood may be similar.

A small number of individuals drew connections between race, blood, and kinship. This included comments about potentially not wanting to receive blood from others of different ethnic or religious backgrounds. There were also references (as in the quote above, although this was somewhat tongue in cheek) to the potential for characteristics of the donor to be transferred in the blood,[3] and by the participant referring to picking donor based on their physical characteristics. However, many of these comments were couched in a rhetoric of "somebody might think", making it hard to tell if these were the views of the participants, who were unwilling to express them directly, or a (potentially incorrectly) perceived notion of what other religions or ethnic groups might think.

M1: Would they be bothered from a racial point of view, is that why they'd be bothered?

M2: I think a lot of people would. A lot of people in northern Europe wouldn't want blood from somebody from somewhere in Africa, even though it might be HIV positive disease free.

M1: What you were saying about northern Europe and Africa, you're probably right, I think this would serve as a good step to eradicate these kinds of differences.

M2: I think it's fine for us to say that but the majority of people, the public who we work for wouldn't want that. I'm not saying that I wouldn't have blood from an African donor or a South American donor, but I think the majority of people in the UK would, the wishes of who we work for, who we serve.

M1: I don't think it's the majority, I think most people would be alright with it. I wouldn't put it as a major concern against this whole thing but we would have to consider it unfortunately, an unpleasant fact.

(Medical Students)

> Apparently even Americans don't want to have our blood because it's impure [due to the possibility of vCJD]. ... It makes you feel devalued as well, and that's something, why on earth if it's good enough for me, it should be good enough for you. (Bioethicist)

The cultured blood product is currently being produced in the UK, with a mainly UK-based research team, and our data did appear to show an attachment to "Scottish", or at least "British" blood. One interviewee (who worked for the SNBTS) referred to previous plasma imports from Germany and the US being discontinued, and that they as an individual felt the public would want a "Scottish" product. This might suggest an attachment specifically to UK blood, rather than simply to blood from developed or European countries, however, more work would be required to clarify this.

The attachment seen both to blood from the UK and to having some choice over the donor of the tissue demonstrates how attitudes towards blood change, as the blood itself moves from being an altruistically donated product to a potentially commercialized one.

Potential Global Production of Cultured RBCs

The ultimate aim of the Novosang team is to produce an infection-free cRBC product that could eliminate the global requirement for RBC donors. Although the cRBC product will not be restricted to the UK, the UK may logistically be unable to produce and distribute the amounts of cRBC product required to replace transfusions on a global scale. The likelihood is that other countries would have to operate some type of central production and distribution system at a country-wide level. Novosang is already gaining intellectual property rights over many of the processes involved in producing the cRBC from iPS cells, which should allow them to retain some future control over the production and distribution of the cRBC product.

Health-care in the UK is paid for by the taxpayer and is free for patients at the point of delivery, a model which does not extend to other countries. Our respondents therefore varied in their views on how the UK should consider exporting this product. There was general consensus that a benefit of the cRBC product was allowing access to safe blood in countries which do not currently have an established transfusion system.

> You were talking about there being countries that don't have transfusion services, this would be a wonderful door to be opened for third world countries for example where people sell blood and damage themselves, this would be in the long term a wonderful thing to happen. (Sports group)

Yet respondents showed a distrust of cRBCs produced overseas, with concern for the iPS cell donors, concern about the potential for inferior products, and concern surrounding a potential monopoly on cRBCs.

Some respondents felt that a single iPS cell donor providing the entire population of cRBCs could be open to abuse in some countries, with donors of iPS cell tissue coerced or encouraged to give tissue without proper consent procedures. At the start of this research process, "The Immortal Life of Henrietta Lacks" (Skloot 2010) had been in the best-seller lists, and some participants raised this as something that could happen with iPS cells.

> It's all ... going to have to come from somebody, it's all consequences of that. The cells that they use in the labs for most research is a cell they call HeLa, and it comes from a black lady's ovarian cancer about 50 years ago and she had her cells taken and she never got any money for it, they did it without her consent, and there's been all sorts of problems from that. I think it would require a lot more control. (Medical students)

Participants were also apprehensive about production of a sub-standard cRBC product in countries with lower safety and quality-control standards than the UK. Concern was not just about the production of inferior cRBCs in other countries, but that these might find their way back into the NHS through procurers who were looking to source the cheapest available product. It would seem impossible to prevent inferior products being produced in other countries, however, one option raised was for regulation of the potential Novosang product, for example through licensing, which would allow it to be produced elsewhere but for the Novosang group to retain a level of control.

Perhaps most importantly for a discussion on the bioeconomy, participants queried the potential mismatch in payment for cRBCs across different global healthcare systems. Respondents felt strongly that cRBCs should be available to all, and highlighted the potential for restrictions on cRBCs which might arise in other countries. For example, countries with privatized health systems might charge patients for cRBCs, and restrict the cRBC product to those who could afford it. As this product is being developed in the UK, and would be available free at the point of the delivery to patients in the UK, it was considered inappropriate that other countries could potentially charge high amounts for cRBCs.[4]

Participants saw global distribution as a benefit, allowing access to safe blood for countries that do not currently have a reliable donation system. There was concern, however, that a two-tiered economy could see cRBCs only available for those who could afford to pay.

Discussion

The development of cRBCs for transfusion seeks to create a supply of donor blood, free from infections and produced under controlled laboratory conditions. Here the views of various publics towards some aspects of cRBCs have been presented, with particular emphasis on how this relates to the potential bioeconomy of cRBCs.

Currently in the UK blood is given altruistically, processed and tested, and then given to patients. The model for cRBCs will involve the donation of tissue by an iPS cell donor (it is expected that this will be an altruistic donation and that the donor would not be compensated), followed by growth, differentiation, and packaging of the cRBC product before it is supplied to recipients. The quantities involved will be significant, with around 2.2 million units a year required for the UK alone. The envisioned aim of the Novosang project is to produce a clinical product that could be used in both the UK and abroad, eventually supplying the world's requirement for RBCs and reducing the need for RBC donors.

Whilst various other technologies have previously been attempted by different research teams, such as chemical alternatives to RBCs, none has so far proved successful. Using in vivo red blood cells as a model to produce laboratory grown cRBCs would appear to be a solution to the forecast deficit in RBC donations, whilst also reducing the potential for transmission infection and making RBCs that offer an improved therapy for multiply-transfused patients. CRBCs offer a significant number of advantages, they offer a potentially limitless source of RBCs suitable for transfusion, whilst adopting the existing infrastructure for storage and delivery of transfusions to patients. Unlike other technologies (such as stripping RBCs of their antigens) which would still require donors, cRBCs require only the initial donation of an iPS cell sample.

The funding model for the scale-up and testing of cRBCs is still unclear, as it seems likely that the high costs for clinical trials will be unable to be met by the NHS or the current public and charitable funding of Novosang. The Novosang consortium are currently pursuing funding from research organizations, but some degree of commercial involvement is likely to be required in order to bring this product to clinical use. The delivery of both current donated RBCs and cRBCs require processes that entail costs, but may differ considerably in the distribution of these costs. We see then that cRBCs are both a new economy and a revealing of a hidden bioeconomy, requiring a re-interpretation of how blood is viewed within an altruistic donation system. In the data presented above the ontology of RBCs is seen to change as these cells move from altruistic donations into a more commercialized space.

The altruistic donation system in the UK appeared to mask the true costs associated with procuring, testing, and delivering RBC transfusions. Within the altruistic donation system of the UK, blood was viewed by the participants as a "free" product, with many unaware that transfusions contain only RBCs and expecting that whole blood is taken from the donor and delivered to the recipient with minimal intervention. This demonstrates that there is currently a "hidden" bioeconomy of blood production which conceals the true cost (both in money and effort) of producing blood products. Participants desired cRBCs to retain the same ontology as donated RBCs, and to be distributed by the NHS or research funders rather than commercial producers. There was a fear that commercialization would create a market in which only some people could afford to have access to blood, which was seen as going against the current nature of blood donations as a "gift" that is available to all.

The desire to know who the donor was and not to have blood developed in certain areas of the world reflected varying levels of trust, in both the donors and the producers of cRBCs. The attachment to "home-grown" blood in the UK largely stemmed from the trust put in the current producers of transfusions and the strict regulations that govern the blood transfusion services, as the view that commercial producers would cut corners in the pursuit of profits also extended to scientists in commercial companies within the UK.

Conversely participants were much more accepting of commercial involvement in pharmaceuticals, seeing pharmaceutical companies as providing a service in developing drugs, which would be beyond the means of the NHS and publicly funded research organizations. It was not the commercial model in itself that appeared to be an antithesis to our participants, but rather the changing ontology of RBCs as they move from an altruistic product to a commercialized one. A similar change in ontology was seen by Stephens and Dimond (2015) during the closure of a biobank, as the nature of the tissue held within changed during the process of removal to other storage options, moving through various changes in "value" from being a potential waste product to finding another, useful, home in other biobanks. The distrust of commercial producers (which was not such a concern regarding pharmaceutical products) stemmed from this desire for cRBCs to retain the same ontology as donated ones, and from trying to fit the altruistic donation model into a potentially commercialized market.

Participants constructed a commercial imperative for cRBCs. Despite the majority of participants preferring that the cRBC product be developed by a not-for profit there was a reluctance to use NHS money to fund clinical research trials. This then presented a challenge to discussions about who should develop the cRBC product, with many feeling that the inclusion of commercial producers became a necessity, with no other obvious source of funding available. This recognition of commercial involvement in pharmaceuticals shows that where blood is seen to sit on the commercial/altruistic continuum is important for acceptance of potentially commercially produced cRBCs. This echoes the earlier work of Ferguson et al. (2008) in calling for positive "message framing" in an attempt to increase how safe the public view blood substitutes to be.

Blood was seen within the current altruistic donation system as a gift, with participants expressing gratitude for the donors and the work of the transfusion services. Participants were attached to the altruistic donation system, despite the potential for TTIs in blood that comes from human donors. This "specialness" of blood has been seen elsewhere (Douglas 1966; Tucker 2011; Titmuss 1970). As in the very early transfusion experiments, some participants articulated the feeling that "elements" (over and above potential communicable diseases) are somehow contained within the blood.

One participant referred to the idea of the laboratory space as a black box, when discussing cRBCs, but felt that the standard donation system was better understood. This is especially interesting given that evidence from the wider public outreach events carried out by Novosang has found that most people do not know that blood transfusions contain only RBCs, so the current transfusion system may not be as well understood as this participant felt. This was in contrast to the view of many of the scientists, who saw the laboratory system as being clear and controlled, whilst the donors themselves represented an unknown space from which it was hard to control the spread of TTIs.

Case studies such as the development of cRBCs help us to explore and study people's attitudes and understanding about the bioeconomy. Where potential clinical uptake could be influenced by public perceptions of the new cRBC product it is increasingly important to understand the views of potential users. The data presented here show that the changing ontology of RBCs is seen as problematic by participants, which suggests that making cRBCs a bioeconomic success may require a marketing strategy that distances cRBCs from donated RBCs by portraying the former more as a pharmaceutical product and removing the "specialness" attested to in current literature around blood donation and altruism.

Conclusion

Stem-cell derived RBCs could replace the established practice of transfusing RBCs from human donors. This project set out to examine the views of various publics towards the use of cRBCs. Findings show that the ontology of RBCs changes as they move from an altruistically donated product to a potentially commercialized product. Whilst respondents were willing to accept an unknown altruistic donor, there was a call to know more about the tissue source for commercially produced cRBCs.

While there was a distrust of commercial producers, and a preference for cRBCs made by NHS or charitable funding, respondents articulated a commercial imperative due to a perceived lack of alternative funding models for clinical trials and development of cRBCs. There was a difference in attitude shown toward, on the one hand, blood, with an attachment to the donated product, and on the other hand, pharmaceuticals, where commercial operators, who already dominate, were accepted. A cRBC bioeconomy, therefore, may require the unmaking of the notion of the "specialness" of blood in order that societal attitudes become compatible with a more pharmaceutical and commercialized model of blood production and distribution.

Notes

1. http://novosang.co.uk/.
2. http://www.nhsbt.nhs.uk/news-and-media/news-articles/news_2015_07_10.asp. It should be noted, however, that if the full infrastructure required within the NHS to administer blood is taken into account, the full cost of a unit of blood is much higher, an estimated £635 per unit in 2001 (Varney and Guest 2003), and more likely to be around £800 per unit today.
3. This harks back to early transfusion experiments, where it was accepted that behavioural characteristics were contained within the blood; many experiments took place transfusing the blood of docile animals (such as lambs) into humans, with the hope of curing madness (Brown 1948; Tucker 2011; Lower 1665; Denis 1667).
4. It is interesting to note that assisted reproduction, with the selling of sperm, eggs, and embryos raises similar arguments (Klitzman and Sauer 2015).

References

Brown, H. (1948). Jean Denis and transfusion of blood, Paris, 1667–1668. *Isis, 39*, 15–29.

Bryman, A. (2004). *Social research methods.* Oxford: Oxford University Press.

Cohn, C. S., & Cushing, M. M. (2009). Oxygen therapeutics: Perfluorocarbons and blood substitute safety. *Critical Care Clinics, 25*(2), 399–414.

Denis, J.-B. (1667). A letter concerning a new way of curing sundry diseases by transfusion of blood. *Royal Society.*

Douay, L., Lapillonne, H., & Turhan, A. G. (2009). Stem cells—A source of adult red blood cells for transfusion purposes: Present and future. *Critical Care Clinics, 25*, 383–398.

Douglas, M. (1966). *Purity and danger.* Abingdon: Routledge.

Ferguson, E., Prowse, C., Townsend, E., Spence, A., Van Hilten, J. A., & Lowe, K. (2008). Acceptability of blood and blood substitutes. *Journal of Internal Medicine, 263*, 244–255.

Fleming, P., Ferguson, E., Townsend, E., & Lowe, K. C. (2007). Perceptions in transfusion medicine: A pilot field study on risk and ethics for blood and blood substitutes. *Artificial Cells, Blood Substitutes, and Immobilization Biotechnology, 35*, 149–156.

Fontana, A., & Frey, J. H. (1994). Interviewing: The art of science. In N. K. Denzin & Y. S. Lincoln (Eds.), *Handbook of qualitative research.* London: Sage.

Goven, J., & Pavone, V. (2015). The bioeconomy as political project: A polanyian analysis. *Science, Technology & Human Values, 40*(3), 302–337.
Grethlein, S. J., & Rajan, A. (2012). Blood substitutes. *Medscape Online*. http://emedicine.medscape.com/article/207801-overview. Accessed April 2017.
Henkel-Honke, T., & Oleck, M. (2007). Artificial oxygen carriers: A current review. *AANA Journal, 75*, 205–211.
Kent, J., Faulkner, A., Geesink, I., & Fitzpatrick, D. (2006). Culturing cells, reproducing and regulating the self. *Body & Society, 12*, 1–23.
King, E. (2013). *Blood, sweat and tears: A case study of the development of cultured red blood cells for transfusion*. Ph.D, Edinburgh: The University of Edinburgh.
King, E. (2015). Biodiversity and perceptions of risk: Reactions to the use of a single donor for stem-cell-derived red blood cell transfusions. *Journal of Medical Law and Ethics, 3*(3), 151–163.
Kitzinger, J. (1995). Qualitative research: Introducing focus groups. *BMJ, 29*, 311.
Klitzman, R., & Sauer, M. V. (2015). Creating and selling embryos for "donation": Ethical challenges. *American Journal of Obstetrics and Gynecology, 212*(1), 167–170.
Krueger, R. A. (1988). *Focus groups: A practical guide for applied research*. London: Sage.
Lower, R. (1665). The success of the experiment of transfusing the blood of one animal into another. *Philosophical Transactions of the Royal Society B: Biological Sciences, 1666*(1), 352.
Lyall, C., & King, E. (2016). Using qualitative research methods in biomedical innovation: The case of cultured red blood cells for transfusion. *BMC Research Notes, 9*, 267.
Martyn, V., Farmer, S. L., Wren, M. N., Towler, S. C. B., Betta, J., Shander, A., et al. (2002). The theory and practice of bloodless surgery. *Transfusion and Apheresis Science, 27*, 29–43.
Mittra, J. (2016). *The new health bioeconomy*. US: Palgrave Macmillan.
Mittra, J., Tait, J., Mastroeni, M., Turner, M. L., Mountford, J. C., & Bruce, K. (2015). Identifying viable regulatory and innovation pathways for regenerative medicine: A case study of cultured red blood cells. *New Biotechnology, 32*, 180–190.
Mountford, J. C. (2008). Human embryonic stem cells: Origins, characteristics and potential for regenerative therapy. *Transfusion Medicine, 18*, 1–12.
Mountford, J. C., & Turner, M. (2011). In vitro production of red blood cells. *Transfus Apheresis Science, 45*, 85–89.
NHS Blood and Transplant. (2010). The future of unrelated donor stem cell transplantation in the UK: Part 1. UK Stem Cell Strategic Forum.
NHS England and Monitor. (2016). *2016/17 National Tariff Payment System*. www.gov.uk/monitor.

Penrose, G. (2015). *The Penrose inquiry final report.* Edinburgh: APS Group Scotland.

Schneider, W. H. (2003). Blood transfusion between the wars. *Journal of the History of Medicine, 58*, 187–244.

Silverman, T. A., & Weiskopf, R. B. (For the planning committee). (2009). Haemoglobin-based oxygen carriers: Current status and future directions. *Anesthesiology, 111*, 946–963.

Skloot, R. (2010). *The immortal life of Henrietta Lacks.* London: Pan Macmillan.

Starr, D. (1999). *Blood: An epic history of medicine and commerce.* Boston, MA: Little, Brown and Company.

Stephens, N., & Dimond, R. (2015). Unexpected tissue and the biobank that closed: An exploration of value and the momentariness of bio-objectification processes. *Life Sciences Society and Policy, 11*, 14.

Strauss, A. L., & Corbin, J. (1994). Grounded theory methodology. In N. K. Denzin & Y. S. Lincoln (Eds.), *Handbook of qualitative research, 17*, 273–285. Thousand Oaks: Sage.

Titmuss, R. M. (1970). *The gift relationship: From human blood to social policy.* London: LSE Books.

Tucker, H. (2011). *Blood Work: A tale of medicine and murder in the scientific revolution.* New York: Norton.

Varney, S. J., & Guest, J. F. (2003). The annual cost of blood transfusions in the UK. *Transfusion Medicine, 13*, 205–218.

Waldby, C. (2002). Stem cells, tissue cultures and the production of biovalue. *Health, 6*, 305–323.

Webster, A., Haddad, C., & Waldby, C. (2011). Experimental heterogeneity and standardisation: Stem cell products and the clinical trial process. *Biosocieties, 6*(4), 401–419.

Wield, D., Hanlin, R., Mittra, J., & Smith, J. (2013). Twenty-first century bioeconomy: Global challenges of biological knowledge for health and agriculture. *Science and Public Policy, 40*(1), 17–24.

Author Biography

Dr. Emma King trained in genetics before moving into the area of Science and Technology Studies. Her PhD, and subsequent research project, focused on the development of cultured red blood cells. Emma is now based in the Nursing, Midwifery, and Allied Health Professions Research Unit (NMAHP-RU) at the University of Stirling, where she works on the delivery of complex health interventions.

CHAPTER 14

Conclusion

Vincenzo Pavone and Joanna Goven

In the introduction to this book, we noted our discomfort with the term bioeconomy. The intended performativity of the concept—as a unique solution to urgent problems, for which we must allegedly make changes to democratic and scientific institutions, expand the role and rights of private capital, and restrict the rights of citizens—is itself an argument against its use. At the same time, it is increasingly adopted in academic studies as a way of pointing to the developing interrelations of life sciences and capital.

We continue to caution against the unquestioning adoption of "the bioeconomy" as if it were a politically innocent designation of a particular set of economic (and/or scientific) activities. At the same time, it is essential to investigate how, and with what effects, existing biosciences and biotechnologies are intertwined with the power of capital and with modes of governance. It is also essential to track the activities to which the bioeconomy concept and its associated assumptions and

V. Pavone (✉)
Institute of Public Goods and Policies (IPP), Consejo Superior Investigaciones Científicas (CSIC), Madrid, Spain

J. Goven
Kukupa Research, Pigeon Bay, New Zealand

V. Pavone and J. Goven (eds.), *Bioeconomies*,
DOI 10.1007/978-3-319-55651-2_14

strategies give rise. In order to signal that this is the terrain explored by this book, we have used the plural term "bioeconomies".

The chapters contained herein have revealed a wide-ranging, context-dependent diversity of activities carried out on this terrain. At the same time, certain themes have emerged repeatedly from these different contexts. Here we draw out these themes and discuss their implications for future research.

Remaking Science for the Bioeconomy

What is the bioeconomy project doing to, as well as with, science? And are scientific practices articulated and operating differently in different bioeconomies? A number of chapters in this book portray a promissory-bioeconomy-inspired reconfiguration of relationships among science, the state, and industry (see especially Chaps. 2, 3, 4, 7, 8, 13).

As these chapters have shown, various government-supported initiatives are reshaping, or attempting to reshape, scientific practices in the name of harnessing scientific research for economic growth and global economic competitiveness. Within major initiatives in the health bioeconomy, science is being reconfigured as innovation, while innovation is equated with commercialization (Gardner, Webster, and Mittra, this volume). In the regenerative medicine space, the UK government has funded an industry-guided and commercialization-oriented innovation agency and an interdisciplinary research platform, both aimed at reconfiguring the biomedical research landscape in the direction of closer collaboration among clinicians, researchers, and industry (Chap. 2). In the Innovative Medicines Initiative, academic-industry collaboration is structured so as to empower pharmaceutical companies to set the research agenda as well as the epistemic standards around which the research is organized (Chap. 4). Scientific aims and practices are coming into conflict with, and being subordinated to, industry expectations: for example, rather than striving for epistemic novelty (scientific innovation), under the banner of innovation(-as-commercialization), scientists are expected to assist industry in generating tangible outputs, through highly standardized systems of production (Chap. 4). Industry practices with regard to intellectual property rights and access to data (which claim more access and ownership for industry) are contesting and sometimes displacing those prevalent among scientists involved in the initiative. More generally, publicly funded researchers are often expected to produce results that will attract industry investment (Chaps. 2, 4, 13).

As Tupasela notes (Chap. 8), state authorities across Europe are seeking new ways in which market forces and commercialization imperatives can be introduced into the scientific research system. With regard to biobanks, it is somewhat paradoxically translated into an open-science, sharing imperative; this, however, is an enabling strategy for future commercialization: such biobanks are viewed as a "gold mine" from which, ideally, many will obtain valuable data that will underpin commercializable innovations.

While it is clear that major bioeconomy initiatives are seeking to reshape science around the imperatives of commercialization and economic competitiveness, the question of how exactly these efforts are shaping scientific practices needs additional investigation. It cannot be assumed that the effects are uniform, or necessarily anticipatable in advance. The studies contained in this book point to the importance of tracking in what directions and to what degree the bioeconomy project is reshaping the focus, methods, career structures, and routine practices of science.

Doezema and Hurlbut provide a different take on the role of science in the bioeconomy. In their account of the AquaAdvantage salmon as an instantiation of the bioeconomy project, they describe an imaginary of innovation-as-governance that calls for a partnership between science and the state, in which "science assumes the authority to envision and realize technological futures" and is granted a privileged agenda-setting role in governance (Chap. 3). This calls our attention to the degree to which developments in science, and their public interpretation and promotion by scientists, underpin and inspire bioeconomy initiatives. At the same time, other studies presented here (Chaps. 2, 4, 13) show "science" as enacted in various bioeconomies to be constrained and directed by the interests of industry; the state's role, rather than as a governing partner with science, is to facilitate the governing role of industry in these "scientific" initiatives. In still others (Chaps. 7, 8), the state directs science to adapt its practices to facilitate state visions of innovation through eventual commercialization. If, as Gardner, Webster, and Mittra argue, innovation is equated with the commercialization of scientific discoveries, then an imaginary of innovation-as-governance will grant a governing role to those to whom commercialization is entrusted. While bioeconomy strategies, as Doezema and Hurlbut note, construe (many) bioeconomy product markets as something to be actively constructed in the future, market actors—in the sense of powerful industry actors and interests—are present from the start, playing an agenda-setting role long before markets for their new products exist. These studies, taken together, raise the question of whether the epistemic authority of science can continue to be a legitimating resource as science is increasingly subordinated to industry interests.

Science as a Political Project

Many of the chapters in this book provide insights into not only how the practices we call science are being remade, but also how that re-making is implicated in a particular imagining of political communities. As a number of chapters demonstrate, in the bioeconomy science becomes a constituent element of national and supranational political projects.

Marelli and Testa provide a detailed analysis of how the bioeconomy project's promise to address "societal challenges" while generating economic growth can be deployed to construct, integrate, and legitimate political communities. Their study of the European Union's Innovative Medicines Initiative (IMI) elucidates it as an instantiation of the European Commission's vision of the bioeconomy, which serves as "the lynchpin around which to consolidate the process of European integration and re-orient the European polity along the lines of competitiveness and innovation" (Chap. 4). In a narrative that locates "Europe" as a competitor in a global economic race, the alignment of European research efforts around the bioeconomy and its priorities becomes the solution to major European societal challenges, which are framed in terms of mutually reinforcing goals of economic competitiveness and sustainability. In the health domain, this translates into "positioning Europe as a leader in the rapidly expanding global markets for health and wellbeing innovations" (IMI, in Chap. 4). This narrative thus constructs "European" problems for which a consolidated, integrated European-Union bioeconomy effort—aligning research, institutions, priorities, and policies—is the solution. What kind of political community is constituted by this co-productive process? In Marelli and Testa's analysis, it is one in which inequalities between center and periphery are exacerbated and which, through the governing role granted to the pharmaceutical industry and its norms, threatens to undermine public accountability.

Given its discourse of present urgency and great future promise and its projection of "global economic competitiveness" as an unquestionable priority for any political community (see Chap. 1), we should not be surprised to find that the bioeconomy project can have anti-democratic effects. Similar dynamics have also been identified and discussed by Gardner, Webster, and Mittra (Chap. 2, this volume). Here it is the UK's identity as a "global leader" that the health bioeconomy promises to secure. Again, industry interests and priorities are given a governing role, not only in the investment of public funds, but also in the restructuring of research, the reform of regulatory frameworks, and the reframing of the NHS from a public health service to a facilitator of industry innovation. The political role of the bioeconomy here can be understood as legitimating and consolidating an on-going neoliberal transformation of the relations between the public and private sectors and between the nominally democratic state and its citizens.

The reconfigured role of democratic citizens is highlighted in Doezema and Hurlbut's account of the AquaAdvantage salmon as part of a broader bioeconomy project (Chap. 3). Here bioscientists, attributed with superior ability "to see ahead into a future where particular technological potentials are realized and specific urgent needs are thereby met", are granted a kind of paternalistic power over a citizenry that, without such an ability, is not able to act in its own best interests. This is played out in the determination of criteria to be applied in regulatory decision-making, in which citizens' knowledge and concerns, if deemed "unscientific", are rendered illegitimate. Thus, in Doezema and Hurlbut's account, the bioeconomy project engenders a polity in which "science" is empowered at the expense of the democratic rights and powers of citizens.

Delvenne's chapter illustrates quite a different instrumentalization of bioeconomic visions for political purposes. He conceptualizes the bioeconomy in Argentina as an act of political imagination through which political elites try to "conclusively transcend a historically entrenched dichotomy between agriculture and industry" (Chap. 10). The expansion of GM soy production, which had occurred partly in response to the crisis-induced economic restructuring of the 1990s, made possible the Peronist Kirchner governments' "export-oriented populism": by taxing soy exports, they were able to finance a range of public services without either unmaking the neoliberal restructurings of previous governments or eroding the living standards of the urban population through consumption taxes. The continual ratcheting up of these export taxes, however, eventually generated a political crisis of a surprising nature: anti-government protesters united across the traditional opposition between rural interests (formerly identified with an agrarian bourgeoisie) and those of an urban industrial working class (the traditional constituency of Peronists). The vision of an agro-industrial bioeconomy enabled Cristina Kirchner's government not only, in a way familiar from other studies, to (appear to) address its economic crisis by enhancing economic competitiveness, but also to address its political crisis by demonstrably "mov[ing] beyond the idea of agriculture as a backward, elite rent-seeking sector" in order to (attempt to) forge a new political coalition. Unsurprisingly, this competitiveness-oriented bioeconomy vision has proved sufficiently plastic to continue to be deployed as a legitimating strategy by the recently-elected conservative successor to the Kirchner government.

In van Wichelen's chapter we are shown how particular bioeconomies, as opposed to an envisioned bioeconomy project, can be deployed in the interest of policing the nation as a particular kind of imagined community (Chap. 9). In this case, an immigration bioeconomy, in conjunction with discriminatory and differentiated legal constructions of kinship, is facilitating an unacknowledged recapitulation of the "White Australia" imaginary, which excludes particular groups of non-whites from membership in the political community. Work in STS has long shown that science and technologies "have politics" (Winner 1980) and van Wichelen's chapter is a forceful illustration of this.

Bioeconomy Tales and the Generation of Value

The generation, distribution, and appropriation of value, especially economic value, in the bioeconomy currently stand at the heart of a flourishing debate. While the debates on economic value usually revolve around the actual sources of value, contributions informed by valuation studies tend to broaden the focus to the complex articulation between economic value and other types of value.

Several chapters in this book cogently address this articulation. They show both that value(s) can be derived from different sources and that the ways in which bioeconomies extract value(s) also vary, according to the type of bioeconomy, the technology at stake, the local regulatory regime, and the local epistemological culture. Value of various kinds can indeed be extracted, in various ways, from bodily tissues and living organisms in, for example, cord-blood banking (Chap. 11) and fertility services (Chaps. 6, 12), as well as in the imagined cultured blood product industry (Chap. 13) and the GM feed and food market (Chaps. 3, 10).

It can also be generated from data extracted from—and paired with—bodily tissues and waste, as in the biobanking sector (Chap. 8) and the Underworlds project (Chap. 4). In the cord-blood and mastitis-treatment bioeconomies (Chap. 11), as well as in the reproductive bioeconomies based on gamete procurement and surrogacy (Chaps. 6, 9, 12), extracting value from bodily tissues is also inextricably linked to clinical and reproductive labor, and to the capitalization of altruism. Indeed, in these bioeconomies the creation and articulation of a sociotechnical value chain, i.e., an organized series of economic, social, normative, and technological steps that are necessary to arrange the operational success of reproductive markets, seems to play a constitutive role (Chap. 6).

However, economic value is not necessarily linked to bodily tissues or cells: it may also proceed from the commercial exploitation of medical and biological data (Chaps. 5, 7, 8). The microbiome story told by Reis Castro is, in this respect, quite instructive. Microbiomes can be subjected to the same sort of process that Waldby and Mitchell (2006) described in relation to tissues, gametes and embryos. Data proceeding from waste become a source of multiple values: medical value in the sense of knowledge that can be used to protect public health; academic and epistemic value resulting from the various research lines pursued through the data; status and economic value to the institution (MIT) hosting a high-profile, media-friendly project that has also attracted a well-funded partnership with Kuwait; and finally, a projected market value, as this freely "donated" data has the potential to be enclosed through intellectual property rights and sold or licensed. While the proprietary status of the waste microbiome remains controversial, such enclosure has become the norm in health bioeconomies: what is configured as waste can and should re-enter the economy as private property, with no benefits shared with the producers of the "waste". In this case, in contrast to tissue bioeconomies, what is being appropriated is already widely regarded as both waste and not-human. As we see also in the biobanking sector, like bodily tissue, data have to be framed and conceptualized as either "waste" or as "donation" in order to pre-empt proprietary claims from original owners and allow for appropriation (and exchange or rent-gathering) by researchers or a commercial partner.

Other chapters have also shown that value can be produced through the setting up of promissory regimes in which financial and scientific speculation jointly operate to attribute value to on-going research on future and emerging technologies and treatments. This is well illustrated in chapters on initiatives framed by the "health and wealth" discourse (Chaps. 2, 4). In highly speculative sectors, such as regenerative medicine, enclosure through intellectual property rights also plays a role in the generation as well as appropriation of value. However, in contrast to perspectives that argue that the economic value in the bioeconomy is derived primarily from these proprietary rights to extracted knowledge, Aarden's focus on social contracts (Chap. 7) shows how the biological itself still matters—precisely because it generates other forms of value than those amenable to exchange in a market.

These chapters highlight the degree to which value generation and appropriation depend on the existence, or creation, of trust: among researchers, between researchers and research subjects, or between researchers and commercial actors. This suggests that the relational work involved in building and maintaining trust ought to be considered by those theorizing the contribution of labor to the production of value in bioeconomies.

Some chapters explicitly address the complexity of the relationship between economic value and other values, as well as the situated, political articulation of the valuation process (Chaps. 2, 7, 8). Recent studies in valuation practices have challenged the distinction between economic value and other types or regimes of value. As a result, valuation studies suggest focusing on how different actors frame and pursue what they count as desirable or valuable. Given that any common project may enact several different values at the same time, there is always the potential for conflict among them. In their chapter, Gardner, Webster, and Mittra argue that the collaboration of diverse actors in a complex project such as the regenerative medicine initiative is brought about through appealing to a number of different types of value: clinical value, path-finding value, safety value, scientific value, global competitiveness value, and commercialization value. How the conflicts among these different values are negotiated becomes a key determinant of the success of such initiatives.

Importantly, Aarden's chapter (Chap. 7) reminds us that the process of value-generation is not value-neutral: it always endorses specific ways of distributing benefits and risks, of prioritizing or marginalizing different interests and needs. Utilizing the framing device of a social contract, his study shows, first, how the expectations of biomedical research go well beyond the domain of biocapital: biomedical research can be expected to be economically beneficial, as in the Singapore Tissue Network, but also to inform the development of new health policies, as in the Million Death Study, or to provide key epidemiological information to inform individual behavior, as in the Framingham Heart Study. Second, the social contract approach forces us to consider that the production of biovalue does not benefit equally all sectors of the population: power and its social priorities are indeed reflected in the organization and effects of biomedical research.

Finally, Tupasela's study (Chap. 8) shows particularly clearly how the drive for economic or commercialization value can clash with other valued things: for example, scientific discovery or reputation and career advancement. Such clashes, shaped by underlying contexts of competition for scarce resources and workplaces governed through New Public Management, reflect contradictory logics that can derail the generation of value; such contradictory positionings of bioeconomy actors thus deserve greater attention in studies of bioeconomies.

Inequality

The studies in this book have shown bioeconomies to be productive of inequality in a number of ways. As part of the political project outlined in bioeconomy strategies, bioeconomy initiatives regressively redistribute income and investment, as well as occluding existing inequality and distracting from the need to address it (Goven and Pavone 2015). Particular bioeconomies examined in this book also mobilize, and sometimes reinforce, existing inequalities in order to extract labor and tissue and to sell products.

As we have seen particularly in relation to major health-bioeconomy initiatives, (some) bioeconomies are likely to have a regressively redistributive effect. As we noted in Chapter One, bioeconomy strategies call for a particular set of policy responses, including various kinds of public subsidies of private bioeconomic activity. Examining the case of government initiatives in the field of regenerative medicine in the UK, Gardner, Webster, and Mittra have shown that their likely redistributive impact will be to increase inequality. The acknowledged role of publicly funded initiatives in this area—whether the Cell and Gene Therapy Catapult or the UK Regenerative Medicine Platform—is to "de-risk future commercial investment". Public investment is expected to be translated into private profit, in this case the profits of large pharmaceutical companies. Moreover, the publicly funded National Health Service (NHS) would be a major purchaser of these privately-owned therapeutic products.

The Novosang project featured in King's chapter appears to be operating under similar expectations of public subsidy, de-risking, and publicly-funded purchase (Chap. 13).

With no, or inadequate, provisions made to recoup this significant public investment from its private beneficiaries, the result is a socialization of risk (and cost) and the privatization of rewards. This has a significant and regressive redistributive effect, converting taxes paid by the population as a whole into profits that will mostly benefit the wealthiest. As the authors of the regenerative medicine study note (echoing Mazzucato), the effect of this is to exacerbate inequality (Chap. 2). Similar dynamics of regressive wealth transfer can be seen at the European level in the Innovative Medicines Initiative, as outlined by Marelli and Testa (Chap. 4). The degree to which these practices have become normalized is perhaps reflected in the acceptance by the participants in King's study, through analogy with existing pharmaceutical-sector practices, of the publicly funded cRBC project being capitalized upon by private biomedical companies (Chap. 13).

Beyond this, large-scale public investments in activities meant to generate private profits entail significant opportunity costs, which are themselves likely to exacerbate inequality. As Gardner, Webster, and Mittra note, these investments are occurring at a time when public funding is being withdrawn from social services and the NHS is grossly under-funded. It is likely that a much greater, and more equitably distributed, health benefit would be generated by investing more in those services; moreover, given the mutually reinforcing effects of poor health and poverty, it is likely that a lack of investment in those services is exacerbating structurally determined health inequalities, while the under-funding of public health care may exacerbate economic inequality. A similar argument can be made at the European level, where massive investment in bioeconomy initiatives is carried out alongside the imposition of austerity policies that are directly and negatively impacting the wellbeing of many Europeans.

The enthusiasm for biomedical innovation and its promissory claims that is characteristic of medical bioeconomies may also occlude and distract from other, more fundamental causes of ill health. As Aarden has observed, equating improvements in health with biomedical innovation can act to "obscure the importance of structural dimensions in health, and of coordinated collective action as a way of improving population health" (Chap. 7). The search for genetic "risk factors"—as in the Framingham Heart Study and many of the biobanking initiatives discussed in this book—shifts focus and resources away from addressing those structural dimensions, that is, away from the health problems generated by social inequality. The Innovative Medicines Initiative, constructed as a way of "innovating Europe out of the [economic] crisis" (Chap. 4), occludes the underlying demand-side causes of that ongoing crisis, argue Marelli and Testa, thus distracting from the need to address wage stagnation, unemployment, and rising inequality.

All of this suggests that those studying bioeconomies associated with promises of "health and wealth" ought routinely to ask: whose health and whose wealth? Indeed, Aarden's use of the social-contract metaphor led him to ask a similar question: for whom does biomedical research produce value, and who is excluded from this process?

A number of chapters illustrate ways in which bioeconomies rely upon, mobilize, and reinforce existing inequalities. Building on theories of clinical labor, Lafuente Funes has clearly shown how the private reproductive bioeconomy in Spain capitalizes on the growing desperation of those badly affected by years of austerity policies, while at the same time working to obscure this dependency by deploying norms that construct women as altruistic and self-sacrificing (Chap. 11). Similar gendered norms, this time representations of the "good mother", are mobilized in another bioeconomy of parenting: the market for new probiotic products to treat mastitis capitalizes upon and reinforces the, often unacknowledged, self-sacrifice expected of mothers in order to ensure their children have the benefits of breastfeeding (Chap. 12). Like the reproductive bioeconomy, cord-blood banking also mobilizes economic inequality, though in a different way: as Santoro and Romero-Bachiller demonstrate, Spanish private cord-blood banks not only activate relations of care (of parents toward children), but also mobilize a logic of distinction and "exclusiveness"—that is, capitalize on inequalities of class—to sell their services (Chap. 12). The effects of this extend to the excluded: the authors also point to how it can lead to distress among those unable to afford what is presented as an expression of care.

In the migration bioeconomy, as analyzed here by van Wichelen (Chap. 9), claims of scientific authority and precision are deployed to enable a superficially neutral but deeply discriminatory border policing, which reinforces global structural inequalities. These inequalities, which underlie both international migration and international surrogacy, are, as in the health domain, obscured by a molecular gaze that sees only the genetic attributes of individuals. Indeed, the prospect of such global inequalities being both exploited and exacerbated by commercial producers of cultured red blood cells was identified by the participants in King's study as a significant concern (Chap. 13).

Relations and Subjectivities

The chapters in this book have revealed a wide variety of ways in which bioeconomies construct and mobilize relations. As described above, bioeconomic endeavors often involve the deliberate creation of (closer) relations between academic scientists and private capital. They also involve the construction and negotiation—or in some cases disentanglement and denial—of relations between scientists and the subjects of their research; indeed, as Aarden's example of the Framingham Heart Study shows, this relationship can materially shape whether and how the research can be capitalized upon. Beyond these and other relationships characterizing research, the studies in this book have revealed a number of ways in which relations, and the subjectivities they entail, are fundamental to a variety of bioeconomies.

The bioeconomies studied by Santoro and Romero-Bachiller depend for their economic success on the mobilization of relations of care: private cord-blood banks mobilize parents' and grandparents' feelings of protection toward (and anxieties regarding) their children and grandchildren; similarly, the market for new mastitis treatments trades on mothers' desire to give their children the healthy start that breastfeeding represents and on their willingness to endure considerable pain to do so (Chap. 12). Santoro and Romero-Bachiller argue that these participants in health bioeconomies are much more ambivalent subjects than either the empowered biological citizen or the exploited clinical laborer suggest; rather, networks of affect and relationality, which are also permeated by social inequalities, produce deeply ambivalent and complex subjectivities.

The reshaping of subjectivities plays a particularly enabling role in reproductive bioeconomies. Lafuente Funes describes the ways in which (potential) egg donors are encouraged, or even required, to see themselves as altruistic, solidaristic subjects (Chap. 11). In contrast, as Pavone shows in in his study of the Spanish reprogenetics economy, clients of pre-implantation genetic screening services are constructed as rationally-individualistic subjects: genetic selection is framed as an act of (reproductive) autonomy by prospective parents addressed as consumers (Chap. 6).

Family relations exist in law as well as in affect; van Wichelen (Chap. 9) demonstrates that the legal constitution of family relatedness is a crucial enabler of transnational surrogacy, helping to create and legitimate new types of families. But when law and biotechnology are deployed at the border, they enable the imposition of much more restrictive criteria, which, rather than producing new families, deny and thereby break apart existing familial relations. Kinship here is subject to the interplay of politico-legal practices and an actively promoted "security" bioeconomy, in a context of extreme inequalities of power. This configuration leads van Wichelen to challenge the notion of biological citizenship. Rather than providing a basis for the formation of new bonds of community and for the extension of rights, the application of genetic technologies here is profoundly individualizing and disempowering. Seen through the lens of migration, "biological citizenship" appears as the domain of the privileged. Rather than cross- (or under-)cutting categories of race, class, and nation and engendering new communities, genetic knowledge becomes a tool to reinforce (while obscuring) discrimination based on those categories, to fracture existing relations, and to forbid membership in national communities: "genetic relatedness, rather than helping to extend rights, is actually restricting rights of citizenship" (Chap. 9). From this perspective, biological citizenship becomes an additional privilege of those with the resources to successfully lobby states for extensions of rights and benefits.

Conceptions of race are also at work in other bioeconomies as well. As Lafuente Funes indicates, fertility-clinic practices are guided by a kind of racial ideal—which may be embraced by clinic staff or simply imputed to potential clients—and by a kind of racial aggregation of the "other" that results in a different kind of service being provided to non-white clients (Chap. 11). The participants in King's study imputed to (white) others an unwillingness to receive blood products derived from someone "racially" unlike themselves (Chap. 13). We can expect health bioeconomies to continue to have to negotiate—accepting, contesting, reinforcing, mobilizing—racial (and racist) identities and subjectivities; these negotiations will have wider social effects.

Reis Castro's study points to a very different process of constituting communities: as a collective microbiome (Chap. 5). As in the case of international migration, this is not an act of self-determination, but rather a construction imposed from outside. The Underworlds project suggests that constituting a community as a collective microbiome will mean constructing it as more or less ill, more or less criminal. This not only renders the community as an object for external intervention; it also raises the question of how such a construction will shape the subjectivities and relations of the community members.

As van Wichelen's study clearly shows, bioeconomies not only create new relations and mobilize (and capitalize upon) existing ones; they also unmake relations. But this is not limited to familial relations, nor is it limited to humans. Delvenne's study (Chap. 10) outlines a financialized agricultural bioeconomy that depends fundamentally on breaking the relationship between the farmer and the land. A soy bioeconomy constituted by sowing pools puts a premium on contracting services and leasing land, and on asset-light, mobile companies at the peak of the industry hierarchy. Ironically for a bioeconomy traceable in part to a movement to improve soil health, the soy bioeconomy in effect disembeds farming from land and from local communities. In this sense, it represents the diametric opposite of the eco-economy (see Chap. 1). In GMO bioeconomies more generally, enclosure (patentability) depends on a partial unmaking of relations (claims of novelty), which then have to be re-established in order for the GM organism to be declared "safe" or "substantially equivalent" to a non-GM organism. As Doezema and Hurlbut's account (Chap. 3) of the AquaAdvantage Salmon reveals, the marketability of AquaAdvantage Salmon was contingent on the deployment of mitochondrial DNA analysis (as a determinant of ancestry and belonging) to legally embed the biotechnologically remade (novel) fish into the community of Atlantic salmon.

Conflict, Contradiction, Resistance

Bioeconomies are complex political, social, scientific, technical, and economic phenomena, which bring together actors with diverse, and sometimes conflicting, expectations. The bioeconomy project and, to varying degrees, the bioeconomies examined in this book are characterized by unacknowledged assumptions, priorities, and values and bring about a range of effects on participants, institutions, and communities. How might we conceptualize resistance to such a phenomenon? The studies in this volume highlight the ways in which this diversity produces conflict in particular contexts, and how that conflict sometimes leads to acts of resistance against the bioeconomy project.

The promissory construct of "health and wealth", which featured in several chapters, contains contradictory imperatives: the channeling of public funds to science geared for health innovation, where innovation is understood to necessarily entail commercialization, is not—as we have noted in our discussion of inequality—the most effective way to improve the health of the population and may even undermine it. Thus it is not surprising that some clinicians have pushed back against the emphasis on commercialization and some scientists have challenged the benefit to health of regulatory changes to promote "accelerated access" to new treatments (Gardner, Webster, and Mittra, Chap. 2). What is perhaps more remarkable is the failure by promoters of these bioeconomic visions to acknowledge the (at least potential) conflict between population health and wealth understood as corporate profits.

Tupasela's study (Chap. 8) vividly illustrates the contradictory logics built into expectations for large prospective cohort (LPC) biobanks. These biobanks, too, are meant to generate "health and wealth" by providing a valuable resource to health researchers. This expectation sits in a larger context that values or requires competitiveness. The governmental expectation that LPC biobanks will adopt an open-science, data-sharing approach is at odds with the system of performance evaluation in which biobankers and researchers operate, which does not reward sharing (particularly with rival researchers) and is generally "toxic to collaboration". This conflict is only exacerbated by policies of austerity that are cutting public funding for biobanks and forcing biobanks to become economically self-supporting. These conflicting forces have generated what Tupasela calls a competition logic, in which biobanks, and research groups associated with them, compete among themselves for research funding and economically and/or reputationally desirable collaborators.

Here we might also point to the legitimating, conflict-management role of "open science". While openness and sharing are argued to be conducive to scientific progress and innovation, they also come into play as a response to growing concerns, and conflict, over the role of intellectual property rights in science and more generally (Goven and Pavone 2015). Areas of science particularly dependent on public cooperation, such as population biobanks, are particularly vulnerable to such concerns. But such legitimatory calls for open science are likely to produce dysfunctional outcomes in a context of austerity-linked cuts in public funding and where the imposition of managerialism-inspired performance-evaluation practices puts a premium on competition for both funding and scientific (and commercial) achievement.

That the bioeconomy project requires such legitimating efforts is suggested by the journalistic investigation of the Innovative Medicines Initiative

(Chap. 4, Marelli and Testa), which highlighted the public subsidization of private profit ("the people pay, corporations cash in") and criticized the Initiative's focus on industrial profits rather than medical innovation. However, in a context where most research scientists are dependent on such bioeconomy initiatives for the funding that makes their research possible, we can expect opposition to the bioeconomy project that comes from within the research domain to more often take passive or adaptive forms (for example, individual strategies for maintaining control over data), such as those identified by Tupasela and Aarden; open, explicit resistance is too costly.

Santoro and Romero-Bachiller pose the question of how we, as researchers, might "collaborate with patients, citizens, and activists to imagine and co-construct a 'more liveable' bioeconomy" (Chap. 12). How might we engage in a research practice of constructive resistance with those who are, in effect, targets of the bioeconomy? How might we "communicate our analyses to, and share our analytical tools with, citizens and collectives concerned or affected by the bioeconomic project?" They urge us to imagine more collaborative approaches to research that recognize both the effects of asymmetrical power and the agency of the less powerful, and that highlight the instrumentalization of care while making the research process more caring.

Beyond this, how might we develop approaches to researching bioeconomies that are grounded in a more respectful attitude to "life" in its diversity and relational embeddedness, in its inherent complexity? How do we bring proper recognition to the integrity and agency of non-human life, which is subjected to novel forms of instrumentalization through bioeconomies? How might we conceptualize life's complexity, its failure to conform to bioeconomic plans for it, as a form of resistance? Such approaches to researching bioeconomies would themselves be acts of resistance against the widespread academic normalizing of the bioeconomy project.

*

It has been nearly ten years since we, the editors of this book, began thinking about how to think about "the bioeconomy". Over that time, it became clear to us that the bioeconomy exists first as a promissory construct that partakes of broader imaginaries and is translatable into a very particular policy framework. In other words, the bioeconomy is a political project, and that political project is currently being enacted at many different sites around the world. One of the aims of this book has been to chronicle what that enactment looks like in practice.

A second aim of the book has been to bring together and into conversation a set of empirical studies of activities often labeled as bioeconomies, or as aspects of a singular (economic-scientific) bioeconomy. In doing so, we also wished to highlight how an awareness of the political-institutional and political-economic dimensions of power can inform such analyses. The chapters of this book speak for themselves, but we hope our attempt above to bring them into conversation with each other also helps to demonstrate how productive this effort has been. It is clear that the bioeconomy project and particular bioeconomies are articulating in highly consequential ways with semi-permanent structures of power and inequality. These are playing out in regulatory agencies and research institutions, in parent support groups and fertility clinics, in "public-private partnerships" and struggling biobanks, in farmers' fields and national health services, in refugee camps and financial markets. The implications, and the opportunities for research and for political engagement, are vast.

References

Goven, J., & Pavone, V. (2015). The bioeconomy as political project: A Polanyian analysis. *Science, Technology & Human Values, 40*, 302–337.

Waldby, C., & Mitchell, R. (2006). *Tissue economies: Blood, organs, and cell lines in late capitalism*. Durham: Duke University Press.

Winner, L. (1980). Do artifacts have politics? *Daedalus, 109*(1), 121–136.

Authors' Biography

Vincenzo Pavone is tenured research fellow the Institute of Public Goods and Policies (IPP) of the Spanish National Research Council (CSIC). His research activities address the relationship between neoliberalism and emerging bioeconomies, with a special focus on reproductive bioeconomies.

Joanna Goven is currently the Director of Kukupa Research, Deputy Director of the Centre for Integrated Research in Biosafety, Evaluation Director for the Engagement Programme of the Deep South National Science Challenge, and a researcher with the Centre for Integrated Biowaste Research (New Zealand). She has published on gender politics, antifeminism, transdisciplinarity, and the politics and political economy of science and technology.

Index

V. Pavone and J. Goven (eds.), *Bioeconomies*,
DOI 10.1007/978-3-319-55651-2

R

S

T

The manufacturer's authorised representative in the EU is Springer Nature Customer Service Centre GmbH, Europaplatz 3, 69115 Heidelberg, Germany. If you have any concerns regarding our products, please contact ProductSafety@springernature.com

Printed and bound by CPI Group (UK) Ltd, Croydon, CR0 4YY

25/09/2026

02222859-0009